Equilibrium Problems:
Nonsmooth Optimization and Variational Inequality Models

Nonconvex Optimization and Its Applications

Volume 58

Managing Editor:
Panos Pardalos

The titles published in this series are listed at the end of this volume.

Equilibrium Problems: Nonsmooth Optimization and Variational Inequality Models

Edited by

Franco Giannessi
University of Pisa,
Italy

Antonino Maugeri
University of Catania,
Italy

and

Panos M. Pardalos
University of Florida,
U.S.A.

KLUWER ACADEMIC PUBLISHERS
DORDRECHT / BOSTON / LONDON

A C.I.P. Catalogue record for this book is available from the Library of Congress.

ISBN 978-1-4419-5208-0   e-ISBN 978-0-306-48026-3

Published by Kluwer Academic Publishers,
P.O. Box 17, 3300 AA Dordrecht, The Netherlands.

Sold and distributed in North, Central and South America
by Kluwer Academic Publishers,
101 Philip Drive, Norwell, MA 02061, U.S.A.

In all other countries, sold and distributed
by Kluwer Academic Publishers,
P.O. Box 322, 3300 AH Dordrecht, The Netherlands.

Printed on acid-free paper

In memory of Werner Oettli

Contents

On the storng solvability of a unilateral boundary value problem for nolinear discontinuos operators in the plane   119

Sofia Giuffrè

Most likely traffic equilibrium route flows analysis and computation   129

T. Larsson, J.T. Lundgren, M. Patriksson and C. Rydergren

An overview on projection-type methods for convex large-scale quadratic programs

Valeria Ruggiero and Luca Zanni

Preface

The book contains a collection of selected and refereed papers on the most recent results about equilibrium problems arising from Mechanics, Network Management and Economic Systems.

Some of these results were presented at an international conference held in Taormina in the period December 3–5, 1998. The aim of the book is to cover the fundamental aspects of the research on equilibrium problems: the statement of the problem and its formulation using mainly Variational Methods, its theoretical solution by means of classic and new variational tools, the calculus of solutions and the applications in concrete cases.

In the book is shown how many equilibrium problems, such as the celebrated Signorini problem, the obstacle problem, the discrete and continuous traffic equilibrium problem, the spatial price equilibrium problem (and many others) follow a general law (the so-called user equilibrium condition). Such a law allows us to express the problem in terms of Variational Inequalities, which provide a powerful methodology; by means of it, existence and calculation of the solutions can be obtained.

Moreover, the Lagrangian Theory connected with Variational Inequalities has revealed itself to be very productive and, in some cases, absolutely necessary as, for example, in order to find the equilibrium conditions for the continuous traffic equilibrium problem.

As was first noted by A. C. Pigou, the classic equilibrium principle follows the user optimized approach, instead of the system optimized one. The latter is useful when it is requested by an "authority" to minimize a global cost (or a potential) for management purposes. This problem is connected with the bilevel problem to which some chapters of the book are devoted. Very recent theoretical results and calculation procedures complete this book, which would achieve the objective of following problems from their formulation to their solution and concrete applications.

Franco Giannessi, University of Pisa

Antonino Maugeri, University of Catania

Panos M. Pardalos, University of Florida

ON THE NUMERICAL SOLUTION OF FINITE-DIMENSIONAL VARIATIONAL INEQUALITIES BY AN INTERIOR POINT METHOD

Stefania Bellavia
Dipartimento di Energetica "S. Stecco",
Università di Firenze
Via C.Lombroso 6/17. 50134 Firenze, Italia

Maria Grazia Gasparo
Dipartimento di Energetica "S. Stecco",
Università di Firenze
Via C.Lombroso 6/17, 50134 Firenze, Italia
e-mail: gasparo@de.unifi.it

Abstract In this paper we give an Inexact Interior Point method to solve finite-dimensional variational inequality problems for monotone functions and polyhedral sets. At each iteration the linear system that determines the search direction is solved inexactly, by using a linear iterative solver with an ad hoc stopping criterion. We discuss algorithmic issues concerning the solution of several subproblems arising in the formulation of the method, including: form of the linear systems to be solved, choice of the accuracy in the solution of these systems, strategy for satisfying centering and descent conditions. Not surprisingly, all these choices can affect the actual performance of the method, both in terms of reliability and efficiency. We describe the practical and theoretical considerations behind the decisions included in our implementation. Results of the numerical experimentation on several well known test problems are given. They confirm the effectiveness of the proposed algorithm.

Keywords: Interior Point methods, Inexact methods, Variational inequalities, backtracking techniques, computational experiments.

F. Giannessi et al (eds.),
Equilibrium Problems: Nonsmooth Optimization and Variational Inequality Models, 1–23.
© 2001 *Kluwer Academic Publishers.*

1. Introduction.

Given a non-empty closed set $X \subseteq \mathbb{R}^n$ and a function $F : \mathbb{R}^r \to \mathbb{R}^n$, the finite-dimensional variational inequality problem, denoted by $VI(X, F)$, is to find a vector $s^* \in X$ such that

$$F(s^*)^T (s^* - s) \geq 0 \quad \text{for each} \quad s \in X. \tag{1.1}$$

Problem (1.1) is quite general and includes as special cases some known classes of problems. For example, if $X = \mathbb{R}^n_+$ (the nonnegative orthant of $\mathbb{R}^n$), $VI(X, F)$ becomes the well known nonlinear complementarity problem, denoted as $NCP(F)$, that is to find $s^* \in \mathbb{R}^n_+$ such that

$$F(s^*)^T s^* = 0 \quad \text{and} \quad F(s^*) \geq 0. \tag{1.2}$$

Further, if X is a convex set and $F(s)$ is the gradient of a function $f : \mathbb{R}^n \to \mathbb{R}$, then $VI(X, F)$ is equivalent to the first-order necessary optimality conditions for the nonlinear programming problem:

$$\min_{s \in X} f(s). \tag{1.3}$$

The finite-dimensional variational inequality problems have many important applications in several areas of engineering, such as structural mechanics problems, nonlinear obstacle problems, contact problems, traffic equilibrium problems. Moreover a large number of economic models can be stated as variational inequalities, for example the Nash-Cournot production models, the spatial price equilibrium models, etc. The applications often yield variational inequalities in infinite-dimensional function spaces; hence, these problems are discretized and very large finite-dimensional variational inequalities are obtained. Detailed descriptions of several applications can be found in the survey papers [13], [11], [21] and in the references there quoted.

Numerical methods for $VI(X, F)$ or $NCP(F)$ have been extensively studied in literature. A growing attention was paid in last years in the development of globally convergent methods with fast local convergence rate as well as in their robust and efficient implementations. An excellent introduction to assessed methods and new trends for solving nonlinear complementarity problems, with some discussions about extensions to variational inequalities, is given in [11]. Many numerical approaches have been proposed, mostly based on suitable reformulations of the given problem. A wide class of methods is based on reformulating the problem as a constrained or unconstrained optimization problem (see [1], [9], [10], [34]). In a second approach the problem is transformed into an equivalent system of nonsmooth equations, which is solved by suitable generalizations of globally convergent methods for smooth nonlinear systems (see [19], [30], [36]) or by smoothing techniques (see [5], [6], [25]). The third approach is the interior point approach (see [2], [3], [32], [35], [39]), based on the reformulation of the problem as a constrained system of smooth equations. The constraints are removed by an interior penalty, yielding to the central path equations, iteratively solved by using Newton's directions. At each iteration a positive steplength α is determined so that the new iterate, obtained by taking a step of length α from the old one in the Newton's search direction, satisfies suitable centering and descent conditions. The interior point approach exhibits some attractive theoretical and practical features. In particular, global polynomial complexity and fast local convergence can be often proved without assuming strong regularity conditions.

Here, we propose an Interior Point method for solving $VI(X, F)$ with polyhedral X defined by

$$X = \{s \in \mathbb{R}^n_+ : As = b\}, \tag{1.4}$$

where $A \in \mathbb{R}^{m \times n}$, $m < n$, and $b \in \mathbb{R}^m$. We assume throughout that A is a full rank matrix and F is continuously differentiable and monotone, i.e. with jacobian $F'(x)$ positive semi-definite for all $x \in \mathbb{R}^n_+$, or, equivalently,

$$(x - y)^T (F(x) - F(y)) \geq 0, \quad \text{for any} \quad x, y \in \mathbb{R}^n_+$$

The proposed method, denoted from now on by IIPVI-method, is an inexact Interior Point method, where the Newton's directions are approximately computed with increasing accuracy as the iterative process goes on and the solution is approached. We remind that inexact methods are widely used in literature in the solution of large unconstrained systems of nonlinear equations ([7],[26]). Further, the inexact Interior Point approach is successfully used to solve linear programming and complementarity problems(see [4], [14], [31]); recently, it was extended to the solution of mixed and nonlinear complementarity problems ([2], [3]).

The IIPVI-method results to be a generalization of the method for monotone nonlinear complementarity problems studied in [3], to which we will be refer as IIPNCP-method. Under standard assumptions, in [3] the IIPNCP-method is proved to be globally convergent with local superlinear convergence; further, under a scaled Lipschitz condition, the feasible version has global polynomial complexity. Such results can be easily extended to the IIPVI-method considered in this paper.

In [3] some practical suggestions are given for an effective implementation of the IIPNCP-method, but no experimental result is reported.

Here, our aim is to discuss in detail various algorithmic issues that are crucial in the practical performance of the IIPVI-method, and consequently for the IIPNCP-meyhod,in terms of reliability and efficiency. We describe the approaches to these issues that we took in our implementation and show the results of the numerical experimentation on well known nonlinear complementarity and programming test problems. Such results confirm the effectiveness of the proposed inexact approach and of our algorithmic choices.

Throughout the paper e denotes the vector of all ones and the shorthand notation $(s, z, v) = (s^T, z^T, v^T)^T$ is used. For any vector $y \in \mathbb{R}^n$, the corresponding capital letter Y denotes the diagonal matrix $Y = \mathrm{diag}(y_1, y_2, \cdots, y_n)$ and the notation $\min(y)$ is used for $\min_i(y_i)$. Further, $\| \cdot \|$ is the Euclidean norm.

2. The IIPVI-method.

The monotone variational inequality problem $VI(X, F)$, with X defined by (1.4), can be reformulated via the Karush-Kuhn-Tucker conditions as the constrained system of nonlinear equations:

$$H(s, z, v) \equiv \begin{pmatrix} As - b \\ z - F(s) + A^T v \\ Sz \end{pmatrix} = 0 \quad s, z \geq 0 \qquad (2.1)$$

where $s, z \in \mathbb{R}^n$ and $v \in \mathbb{R}^m$. A point (s, z, v) is called *feasible point* if $As = b$ and $z = F(s) - A^T v$. In the following we will use

$$\rho(s, z, v) = \sqrt{\|As - b\|^2 + \|z - F(s) + A^T v\|^2}$$

as a measure of the distance of (s, z, v) from the set of the feasible points.

A vital role in the definition of Interior Point methods is played by the central path, an arc of feasible points parametrized by a scalar $\tau > 0$. For problem (2.1) the points $(s, z, v) = (s(\tau), z(\tau), v(\tau))$ of the central path solve the following system:

$$H(s, z, v) = \begin{pmatrix} 0 \\ 0 \\ \tau e \end{pmatrix} \quad s, z > 0. \qquad (2.2)$$

Given a point (s_0, z_0, v_0), with $s_0, z_0 > 0$, the IIPVI-method generates a sequence of points (s_k, z_k, v_k), with $s_k, z_k > 0$. The iterates are computed by using inexact Newton's directions on the equation of the central path. Two well known centering conditions are used to avoid the premature approaching to the boundary of $\mathbb{R}^n_+$ and an Armijo-like condition is incorporated to ensurea sufficient decrease in the merit function $\psi(s, z) \equiv s^T z$. More precisely, given (s_k, z_k, v_k) and the movement direction $(\Delta s, \Delta z, \Delta v)$, the next iterate is computed as follows:

$$
\begin{aligned}
s_{k+1} &= s_k(\alpha) \equiv s_k + \alpha \Delta s \\
z_{k+1} &= z_k(\alpha) \equiv z_k + \alpha \Delta z + g_k(\alpha) \\
v_{k+1} &= v_k(\alpha) \equiv v_k + \alpha \Delta v
\end{aligned}
\tag{2.3}
$$

with $g_k(\alpha) = F(s_k(\alpha)) - F(s_k) - \alpha F'(s_k)\Delta s$ and $\alpha \in (0, 1]$ chosen in such a way that:

$$
\min(S_k(\alpha) z_k(\alpha)) \geq \tau_1 \mu_k(\alpha),
\tag{2.4}
$$

$$
\mu_k(\alpha) \geq \tau_2 \rho_k(\alpha),
\tag{2.5}
$$

$$
\psi(s_k(\alpha), z_k(\alpha)) \leq \psi(s_k, z_k) + \alpha\beta[\nabla\psi(s_k, z_k)]^T \begin{pmatrix} \Delta s \\ \Delta z \end{pmatrix},
\tag{2.6}
$$

where:

$$
\mu_k(\alpha) = s_k(\alpha)^T z_k(\alpha)/n,
$$

$$
\rho_k(\alpha) = \rho(s_k(\alpha), z_k(\alpha), v_k(\alpha)),
$$

$$
\tau_1 = \nu \frac{\min(S_0 z_0)}{s_0^T z_0}, \qquad \tau_2 = \nu \frac{s_0^T z_0}{\rho(s_0, z_0, v_0)}
$$

and $\nu \in [0.5, 1)$, $\beta \in (0, 1)$ are given constants.

Condition (2.4) ensures that the pairwise products in $s^T z$ are reduced to zero at more or less the same rate. Condition (2.5) prevents $s^T z$ from becoming too small with respect to the distance of the point from the set of the feasible points. We remark that τ_2 is not well defined if the initial guess (s_0, z_0, v_0) is a feasible point; as we will see later, in this case all the iterates are feasible points and condition (2.5) is not used. Condition (2.6) is the usual Armijo's rule used in the line-search strategies in the framework of global iterative methods for solving optimization problems and nonlinear systems. In the sequel we will refer to a point $(s_k(\alpha), z_k(\alpha), v_k(\alpha))$ satisfying (2.4),(2.5) and (2.6) as an *acceptable point*.

Below we sketch a generic iteration of the method:

0. Given $\nu \in [0.5, 1)$, $\beta \in (0, 1)$ and (s_k, z_k, v_k) with $s_k, z_k > 0$.
1. Compute $\mu_k = (s_k^T z_k)/n$.

2. Choose $\sigma_k \in (0,1)$, $\omega_k \in [0,1)$ and $\theta_k \in (0,1]$.
3. Compute $(\Delta s, \Delta z, \Delta v)$ by solving the linear algebraic system:

$$\begin{cases} A\Delta s = \theta_k(b - As_k) \\ F'(s_k)\Delta s - \Delta z - A^T\Delta v = \theta_k(z_k - F(s_k) + A^T v_k) \\ S_k\Delta z + Z_k\Delta s = -S_k z_k + \sigma_k \mu_k e + r_k, \quad \|r_k\| \le \omega_k \mu_k \end{cases} \quad (2.7)$$

4. Find α_k s. t. $(s_k(\alpha_k), z_k(\alpha_k), v_k(\alpha_k))$ is an acceptable point.
5. Put $(s_{k+1}, z_{k+1}, v_{k+1}) = (s_k(\alpha_k), z_k(\alpha_k), v_k(\alpha_k))$
6. End of iteration.

We point out that, if $\theta_k = 1$ is chosen, the Step 3 in the above algorithm corresponds to the inexact solution of the Newton's linear system for approximating the point on the central path $(s(\tau), z(\tau), v(\tau))$, $\tau = \sigma_k \mu_k$. If $\theta_k < 1$, the system (2.7) yields to approximate a point on the surface of analytic centers [27]. From a practical point of view, $\theta_k < 1$ acts as a step control parameter to avoid the use of excessively small steplengths α_k. We recall that a step control strategy is often used to stabilize global iterative algorithms based on line-search strategies [29].

The parameter $\omega_k \in [0,1)$ is used to control the level of accuracy in the solution of the linear system and it is called *forcing term*; if $\omega_k = 0$ the method becomes an exact Interior point method. We emphasize that the first $m+n$ equations of the linear system (2.7) are always solved exactly, while the last n equations are approximately solved with a residual r_k satisfying $\|r_k\| \le \omega_k \mu_k$. It is simple to verify that, with this choice, the following relation holds:

$$\rho(s_{k+1}, z_{k+1}, v_{k+1}) = (1 - \theta_k \alpha_k)\rho(s_k, z_k, v_k). \quad (2.8)$$

This implies that in the infeasible case $(\rho(s_0, z_0, v_0) \ne 0)$ the sequence $\{\rho(s_k, z_k, v_k)\}$ monotonically decreases with rate depending on both the sequences $\{\theta_k\}$ and $\{\alpha_k\}$. Further, if $\rho(s_{k_1}, z_{k_1}, v_{k_1}) = 0$ for some $k_1 \ge 0$, the same holds for all $k > k_1$, i. e. the method maintains feasibility.

3. Algorithmic issues.

In this section we discuss the following issues of algorithmic nature involved in the implementation of the proposed method: definition of the parameters σ_k, ω_k and θ_k at Step 2, solution of the linear system (2.7) at Step 3, computation of the steplength α_k at Step 4. Different strategies can be designed for each of these issues, yielding to algorithms with different practical behaviours in terms of reliability, robustness and efficiency.

Choosing $\sigma_k, \omega_k, \theta_k$

It must be stressed that the choice of the parameters $\sigma_k, \omega_k, \theta_k$ is critical to the efficiency of the algorithm and can affect robustness as well. A key point is to avoid that the line-search procedure at Step 4 breaks down.

Let be σ_k fixed. We know that for exact Interior point methods ($\omega_k = 0$), the existence of an $\bar{\alpha}_k > 0$ such that the centering condition (2.4) holds for all $\alpha \in (0, \bar{\alpha}_k]$ is ensured [27]. In the inexact context, a condition ensuring the existence of such an $\bar{\alpha}_k$ is given by:

$$0 \leq \omega_k \leq \frac{1 - \tau_1 \nu}{1 + \frac{\tau_1 \nu}{\sqrt{n}}} \sigma_k; \tag{3.1}$$

This is proved in [3] for the IIPNCP-method and can be easily extended to the IIPVI-method.

Taking into account (2.8), it can be proved as in [3] that there exists α_k satisfying the centering condition (2.5) if

$$1 - \sigma_k + \frac{\omega_k}{\sqrt{n}} < \theta_k \tag{3.2}$$

holds.

Finally, it can be easily verified that for $\omega_k = 0$ the direction $(\Delta s, \Delta z)$ is a descent direction for the merit function $\psi(s, z) = s^T z$ and the Armijo condition (2.6) is satisfied for some $\alpha_k > 0$. It can be proved as in [3] that this still holds for ω_k satisfying

$$\omega_k < \sqrt{n}(1 - \sigma_k). \tag{3.3}$$

As regards the choice of σ_k, let us observe that for $\theta_k = 1$, (2.7) can be viewed as the linear system generated at the k-th iteration of an Inexact Newton method for the solution of the nonlinear system $H(s, z, v) = 0$, with residual $(0, 0, \hat{r}_k) = (0, 0, \sigma_k \mu_k e + r_k)$ and

$$\|\hat{r}_k\| \leq (\sqrt{n}\sigma_k + \omega_k)\mu_k \leq (\sigma_k + \omega_k)\|H(s_k, z_k, v_k)\|$$

. Hence, the term $\sigma_k + \omega_k$ can be interpreted as a forcing term and techniques used in the analysis of Inexact Newton methods [7] can be applied to study the convergence of the sequence $\{(s_k, z_k, v_k)\}$. The global convergence of the IIPVI-method can be proved as in [3] under the assumption

$$\sigma_k + \omega_k \leq \eta_{max} < 1. \tag{3.4}$$

Summarizing, a robust algorithm based on the IIPVI-method can be obtained if σ_k, ω_k and θ_k satisfying (3.1), (3.2), (3.3) and (3.4) are chosen. Further, the efficiency of the algorithm will be enhanced by choosing the parameters in such a way that local fast convergence is ensured and oversolving in the solution of the linear systems is avoided in the early iterations. These requirements can be satisfied if $\omega_k = O(\|H(s_k, z_k, v_k)\|)$ and $\sigma_k = O(\|H(s_k, z_k, v_k)\|^p)$, $0 < p < 1$. Such choice yields local superlinear convergence [3] and gives large forcing terms away from the solution. In view of all above considerations, some possible choices are the following:

Choice 1

$$\text{Given:} \quad \eta_{max} < 1, C = \frac{1 - \tau_1 \nu}{1 + \tau_1 \nu}, M > 2n;$$

$$\text{Compute:} \quad \sigma_k = \min\{\sqrt{\|H(s_k, z_k, v_k)\|}, 0.5\eta_{max}\},$$
$$\omega_k = \min\{\|H(s_k, z_k, v_k)\|, C\sigma_k\},$$
$$\theta_k = 1 - \frac{\omega_k}{M}.$$

This choice is suggested in [3], but was not completely satisfactory in our numerical experiments.

Choice 2

$$\text{Given:} \quad \eta_{max} < 1, C = \frac{1 - \tau_1 \nu}{1 + \tau_1 \nu}, \hat{C} = 0.5(1 + \tau_1 \nu);$$

$$\text{Compute:} \quad \sigma_k = \min\{\sqrt{\|H(s_k, z_k, v_k)\|}, \hat{C}\eta_{max}\},$$
$$\omega_k = \min\{\|H(s_k, z_k, v_k)\|, C\sigma_k\},$$
$$\theta_k = 1 - (1 - C)\sigma_k.$$

This choice gives σ_k-values, and consequently ω_k-values, larger than those obtained by using Choice 1. Further, regarding θ_k, this choice is more flexible than Choice 1 and avoids too fast convergence of $\{\theta_k\}$ to 1 as ω_k goes to zero.

In order to test the behaviour of the method with respect to the choice of θ_k, we consider also a third possibility, named Choice 3, where σ_k and ω_k are as in Choice 2 while $\theta_k = 1$ is fixed.

Solving the linear system

Note that the linear system (2.7) has dimension $2n + m$ and the associated computational cost might be viewed as excessive for large dimensional problems. On the other hand, the structure of the system can be fruitfully exploited, also taking into account that only the last n equations are solved approximately. An efficient strategy consists in solving the system in three steps.

Since $A\Delta s = \theta_k(b - As_k)$ and A has full rank m, there exists $y \in \mathbb{R}^{n-m}$ such that
$$\Delta s = Ny + \hat{w} \qquad (3.5)$$

where the columns of $N \in \mathbb{R}^{n \times (n-m)}$ span the null space of A and $\hat{w} \in \mathbb{R}^n$ is the minimum-norm solution of $Aw = \theta_k(b - As_k)$. Note that both N and $\hat{w}$ are easily computed in a stable manner by using the SVD-decomposition of A [17] which may be carried out once and for all before starting the iterative process.

From the second set of equations in (2.7), Δz can be written as
$$\Delta z = F'(s_k)\Delta s - A^T \Delta v - \theta_k(z_k - F(s_k) + A^T v_k). \qquad (3.6)$$

Combining (3.5), (3.6) and the last block of equations in (2.7), we obtain the following linear system of n equations in the unknowns $y \in \mathbb{R}^{n-m}$ and $\Delta v \in \mathbb{R}^m$:

$$(S_k F'(s_k) + Z_k)Ny - S_k A^T \Delta v = d_k + r_k \quad , \quad \|r_k\| \leq \omega_k \mu_k \qquad (3.7)$$

with

$$d_k = -S_k z_k + \sigma_k \mu_k e -$$
$$- (S_k F'(s_k) + Z_k)\hat{w} + \theta_k S_k(z_k - F(s_k) + A^T v_k). \qquad (3.8)$$

Hence, first $\hat{w}$ is computed and the linear system

$$((S_k F'(s_k) + Z_k)N \quad | \quad -S_k A^T) \begin{pmatrix} y \\ \Delta v \end{pmatrix} = d_k \qquad (3.9)$$

is solved by an iterative linear solver with the stopping criterion $\|r_k\| \leq \omega_k \mu_k$. Then, (3.5) and (3.6) are used to compute Δs and Δz.

Line-search strategy

In order to take advantage of local fast convergence of the IIPVI-method, unity steplengths must be taken as soon as possible. Hence, at each iteration the value $\alpha_k = 1$ is firstly tried. If $\alpha_k = 1$ fails to satisfy

(2.4), (2.5) and (2.6), some backtracking strategy must be implemented. The simplest strategy consists in successive reductions of α_k by a factor $\lambda \in (0,1)$ until an acceptable point is found [38]. Numerical experiments performed with several values of λ showed that often this strategy fails to find α_k in a reasonably small number of trials. We observed that failures are mostky due to difficulty in satisfying the condition (2.4). This fact suggested us the following polynomial interpolation technique, where the most current information about the function $\phi(\alpha) = S_k(\alpha)z_k(\alpha) - \tau_1\mu_k(\alpha)e$ is used to model it and find a good trial value $\hat{\alpha} < 1$.

Consider the functions $\phi_i(\alpha) = e_i^T S_k(\alpha)z_k(\alpha) - \tau_1\mu_k(\alpha)$, for $i = 1,...,n$, where e_i denotes the i-th unit vector in $\mathbb{R}^n$. For each i, we interpolate $\phi_i(\alpha)$ by the quadratic polynomial $p_i(\alpha)$ such that $p_i(0) = \phi_i(0)$, $p_i(1) = \phi_i(1)$ and $p_i'(0) = \phi_i'(0)$, i.e. $p_i(\alpha) = a_i\alpha^2 + b_i\alpha + c_i$, with

$$a_i = \phi_i(1) - \phi_i(0) - \phi_i'(0)$$
$$b_i = \phi_i'(0)$$
$$c_i = \phi_i(0).$$

Then, for each i, we compute a value $\hat{\alpha}_i$ such that $p_i(\alpha) \geq 0$ for $\alpha \in (0, \hat{\alpha}_i]$. Taking into account that $c_i \geq 0$ for each i, and denoting by $r_{i,1}$ and $r_{i,2}$ the roots of $p_i(\alpha)$, the computation of $\hat{\alpha}_i$ is carried out in the following way:

- If $r_{i,1}$ and $r_{i,2}$ are complex, then $\hat{\alpha}_i = 1$;
- . If $a_i > 0$ and $r_{i,1}, \quad r_{i,2} > 0$, then $\hat{\alpha}_i = \min\{r_{i,1}, r_{i,2}\}$;
- . If $a_i > 0$ and $r_{i,1}, \quad r_{i,2} < 0$, then $\hat{\alpha}_i = 1$;
- . If $a_i < 0$, then $\hat{\alpha}_i = \max\{r_{i,1}, r_{i,2}\}$.

Finally, we compute $\hat{\alpha} = \min\{\hat{\alpha}_1, ...\hat{\alpha}_n\}$ and, if $(s(\hat{\alpha}), z(\hat{\alpha}), v(\hat{\alpha}))$ satisfy (2.4), (2.5) and (2.6), we put $\alpha_k = \hat{\alpha}$. If we need to backtrack again, two different strategies could be followed. In the first one, new interpolating polynomials are constructed by exploiting the information at $\hat{\alpha}$; in the second one, successive reductions of $\hat{\alpha}$ are made until an acceptable point is found. We implemented the second strategy, using two different reduction factors. When condition (2.4) is not satisfied we reduce the steplength of a factor $\lambda = 0.9$, while we use $\lambda = 0.5$ when condition (2.4) is met and condition (2.5) or (2.6) does not hold.

To sum up, we sketch the line-search algorithm we implemented: a maximum number R_{max} of reductions and a lower bound α_{min} for α_k are provided as safeguard parameters.

0. Given: $s_k > 0$, $z_k > 0$, v_k, Δs, Δz, Δv, $R_{max} > 0$, $\alpha_{min} > 0$
1. Put $\alpha_k = 1$.

2. If $(s_k(\alpha_k), z_k(\alpha_k), v_k(\alpha_k))$ is an acceptable point, then stop (success).
3. Compute $\hat{\alpha}$ and put $\alpha = \hat{\alpha}$.
4. For $j = 1, ..., R_{max}$, do
    4.1. If (2.4) is satisfied, then
         if (2.5) or (2.6) is not satisfied, then
            put $\alpha = 0.5\alpha$
        else
            put $\alpha_k = \alpha$ and stop (success)
      else
        put $\alpha = 0.9\alpha$
      end if
    4.2. If $\alpha < \alpha_{min}$, then stop (failure)
5. Stop (failure).

This algorithm turned out to be reliable and efficient, particularly for the feasible case. Due to the modelling of $\phi(\alpha)$, condition (2.4) is generally satisfied within a very small number of trials. Further unnecessary small steplengths are avoided and values near to the unity are produced whenever the current iterate is not too away from the solution.

4. Numerical experiments.

In this section we report on results obtained in solving from several starting points some well known nonlinear complementarity and programming problems appearing in recent numerical literature ([8],[34], [?],[38]).

The first three problems are nonlinear complementarity problems and the fourth one is obtained by reformulating a well known nonlinear programming problem from [22]. These four problems have small dimensions ($n \leq 10$) and are used to test the robustness of the proposed algorithm. The effectiveness of the inexact approach is showed in the results for the last problem, an NCP reformulation of the Bratu obstacle problem, where $n \leq 1600$ is used. For sake of brevity we omit the mathematical formulation of the used test problems, referring to a source paper for each of them.

The algorithm was implemented in a double precision Fortran 77 code running on a RISC 6000 workstation 3CT (machine precision $\epsilon_m \cong 10^{-16}$). The system of linear equations (3.9) generated at each iteration is solved by $GMRES(m)$ [33] with the modified stopping criterion

$$\|r_k\| \leq \max\{8\epsilon_m, \omega_k\mu_k\}. \tag{4.1}$$

No scaling or preconditioning is used and default data are used (maximum dimension of the Krylov subspace $m = 10$, maximum number of restarts of the Krylov iterations $mr = 10$), with an exception for the Bratu problem, where $mr = n$ is used. The SVD of the matrix A is computed by the LAPACK routine DGESVD.

The numerical experiments showed that the performance of the algorithm is essentially independent of the values of the parameters ν and β. Moreover, a good convergence behaviour and high efficiency are achieved when large values of η_{max} are used. Here, we give results obtained by using $\nu = 0.5$, $\beta = 10^{-5}$ and $\eta_{max} = 0.7$ or $\eta_{max} = 0.8$. The used stopping criterion is $\|H(s_k, z_k, v_k)\| \leq 10^{-8}$. Failure is declared when the stopping criterion is not satisfied after 300 iterations, when at some iteration GMRES fails to solve the linear system within the required accuracy, or when the line-search algorithm is not successfull (we use $R_{max} = 15$ and $\alpha_{min} = 10^{-8}$).

The results are summarized in several tables; the column headings mean the following:

sp : starting point,
I_N : total number of nonlinear iterations,
I_L : total number of linear iterations,
N_r : total number of reductions from $\alpha = 1$ in the line-search procedure,
$\|H\|$: norm of H in the last iterate,
f_{ls} : failure in the line-search procedure,
f_g : failure of GMRES,
f_{it} : failure due to excessive number of iterations,

We remark that I_N corresponds to the total number of evaluations of $F'(s)$ and $I_N + N_r$ gives the total number of evaluations of $F(s)$. From the experiments it results that almost all reductions from $\alpha = 1$ are due to the centering condition (2.4). Indeed, the other two conditions are always satisfied for $\alpha_k = 1$ or $\alpha_k = \hat{\alpha}$, with a few exceptions for the Armijo condition.

Problem 1.   Mathiesen's Walrasian equilibrium [28]

This is a nonlinear complementarity problem of dimension $n = 4$. It depends on three parameters (a, b_2, b_3); we consider the triplets (0.75, 1.0, 0.5), (0.75, 1.0, 2.0) and (0.9, 5.0, 3.0). The used starting points, all infeasible, are the following:

$$sp_1 : \quad s_0 = (0.1, \cdots, 0.1), \quad z_0 = (0.1, \cdots, 0.1)$$
$$sp_2 : \quad s_0 = (1, \cdots, 1), \quad z_0 = (1, \cdots, 1)$$
$$sp_3 : \quad s_0 = (5, \cdots, 5), \quad z_0 = (5, \cdots, 5)$$
$$sp_4 : \quad s_0 = (10, \cdots, 10), \quad z_0 = (10, \cdots, 10)$$

In Table 1 we show the results obtained with $\eta_{max} = 0.8$ and θ_k chosen according to both Choice 2 ($\theta_k < 1$) and Choice 3 ($\theta_k = 1$). The results confirm the role of θ_k as a step control parameter: Choice 2 yields α_k-values larger than Choice 3 and $\alpha_k = 1$ is taken after a few initial iterations. In this way fast convergence and great saving both in linear iterations and in function evaluations are achieved. This behaviour was observed for all test problems, and in the following only results obtained by using Choice 2 will be given.

	$\theta_k < 1$				$\theta_k = 1$			
	sp_1	sp_2	sp_3	sp_4	sp_1	sp_2	sp_3	sp_4
	(a, b_1, b_2)=(0.75,1.0,0.5)							
I_N	16	20	22	22	63	80	93	98
I_L	59	71	79	78	246	309	391	479
N_r	22	27	25	20	85	113	153	158
$\|H\|$	5(-9)	4(-9)	3(-9)	8(-9)	8(-9)	8(-9)	7(-9)	8(-9)
	(a, b_1, b_2)=(0.75,1.0,2.0)							
I_N	22	18	23	25	67	82	102	f_{ls}
I_L	84	64	84	92	264	320	451	
N_r	25	22	26	25	105	129	162	
$\|H\|$	5(-9)	1(-8)	5(-9)	4(-9)	8(-9)	8(-9)	8(-9)	
	(a, b_1, b_2)=(0.9,5.0,3.0)							
I_N	36	23	27	28	101	89	116	f_g
I_L	140	87	99	103	400	350	507	
N_r	50	19	33	36	154	136	186	
$\|H\|$	6(-9)	7(-9)	4(-9)	9(-9)	8(-9)	8(-9)	7(-9)	

Table 1. Results for Problem 1.

Stefania Bellavia – Maria Grazia Gasparo

Problem 2.   Nash equilibrium non cooperative games [20]

This is a small nonlinear complementarity problem, depending on a parameter γ. We consider $n = 5$ with $\gamma = 1.1$ and $\gamma = 2.0$, and $n = 10$ with $\gamma = 1.2$. The starting points are given by:

	sp_1	sp_2	sp_3	sp_4	sp_5	sp_6	sp_7	sp_8	sp_9
				$n = 5$	$\gamma = 1.1$				
I_N	19	21	22	24	27	19	20	22	24
I_L	41	39	41	47	56	38	37	44	48
N_r	2	0	0	0	0	0	0	0	0
$\|H\|$	2(-11)	2(-10)	2(-9)	7(-10)	2(-11)	2(-10)	4(-9)	3(-11)	4(-12)
				$n = 5$	$\gamma = 2.0$				
I_N	19	21	23	25	28	21	22	23	25
I_L	51	54	60	64	71	52	54	56	60
N_r	0	0	0	0	0	0	0	0	0
$\|H\|$	9(-11)	9(-9)	3(-10)	2(-10)	5(-12)	2(-11)	1(-11)	1(-10)	3(-12)
				$n = 5$	$\gamma = 1.2$				
I_N	25	33	37	46	62	22	23	25	27
I_L	89	129	153	217	328	64	69	81	89
N_r	15	45	56	101	177	0	0	0	0
$\|H\|$	2(-9)	1(-9)	2(-11)	8(-10)	1(-9)	2(-10)	3(-9)	4(-10)	3(-9)

Table 2. Results for Problem 2.

$$
\begin{aligned}
sp_1: \quad & s_0 = (10, \cdots, 10), & z_0 = (10, \cdots, 10) \\
sp_2: \quad & s_0 = (20, \cdots, 20), & z_0 = (20, \cdots, 20) \\
sp_3: \quad & s_0 = (30, \cdots, 30), & z_0 = (30, \cdots, 30) \\
sp_4: \quad & s_0 = (50, \cdots, 50), & z_0 = (50, \cdots, 50) \\
sp_5: \quad & s_0 = (100, \cdots, 100), & z_0 = (100, \cdots, 100) \\
sp_6: \quad & s_0 = (20, \cdots, 20), & z_0 = F(s_0) \\
sp_7: \quad & s_0 = (30, \cdots, 30), & z_0 = F(s_0) \\
sp_8: \quad & s_0 = (50, \cdots, 50), & z_0 = F(s_0) \\
sp_9: \quad & s_0 = (100, \cdots, 100), & z_0 = F(s_0)
\end{aligned}
$$

Some results obtained with $\eta_{max} = 0.7$ are given in Table 2. We observe that for $n = 5$ the problem is easily solved from both feasible and infeasible starting points. The value $\alpha_k = 1$ is practically always accepted and 2 or 3 linear iterations are enough to satisfy the stopping criterion (4.1) at each nonlinear iteration. More computational work is required to solve the problem for $n = 10$ from infeasible starting points.

Problem 3.   Josephy's problem [24]

This small nonlinear complementarity problem ($n = 4$) was solved with the following starting points: sp_1 : $s_0 = (2, \cdots, 2)$, $z_0 = F(s_0)$

sp_2 : $s_0 = (3, \cdots, 3)$, $z_0 = F(s_0)$
sp_3 : $s_0 = (4, \cdots, 4)$, $z_0 = F(s_0)$
sp_4 : $s_0 = (5, \cdots, 5)$, $z_0 = F(s_0)$
sp_5 : $s_0 = (3, \cdots, 3)$, $z_0 = (3, \cdots, 3)$
sp_6 : $s_0 = (6, \cdots, 6)$, $z_0 = (6, \cdots, 6)$
sp_7 : $s_0 = (10, \cdots, 10)$, $z_0 = (10, \cdots, 10)$

In the Table 3 we give the results obtained by using $\eta_{max} = 0.8$ and $\eta_{max} = 0.7$. The convergence from feasible points is easily obtained for both the used values of η_{max}, with a slight saving of computational work for $\eta_{max} = 0.7$. The solution from infeasible starting points is more laborious, and in one case the value $\eta_{max} = 0.8$ results to be too large so that the algorithm fails to converge.

	sp_1	sp_2	sp_3	sp_4	sp_5	sp_6	sp_7
				$\eta_{max} = 0.8$			
I_N	20	22	23	24	f_{it}	28	36
I_L	46	48	47	48		72	86
N_r	0	0	0	0		23	35
$\|H\|$	2(-10)	1(-10)	4(-9)	1(-8)		7(-9)	1(-11)
				$\eta_{max} = 0.7$			
I_N	18	20	21	22	23	25	33
I_L	43	45	47	48	71	68	84
N_r	0	0	0	0	13	30	33
$\|H\|$	3(-10)	4(-11)	3(-10)	2(-10)	1(-11)	1(-9)	2(-11)

Table 3. Results for Problem 3.

Problem 4.   Chemical equilibrium problem [22]

This problem is given by the KKT conditions for the nonlinear programming problem 4 in [22]. This is a problem in the chemical equilibrium at constant temperature and pressure. The number of independent variables is $n = 10$ and the number of linear equality constraints is $m = 3$. The used starting points are the following, all infeasible:

$$sp_1: \quad s_0 = (0.1, \cdots, 0.1), \qquad\qquad z_0 = (0.1, \cdots, 0.1),$$
$$sp_2: \quad s_0 = (1, \cdots, 1), \qquad\qquad\qquad z_0 = (1, \cdots, 1),$$
$$sp_3: \quad s_0 = (10, \cdots, 10), \qquad\qquad\; z_0 = (10, \cdots, 10),)$$
$$sp_4: \quad s_0 = (1.4, 0.1, 0.1, 0.6, 0.1, 0.1, 0.1, 0.5), \quad z_0 = (1, \cdots, 1),$$
$$sp_5: \quad s_0 = (1.4, 0.1, 0.1, 0.6, 0.1, 0.1, 0.1, 0.5), \quad z_0 = (0.1, \cdots, 0.1),$$
$$sp_6: \quad s_0 = (1.4, 0.1, 0.1, 0.6, 0.1, 0.1, 0.1, 0.5), \quad z_0 = (1, \cdots, 1),$$
$$sp_7: \quad s_0 = (1.4, 0.1, 0.1, 0.6, 0.1, 0.1, 0.1, 0.5), \quad z_0 = (1, \cdots, 1),$$

$$v_0 = (0.1, \cdots, 0.1) \; v_0 = (1, \cdots, 1)$$
$$v_0 = (10, \cdots, 10)$$
$$v_0 = (1, \cdots, 1)$$
$$v_0 = (0.1, \cdots, 0.1)$$
$$v_0 = (-1, \cdots, -1)$$
$$0 = (-10, \cdots, -10)$$

The results obtained by using $\eta_{max} = 0.8$ are shown in Table 4. Note that the large number of linear iterations for this problem depends on the behaviour of GMRES. For each linear system, GMRES stagnates during about 8 or 9 iterations with large residuals and then goes to satisfy the stopping criterion (4.1) in one iteration.

	sp_1	sp_2	sp_3	sp_4	sp_5	sp_6	sp_7
I_N	100	24	74	15	52	17	19
I_L	990	230	729	139	510	160	180
N_r	91	17	66	10	193	12	15
$\|H\|$	7(-10)	1(-10)	9(-13)	9(-9)	8(-9)	6(-9)	9(-9)

Table 4. Results for Problem 4.

Problem 5   Obstacle Bratu problem [23]

This problem consists in finding the equilibrium position of an elastic membrane subject to a nonconstant vertical force pushing upwards, with homogeneous Dirichlet conditions on the boundary of the domain Ω and with an upper constraint (obstacle). The vertical force depends on a real parameter $\lambda > 0$. The problem can be formulated as an infinite dimensional λ-dependent nonlinear elliptic variational inequality [23]. We consider $\Omega = (0,1) \times (0,1)$. By a finite-difference discretization with step $h = 1/(N+1)$ and a simple change of variables, we obtain a finite dimensional NCP formulation of the problem in $\mathbb{R}^n$, with $n = N^2$. Using the two-indices notation, the function F is defined by

$$F_{i,j}(s) = 4(s_{i,j} - u_{i,j}) - (s_{i+1,j} - u_{i+1,j}) - (s_{i-1,j} - u_{i-1,j})$$
$$(s_{i,j+1} - u_{i,j+1}) - (s_{i,j-1} - u_{i,j-1}) + \lambda h^2 \exp(s_{i,j} - u_{i,j})$$

for $i,j = 1, \cdots, N$, where $s_{i,j}$ approximates the solution at the interior grid point $P_{i,j}$ and $u_{i,j}$ is the value of the obstacle at the same point. Of course, $s_{i,j} = u_{i,j} = 0$ for boundary points. We consider obstacles of the form

$$u_{i,j} = A + B\sqrt{(i - 0.5)^2 + (j - 0.5)^2},$$

with: $A = 4$ and $B = -1$ (concave obstacle), $A = 4$ and $B = 1$ (convex obstacle) and $A = 4$ and $B = 0$ (constant obstacle). In all cases, the numerical difficulty in the solution increases for increasing λ-values, due to the presence of many fold points. Typically, near to such points the problem may have one, several or no solutions.

We give results obtained for $\lambda \in [1, 20]$ and $n = 400, 900, 1600$. In all cases we used $\eta_{max} = 0.7$. One of our aims is to verify the real advantages of the inexact approach in terms of overall computational cost with respect to the exact approach. We simulate the exact interior point method (EIPVI-method) by using $\omega_k = 0$, while σ_k and θ_k are chosen following Choice 2. In this case, GMRES is stopped by the following criterion

$$\|r_k\| \le \max\{8\epsilon_m, 500\epsilon_m \|b_k\|\}, \tag{4.2}$$

where b_k is defined in (3.8). The criterion (4.2) corresponds to require to GMRES the maximum accuracy, compatibly with the machine precision.

In the Table 5 we show some results for $\lambda = 1, 4, 6$ and $n = 400, 900, 1600$, obtained by using the feasible starting point $s_0 = (10, \ldots, 10)$, $z_0 = F(s_0)$ and the constant obstacle ($B = 0$).

	IPVI-method			EIPVI-method		
	$\lambda = 1$	$\lambda = 4$	$\lambda = 6$	$\lambda = 1$	$\lambda = 4$	$\lambda = 6$
$n = 400$						
I_N	17	17	17	17	17	17
I_L	918	1019	1080	3929	4172	3973
N_r	0	0	0	0	0	0
$\|H\|$	1(-11)	7(-11)	7(-10)	2(-12)	1(-11)	2(-10)
$n = 900$						
I_N	17	17	18	17	17	17
I_L	1917	2264	3983	8262	9506	11272
N_r	0	0	0	0	0	0
$\|H\|$	4(-10)	1(-9)	1(-12)	1(-10)	4(-10)	4(-9)
$n = 1600$						
I_N	17	17	17	17	17	17
I_L	3636	4325	5925	14154	16474	21059
N_r	2	2	2	2	2	2
$\|H\|$	2(-11)	9(-11)	9(-10)	4(-12)	3(-11)	3(-10)

Table 5. Results for Problem 5 with a feasible starting point.

In Table 6 and Table 7 we show the results obtained for $n = 400$ and

$n = 900$, respectively, with $\lambda = 1, 4, 6, 10, 12, 16, 20$. In these cases we used the infeasible starting point $s_0 = (1, \cdots, 1)$, $z_0 = (1, \cdots, 1)$ and the concave obstacle ($B = -1$). In these tables the symbol * denotes that the value $\theta_k = 1$ was used, due to the failure of the choice $\theta_k < 1$ for both IIPVI-method and EIPVI-method.

From the given results, it can be observed that the number of nonlinear iterations performed by the IIPVI-method and the EIPVI-method is almost the same in all the cases, i.e. the inexactness does not slow down the convergence of the method. On the other hand, large saving in the total number of linear iterations is achieved in all cases, whereas the cost in terms of jacobian and function evaluations is practically the same. We remark that the greatest difficulties happen for $\lambda = 4$ and $\lambda = 6$, near a fold point corresponding to a turning point [23].

	$\lambda = 1$	$\lambda = 4$	$\lambda = 6$	$\lambda = 10$	$\lambda = 12$	$\lambda = 16$	$\lambda = 20$
				IIPVI-method			
I_N	14	16	17	16	16	15	16
I_L	755	3965	968	649	816	392	889
N_r	8	9	17	4	5	4	3
$\|H\|$	4(-10)	7(-10)	9(-9)	7(-12)	3(-12)	6(-9)	3(-12)
				EIPVI-method			
I_N	14	f_{it}	17	16	16	15	16
I_L	2818		3955	2479	2704	1884	2469
N_r	8		13	4	5	4	3
$\|H\|$	7(-10)		3(-11)	4(-12)	3(-11)	3(-9)	2(-12)

Table 6. Results for Problem 5, $n = 400$, with an infeasible starting point.

	$\lambda = 1$	$\lambda = 4$	$\lambda = 6$	$\lambda = 10$	$\lambda = 12$	$\lambda = 16$	$\lambda = 20$
				IIPVI-method			
I_N	15	16*	16*	20	18	18	16
I_L	1755	2527	3204	3056	1558	1502	819
N_r	19	8	8	16	11	13	5
$\|H\|$	5(-11)	3(-12)	9(-10)	2(-11)	1(-12)	2(-10)	3(-9)
				EIPVI-method			
I_N	15	16*	16*	20	18	18	16
I_L	5887	8446	11404	7717	4777	4820	3071
N_r	10	8	8	24	11	12	5
$\|H\|$	8(-11)	2(-12)	4(-10)	3(-11)	2(-12)	8(-11)	2(-9)

Table 7. Results for Problem 5, $n = 900$, with an infeasible starting point.

5. Conclusions and perspectives.

The computational results presented in section 4 show that the inexact approach can be used to design good interior point algorithms for the solution of finite dimensional variational inequality problems. In particular, by computing the search directions with low accuracy during early iterations, *oversolving* can be avoided and the overall computational cost can be considerably reduced, while the practical convergence rate of the exact method is maintained. The reasonable performance of the proposed method indicate that the inexact approach can be fruitfully extended to variational inequality problems with more general equality and inequality constraints.

In this paper several algorithmic details crucial in practical performance of the IIPVI-method were discussed and our decisions about such issues were justified. All of the considered issues deserve further research and experimentation to the aim of developing robust and efficient general-purpose interior point algorithms. As an example, from the experiments some relationship appears between efficiency of the line-search strategy and feasibility or infeasibility of the iterates. We are currently investigating this relationship and the possibility of modifying the line-search strategy in order to improve it. At this purpose, we believe that it may be interesting investigate in the direction of nonmonotone line-search techniques, where the descent condition is substituted by a less restrictive condition that does not enforce the monotonic decrease of the chosen merit function. Such strategies, often used to stabilize Newton-like methods in the numerical solution of unconstrained nonlinear systems and optimization problems, allow efficient solution of very difficult problems where methods based on traditional descent conditions converge very slowly or fail to converge (see [12], [15], [16], [18]).

Acknowledgements We are grateful to Alessandro Armellini for numerous discussions and for carrying out most of the computations.

References

[1] R. Andreani, A. Friedlander, J.M. Martinez, *Solution of finite-dimensional variational inequalities using nonsmooth optimization with simple bounds*, J. Opt. Theory and Appl. 94 (1997), 635–657.

[2] S. Bellavia, *Inexact Interior-point method*, J. Opt. Theory and Appl. 96 (1998), 109–121.

[3] S. Bellavia, M. Macconi, *An Inexact Interior-point method for monotone NCP*, Optim. Methods and Software, Special Issue on Interior

Point methods, 11/12 (1999), 211–241.

[4] J.F. Bonnans, C. Pola, R. Rébaï, *Perturbed Path Following Interior Point Algorithms,* Optim. Methods and Software, Special Issue on Interior Point methods, 11/12 (1999), 183–210.

[5] C. Chen, O.L. Mangasarian, *A class of smoothing functions for non-linear and mixed complementarity problems,* Comput. Optim. and Appl. 5 (1996), 97–138.

[6] X. Chen, L. Qi, D. Sun, *Global and superlinear convergence of the smoothing Newton methods and its application to general box constrained variational inequalities,* Math. of Comput. 67 (1998), 519–540.

[7] R.S. Dembo, S.C. Eisenstat, T. Steihaug, *Inexact Newton methods,* SIAM J. Numer. Anal. 19 (1982), 400–408.

[8] S. P. Dirkse, M. C. Ferris, *MCPLIB: A Collection of Nonlinear Mixed Complementary Problems,* Optimization Methods and Software 5 (1995), 319–345.

[9] F. Facchinei, A. Fischer, C. Kanzow, *Regularity properties of a semis-mooth reformulation of variational inequalities,* SIAM J. Optim. 8 (1998), 850–869.

[10] F. Facchinei, A. Fischer, C. Kanzow, *A semismooth newton method for variational inequalities: the case of box constraints,* in: M.C. Ferris and J.S. Pang (eds.) *Complementarity and variational problems: state of the art* SIAM, Philadelphia (1997), 76–90.

[11] M.C. Ferris, C. Kanzow, *Complementarity and related problems: a survey,* preprint 1998

[12] M.C. Ferris, S. Lucidi, *Nonmonotone stabilization methods for non-linear equations,* J. Opt. Theory and Appl. 81 (1994), 53–71.

[13] M.C. Ferris, J.S. Pang, *Engineering and economic applications of complementarity problems,* SIAM Rev. 39 (1997), 669–713.

[14] R.W. Freund, F. Jarre, *A QMR- based interior-point algorithm for solving linear programs,* Math. Programming, Series B 76 (1997), 183–210.

[15] A. Friedlander, M.A. Gomes-Ruggiero, D.N. Kozakevich, J.M. Mar-tinez, S.A. Santos, *Solving nonlinear systems of equations by means quasi-Newton methods with a nonmonotone strategy,* Optim. Methods and Software 8 (1997), 25–51.

[16] M.G. Gasparo, *A nonmonotone hybrid method for nonlinear systems,* Optim. Methods and Software 13 (2000), 79–92.

[17] P.E. Gill, W. Murray, M.H. Wright, *Numerical Linear Algebra and Optimization, vol. 1,* Addison-Wesley Pub. Company, 1991

[18] L. Grippo, F. Lampariello, S. Lucidi, *A nonmonotone line-search technique for newton's method,* SIAM J. Numer. Anal. 23 (1986), 707–716.

[19] J. Han, D. Sun, *Newton-type methods for variational inequalities,*in: Ya-xiang Yuan (ed.),*Advances in Nonlinear Programming,* Kluwer Ac. Pub. (1998), 105–118.

[20] P.T. Harker, *Accelerating the convergence of the diagonalization and projection algorithms for Finite-dimensional Variational Inequalities,* Math. Progr. 41 (1988), 29–59.

[21] P.T. Harker, J.S. Pang, *Finite-dimensional variational inequality and nonlinear complementarity problems: a survey of theory, algorithms and applications,* Math. Progr. 48 (1990), 161–220.

[22] D. M. Himmelblau, *Applied Nonlinear Programming,* Mc.Graw-Hill, Inc., 1972

[23] R.H.W. Hoppe, H.D. Mittelmann, *A multigrid continuation strategy for parameter-dependent variational inequalities,* J. Comput. and Appl. Math. 26 (1989), 35–46.

[24] N.H. Josephy, *Newton's methods for generalized equations,* Tech. Summary Report 1965, Math. res. center, Univ. of Wisconsin-Madison, Madison, Wisconsin, 1979

[25] C. Kanzow, H. Pieper, *Jacobian smoothing methods for nonlinear complementarity problems,* SIAM J. Optim. 9 (1999), 342–373.

[26] C.T. Kelley, *Iterative methods for linear and nonlinear equations,* Frontiers in Applied Mathematics 16, SIAM ,1995

[27] M. Kojima, T. Noma, A. Yoshise, *Global convergence in infeasible interior point algorithms,* Math. Progr. 65 (1994), 43–72.

[28] L. Mathiesen, *An algorithm based on a sequence of linear complementarity problems applied to a Walrasian Equilibrium model: an example,* Math Progr. 37 (1987), 1–18.

[29] J.J. More', M.Y. Cosnard, *Numerical solution of nonlinear equations,* ACM Trans. on Math. Software 5 (1979), 64–85.

[30] S. Pieraccini,*A hybrid Newton-type method for a class of semismooth equations,* preprint 1999

[31] L. Portugal, M. Resende, G. Veiga, J. Judice, *A truncated primal infeasible dual feasible network interior point method,* Networks 35 (2000), 91–108.

[32] D. Ralph, S. Wright, *Superlinear convergence of an Interior-Point Method for monotone variational inequalities*, Complementarity and Variational Problems: State of the Art, SIAM Publications (1997), 345–385.

[33] Y. Saad, M.H. Schultz, *GMRES: a generalized minimal residual method for solving nonsymmetric linear systems*, SIAM J. Sci. Stat. Comput. 6 (1985), 856–869.

[34] D. Sun, R.S. Womersley, *A new unconstrained differentiable merit function for box constrained variational inequality problems and a damped Gauss-newton method*, SIAM J. Optim. 9 (1999), 388–413.

[35] J. Sun, G. Zhao, *Global linear and local quadratic convergence of a long-step adaptive-mode interior-point method for some monotone variational inequality problems*, SIAM J. Optim. 8 (1998), 123–139.

[36] B. Xiao, P.T. Harker, *A nonsmooth Newton method for variational inequalities, I: theory*, Math. Progr. 65 (1994), 151–194.

[37] B. Xiao, P.T. Harker, *A nonsmooth Newton method for variational inequalities, II: numerical results*, Math. Progr. 65 (1994), 195–216.

[38] S. Wright, D. Ralph, *A Superlinear Infeasible-Interior Point Algorithm for Monotone Complementarity Problems*, Mathematics of Operation Research 21 (1996), 815–838.

[39] Y. Ye, *Interior-point algorithms: Theory and analysis*, John Wiley and Sons, New York, NY, 1997

FIXED POINTS IN ORDERED BANACH SPACES AND APPLICATIONS TO ELLIPTIC BOUNDARY–VALUE PROBLEMS

Gabriele Bonanno

D.I.M.E.T., University of Reggio Calabria
via Graziella – Località Feo di Vito,
89100 Reggio Calabria, Italy

Salvatore Marano

D.P.A.U., University of Reggio Calabria
via Melissari – Località Feo di Vito,
89100 Reggio Calabria, Italy
e-mail:marano@dipmat.unict.it

Abstract   The existence of extremal fixed points for a family of increasing, not necessarily continuous, operators in ordered Banach spaces is achieved. The result is then employed to solve elliptic equations on the whole space with discontinuous nonlinearities.

Keywords: Increasing, possibly discontinuous, operators, Minimal and maximal fixed points, Elliptic equations in the whole space, Discontinuous nonlinearities, Strong solutions.

1991 Mathematics Subject Classification. Primary 47H10, 35J60.

1. Introduction.

Many equilibrium problems arising in specific models involve elliptic equations of the second order. As a rule, equations of this type satisfy a maximum principle. From a theoretical point of view, it means

F. Giannessi et al (eds.),
Equilibrium Problems: Nonsmooth Optimization and Variational Inequality Models, 25–31.
© *2001 Kluwer Academic Publishers.*

that the abstract operators, which are in a standard way induced by the above-mentioned problems, come compatible with the natural ordering of the underlying function spaces. Thus, to exploit this additional information, one is lead to the study of (nonlinear) equations in ordered Banach spaces. Once we adopt such a framework, a well-known technique for establishing existence or qualitative results is the method of upper and lower solutions, sometimes combined with the monotone iterative technique; let us mention [1,2,3] as a general reference. Usually, it is required that the relevant operators are increasing and, moreover, completely continuous or at least continuous [1,2,4,5]. However, in the recent monograph [3] the case of discontinuous operators has been extensively investigated and numerous examples discussed in detail.

The main purpose of the present paper is to point out both a fixed point result concerning a class of increasing, possibly discontinuous, operators in ordered Banach spaces and an application to semilinear elliptic equations on the whole space with discontinuous nonlinear terms; see Theorems 2.1 and 3.1, respectively, below. The first of them, although not explicitly stated in [3], follows immediately from the results of [3, Section 1.2]; we also cite [6, Corollary 3.2]. Nevertheless, it improves Theorem 1.1 a) in [4] and contains Theorem 3.1 of [5] as a special case. On the contrary, to the best of our knowledge, the second result is new. An elliptic problem of the type

$$-\Delta u + u = f(x,u) \quad \text{in } \mathbb{R}^n, \qquad \operatorname*{ess\,inf}_{x \in B(0,\rho)} u(x) > 0 \quad \forall \rho > 0,$$

where $n \geq 3$ while $f : \mathbb{R}^n \times \mathbb{R} \to \mathbb{R}$ may be discontinuous even with respect to u, is considered here and solutions $u \in W^{2,p}(\mathbb{R}^n)$, $p > n/2$, are obtained. For a previous, but essentially different, contribution in this setting we refer the reader to [7, Theorem 2.1].

2. Fixed points of increasing functions.

Let $(E, \|\cdot\|)$ be a real Banach space. A set $K \subseteq E$ is called a (positive) cone of E provided

$$K + K \subseteq K, \qquad \alpha K \subseteq K \ \forall \alpha \in \mathbb{R}_0^+, \qquad K \cap (-K) = \{0\}.$$

In this case, for $v, w \in E$, we write $v \leq w$ if and only if $w - v \in K$. When K is also closed, then we say that $(E, \|\cdot\|, K)$ is an ordered Banach space. The cone K is called regular if every sequence $\{v_j\} \subseteq K$ having the properties

$$v_j \leq v_{j+1} \ \forall j \in \mathbb{N}, \qquad v_j \leq w \ \forall j \in \mathbb{N}$$

(for some $w \in E$) is convergent. Evidently, owing to the Monotone Convergence Theorem, each of the usual Lebesgue spaces $(L^p(\Omega), \|\cdot\|_p)$, $1 \leq p < +\infty$, equipped with the positive cone

$$K_p := \{u \in L^p(\Omega) : u(x) \geq 0 \text{ for almost all } x \in \Omega\}, \qquad (1)$$

is an ordered Banach space and the cone K_p comes regular.

Henceforth the symbol $(E, \|\cdot\|, K)$ will denote an ordered Banach space with a regular cone.

Let X be a nonempty subset of E and let $F : X \to E$. We say that the function F is increasing if $v, w \in X$ and $v \leq w$ imply $F(v) \leq F(w)$. A fixed point v of F is called minimal (respectively, maximal) when for every fixed point w of F one has $v \leq w$ (respectively, $w \leq v$). Finally, if $a, b \in E$ and $a \leq b$, we set

$$[a, b] := \{v \in E : a \leq v \leq b\}.$$

Using Theorem 1.2.2 and Corollary 1.2.2 of [3] (see also [6, Corollary 3.2]) produces the following

Theorem 2.1. *Let* $F : [a, b] \to [a, b]$ *be increasing. Then:*

(i_1) *The function* F *has a minimal fixed point* v_* *and a maximal fixed point* v^*.

(i_2) $v_* = \min\{v \in [a, b] : F(v) \leq v\}$ *and* $v^* = \max\{v \in [a, b] : v \leq F(v)\}$.

(i_3) *For continuous* F *one has* $v_* = \lim_{j \to \infty} F^j(a)$ *as well as* $v^* = \lim_{j \to \infty} F^j(b)$.

Proof. If $\{v_j\} \subseteq [a, b]$ is a monotone sequence then, by the assumptions, the sequence $\{F(v_j)\}$ converges. Hence, through Theorem 1.2.2 in [3] we achieve (i_1) and (i_2). Assertion (i_3) is a direct consequence of [3, Corollary 1.2.2]. $\square$

Remark 2.1. The above result improves Theorem 1.1 a) in [4] and contains Theorem 3.1 of [5] as a special case. Moreover, it essentially has the same conclusion of [1, Corollary 6.2], but the hypothesis "F compact" (namely, F continuous and $F([a, b])$ relatively compact) appearing in that result is replaced here by "K regular" alone.

A significant variant of Theorem 2.1 is the following.

Theorem 2.2. *Let* $F : [a, b] \to [a, b]$ *be weakly sequentially continuous besides increasing. Then the minimal and the maximal fixed points of* F *are respectively given by* $v_* = \lim_{j \to \infty} F^j(a)$ *and* $v^* = \lim_{j \to \infty} F^j(b)$.

Proof. Owing to the preceding result, the function F possesses the minimal fixed point v_* and the maximal one v^*. Set $v_j := F^j(a)$, $j \in \mathbb{N}$. One clearly has $\{v_j\} \subseteq [a, b]$. Arguing by induction on j we then infer that the sequence $\{v_j\}$ fulfils the condition $v_j \leq v_{j+1}$ for all $j \in \mathbb{N}$. Therefore, there exists $v \in [a, b]$ such that $\lim_{j \to \infty} v_j = v$. Since $v_{j+1} = F(v_j)$, $j \in \mathbb{N}$, and the function F is weakly sequentially continuous, we obtain $v = F(v)$. Now, applying Corollary 1.2.2 of [3] produces $v_* = \lim_{j \to \infty} F^j(a)$, as claimed. A similar argument leads to $v^* = \lim_{j \to \infty} F^j(b)$. $\square$

Remark 2.2. Simple examples show that for a given possibly increasing function F from $[a, b]$ into itself neither of the following properties implies the other: F is continuous; F is weakly sequentially continuous. Moreover, we feel useful to point out that the relationships between these conditions are fully studied in [8].

Remark 2.3. A careful reading of the above proof reveals that the equality $v_* = \lim_{j \to \infty} F^j(a)$ holds whenever one has $\lim_{j \to \infty} F(v_j) = F(v)$ for every sequence $\{v_j\} \subseteq [a, b]$ satisfying $v_j \leq v_{j+1}$, $j \in \mathbb{N}$, as well as $\lim_{j \to \infty} v_j = v$. By Corollary 5.8.5 in [3], this condition is clearly less restricitive than the weak sequential continuity of F.

Remark 2.4. If E is reflexive and $F : [a, b] \to [a, b]$ comes only weakly sequentially continuous then it still possesses a fixed point; see for instance [9, Theorem 1].

3. Elliptic problems with discontinuous nonlinearities.

Let n be a positive integer, $n \geq 3$, and let $p \in]n/2, +\infty[$. If $k > 0$, we define

$$Lu := -\Delta u + ku, \quad u \in W^{2,p}(\mathbb{R}^n).$$

Some meaningful properties of the operator L are collected in the following

Proposition 3.1.

(j$_1$) $L : W^{2,p}(\mathbb{R}^n) \to L^p(\mathbb{R}^n)$ *is bijective and its inverse* L^{-1} *is continuous.*

(j$_2$) *There exists* $\beta > 0$ *such that* $\|u\|_\infty \leq \beta \|Lu\|_p$ *for all* $u \in W^{2,p}(\mathbb{R}^n)$.

(j₃) *If $v \in L^p(\mathbb{R}^n) \setminus \{0\}$ and $v(x) \geq 0$ almost everywhere in $\mathbb{R}^n$ then*

$$\operatorname*{ess\,inf}_{x \in B(0,\rho)} L^{-1}v(x) > 0 \quad \forall \rho > 0.$$

Proof. Assertions (j₁)–(j₃) are easy consequences of known results. Precisely, (j₁) follows from [10, Proposition 4.3], regarding (j₂) we mention for instance [7, Lemma 2.1] (which clearly holds, with a different constant, provided $p > n/2$), while (j₃) is contained in [11, Proposition 27, p. 635]. □

Next, let $f : \mathbb{R}^n \times \mathbb{R} \to \mathbb{R}$. We say that the function f is sup-measurable when for any measurable $u : \mathbb{R}^n \to \mathbb{R}$ the function $x \mapsto f(x, u(x))$, $x \in \mathbb{R}^n$, enjoys the same property.

Remark 3.1. It is worth pointing out that the conditions $x \mapsto f(x, z)$ measurable for each $z \in \mathbb{R}$ and $z \mapsto f(x, z)$ monotone for almost every $x \in \mathbb{R}^n$ do not guarantee the sup-measurability of f; see [12, p. 218].

We are in a position now to establish the main result of this paper, which gives another contribution to the problem previously investigated by the authors in [7].

Theorem 3.1. *Let $f : \mathbb{R}^n \times \mathbb{R} \to \mathbb{R}$ be sup-measurable. Assume that:*
(a₁) *The function $z \mapsto f(x, z)$, $z \in \mathbb{R}$, is increasing for almost all $x \in \mathbb{R}^n$.*
(a₂) *Setting $a(x) := f(x, 0)$, $x \in \mathbb{R}^n$, one has $a(x) \geq 0$ almost everywhere in $\mathbb{R}^n$ besides $a \neq 0$.*
(a₃) *There exists $r > 0$ such that the function $b(x) := f(x, \beta r)$, $x \in \mathbb{R}^n$, belongs to $L^p(\mathbb{R}^n)$ and its norm in this space is less than or equal to r. Then the equation $Lu = f(x, u)$ in $\mathbb{R}^n$ possesses at least one solution $u \in W^{2,p}(\mathbb{R}^n)$ with the following property:*

$$\operatorname*{ess\,inf}_{x \in B(0,\rho)} u(x) > 0 \quad \forall \rho > 0.$$

Proof. Let $(E, \|\cdot\|, K)$ denote here the ordered Banach space $(L^p(\mathbb{R}^n), \|\cdot\|_p, K_p)$, where the positive cone K_p is given by (1) for $\Omega := \mathbb{R}^n$. Since $a, b \in E$, $a \leq b$, and the operator L is bijective, it makes sense to define

$$F(v)(x) := f(x, L^{-1}(v)(x)), \quad v \in [a, b], \quad x \in \mathbb{R}^n. \tag{2}$$

Obviously, the function $x \mapsto F(v)(x)$, $x \in \mathbb{R}^n$, is measurable. Moreover, using (j₃) of Proposition 3.1 and (a₁) yields $F(v)(x) \le F(w)(x)$ almost everywhere in $\mathbb{R}^n$ whenever $v, w \in [a, b]$ and $v \le w$.

We assert that $a(x) \le F(a)(x)$ for almost all $x \in \mathbb{R}^n$. Indeed, if $\underline{u} := L^{-1}(a)$ then assumption (a₂) and (j₃) of Proposition 3.1 lead to $0 \le \underline{u}$. Hence, through (a₁) we achieve the desired conclusion.

Let us next verify that $F(b)(x) \le b(x)$ almost everywhere in $\mathbb{R}^n$. If $\overline{u} := L^{-1}(b)$ then, combining Proposition 3.1 with (a₃) produces

$$0 \le \overline{u}(x) \le \beta\|L\overline{u}\|_p \le \beta r \quad \text{for almost all } x \in \mathbb{R}^n.$$

In view of (a₁), this implies

$$F(b)(x) = f(x, \overline{u}(x)) \le f(x, \beta r) = b(x) \quad \text{almost everywhere in } \mathbb{R}^n,$$

as claimed.

Summing up, the function $F : [a, b] \to [a, b]$ defined by (2) fulfils all the hypotheses of Theorem 2.1. Therefore, bearing in mind Proposition 3.1, the conclusion follows at once. □

Remark 3.1. Condition (a₁) of the above result can evidently be replaced by the more general one:

(a₁)′ *There exists $\gamma > -k$ such that the function $z \mapsto f(x, z) + \gamma z$, $z \in \mathbb{R}$, is increasing for almost every $x \in \mathbb{R}^n$,*

although in such a case we have to put, in (a₃), $b(x) := f(x, \beta r) + \gamma \beta r$, $x \in \mathbb{R}^n$.

Remark 3.2. Clearly, the function $\underline{u}(x) \equiv 0$ is a lower solution of the problem

$$u \in W^{2,p}(\mathbb{R}^n), \qquad Lu = f(x, u) \quad \text{in } \mathbb{R}^n, \tag{3}$$

whereas the function $\overline{u}(x) \equiv \beta r$ is not an upper solution to (3). However, defining (like in the proof of Theorem 3.1)

$$\underline{u}(x) = L^{-1}(a)(x) \quad \text{besides} \quad \overline{u}(x) = L^{-1}(b)(x), \quad x \in \mathbb{R}^n,$$

gives a lower solution and an upper solution, respectively, of (3) such that $\underline{u} \le \overline{u}$. Thus, concerning existence, the conclusion of Theorem 3.1 could be improved as follows:

Problem (3) has the minimal solution and the maximal solution in the ordered interval $[\underline{u}, \overline{u}]$.

We conclude the paper with the example below, which represents a typical application of Theorem 3.1.

Example 3.1. Let $\alpha > 0$ and let $m, q \in L^p(\mathbb{R}^n)$ be nonnegative, with $q \neq 0$. Denote by $h : \mathbb{R} \to \mathbb{R}$ the Heaviside function. Then there is $u \in W^{2,p}(\mathbb{R}^n)$ such that $Lu(x) = m(x)h(u(x) - \alpha) + q(x)$ almost everywhere in $\mathbb{R}^n$ and, moreover,

$$\operatorname*{ess\,inf}_{x \in B(0,\rho)} u(x) > 0 \quad \forall \rho > 0.$$

Acknowledgements. Work performed under the auspices of G.N.A.F.A. of C.N.R. and partially supported by M.U.R.S.T. of Italy, 1999.

References

[1] H. Amann, *Fixed points equations and nonlinear eigenvalue problems in ordered Banach spaces*, Siam Rev. 18 (1976), 620–709.

[2] E. Zeidler, *Nonlinear Functional Analysis and its Applications I: Fixed-Point Theorems*, Springer-Verlag, New York, 1986.

[3] S. Heikkilä and V. Lakshmikantham, *Monotone Iterative Techniques for Discontinuous Nonlinear Differential Equations*, Monographs Textbooks Pure Appl. Math. 181, Dekker, New York, 1994.

[4] E. Liz, *Monotone iterative techniques in ordered Banach spaces*, Nonlinear Anal. 30 (1997), 5179–5190.

[5] J.J. Nieto, *An abstract monotone iterative technique*, Nonlinear Anal. 28 (1997), 1923–1933.

[6] S. Heikkilä and S. Hu, *On fixed points of multifunctions in ordered spaces*, Appl. Anal. 51 (1993), 115–127.

[7] G. Bonanno and S.A. Marano, *Elliptic problems in $\mathbb{R}^n$ with discontinuous nonlinearities*, Proc. Edinburgh Math. Soc., to appear.

[8] J.M. Ball, *Weak continuity properties of mappings and semigroups*, Proc. Royal Soc. Edinburgh Sect. A 72 (1973/74), 275–280.

[9] O. Arino, S. Gauthier, and J.P. Penot, *A fixed point theorem for sequentially continuous mappings with applications to ordinary differential equations*, Funkcial. Ekvac. 27 (1984), 273–279.

[10] C.A. Stuart, *Bifurcation in $L^p(\mathbb{R}^n)$ for a semilinear elliptic equation*, Proc. London Math. Soc. 57 (1988), 511–541.

[11] R. Dautray and J.L. Lions, *Mathematical Analysis and Numerical Methods for Science and Technology*, vol. 1, Springer-Verlag, Berlin, 1990.

[12] J. Appell, *The superposition operator in function spaces - A survey*, Expo. Math. 6 (1988), 209–270.

A THEOREM OF THE ALTERNATIVE FOR LINEAR CONTROL SYSTEMS

Paolo Cubiotti

Department of Mathematics
University of Messina
Contrada Papardo, Salita Sperone 31
98166 Messina, Italy

Abstract   In this paper we deal with the existence of periodic extremal solutions of a linear control system. We establish an alternative theorem and we present counterexamples to possible improvements.

Keywords: Linear control systems, periodic extremal solutions, tangent cone, relative boundary, fixed points.

1. Introduction.

Consider the linear control system

$$\dot{x} = A(t)\,x + B(t)\,u, \quad u \in U, \tag{1}$$

where $A : [0,T] \to \mathbf{R}^{n \times n}$, $B : [0,T] \to \mathbf{R}^{n \times m}$ are matrix functions (with fixed $T > 0$ and $n, m \in \mathbf{N}$) and $U \subseteq \mathbf{R}^m$ is a nonempty compact set (the restraint set). As usual, we say that an absolutely continuous $x(\cdot) : [0,T] \to \mathbf{R}^n$ is a solution (or a trajectory) of (1) in $[0,T]$ if there exists a measurable $u : [0,T] \to U$ (the controller) such that

$$\dot{x}(t) = A(t)\,x(t) + B(t)\,u(t) \quad \text{a.e. in } [0,T].$$

For $v \in \mathbf{R}^n$, we denote by $K_T(v)$ the *attainable set* at time T starting from v at time 0, namely we put

$$K_T(v) = \{x(T) : x(\cdot) \text{ is a solution of (1) in } [0,T] \text{ and } x(0) = v\}.$$

F. Giannessi et al (eds.),
Equilibrium Problems: Nonsmooth Optimization and Variational Inequality Models, 33–42.
© 2001 *Kluwer Academic Publishers.*

If A and B have entries in $L^1([0,T])$ (henceforth, we shall assume this assumption is satisfied), then (see [1, 2]) for each $v \in \mathbf{R}^n$ the set $K_T(v)$ is nonempty, compact, convex and

$$K_T^*(v) = K_T(v) = J(T)\,v + K_T(0). \tag{2}$$

Here, $J(t)$ is the fundamental matrix solution of the homogeneous system $\dot{x} = A(t)\,x$ (with $J(0) = I$, the identity matrix), and $K_T^*(v)$ denotes the attainable set at time T (starting from v at time 0) for the relaxed problem

$$\dot{x} = A(t)\,x + B(t)\,u, \quad u \in \mathrm{conv}(U) \tag{3}$$

("conv" standing for "convex hull"). By (2), we note that $K_T(v)$ is merely a rigid translate of $K_T(0)$ by the vector $J(T)\,v$. Consequently, the initial point $v \in \mathbf{R}^n$ does not affect the geometry of the attainable set $K_T(v)$, except for the location of this set in $\mathbf{R}^n$. We recall that if the matrix $A(t) = A$ is constant, then $J(t) = e^{At}$ (see [2]).

If $x(\cdot)$ is a trajectory of (1) in $[0,T]$ satisfying $x(T) \in \partial K_T(x(0))$, then $x(\cdot)$ is said an *extremal trajectory*, and the corresponding controller $u : [0,T] \to U$ is said an *extremal controller* [2]. As usual, we say that a solution $x(\cdot)$ of (1) in $[0,T]$ is *periodic* if $x(0) = x(T)$. Of course, the system (1) admits a periodic (resp., extremal and periodic) solution in $[0,T]$ if and only if there is a point $v \in \mathbf{R}^n$ satisfying $v \in K_T(v)$ (resp., $v \in \partial K_T(v)$).

Very recently, in [3], the following existence result for finite-dimensional generalized quasi-variational inequalities was proved, which improved the main result of [4] by a weaker coercivity condition. (Henceforth, we shall write B_r to denote the closed ball of $\mathbf{R}^n$ centered at the origin with radius r, with respect to the Euclidean norm $\|\cdot\|$ of $\mathbf{R}^n$ induced by the usual scalar product $\langle \cdot, \cdot \rangle$).

Theorem 1.1 ([3], Theorem 3.2). *Let $X \subseteq \mathbf{R}^n$ be closed and convex, and let $K \subseteq X$ a non-empty compact set, $\Phi : X \to 2^{\mathbf{R}^n}$ and $\Gamma : X \to 2^X$ two multifunctions. Assume that:*

(i) the set $\Phi(x)$ is non-empty and compact for each $x \in X$, and convex for each $x \in K$, with $x \in \Gamma(x)$; (ii) for each $y \in X - X$, the set $\{x \in X : \inf_{z \in \Phi(x)} \langle z, y \rangle \le 0\}$ is closed; (iii) Γ is a lower semicontinuous multifunction with closed graph and convex values.

Moreover, assume that there exists an increasing unbounded sequence $\{r_k\}$ of positive real numbers, with $X \cap B_{r_1} \ne \emptyset$, such that for each $k \in \mathbf{N}$ one has: (iv) $\Gamma(x) \cap B_{r_k} \ne \emptyset$ for all $x \in X \cap B_{r_k}$; (v) for each $x \in (X \cap B_{r_k}) \setminus K$, with $x \in \Gamma(x)$, one has

$$\sup_{y \in \Gamma(x) \cap B_{r_k}} \ \inf_{z \in \Phi(x)} \ \langle z, x - y \rangle > 0.$$

Then there exists a pair $(\hat{x}, \hat{z}) \in K \times \mathbf{R}^n$ which solves the generalized quasi-variational inequality

$$\hat{x} \in \Gamma(\hat{x}), \quad \hat{z} \in \Phi(\hat{x}) \quad and \quad \langle \hat{z}, \hat{x} - y \rangle \leq 0 \quad for\ all \quad y \in \Gamma(\hat{x}).$$

The peculiarity of Theorem 1.1 resides in the absence of compactness for the set X and in the assumption (ii) for the multifunction Φ. Such an assumption is weaker than the upper semicontinuity condition usually required in the literature. For more details on multifunctions Φ satisfying (ii), we refer to [5, 6, 7, 8] and to the references therein.

By using Theorem 1.1 and a continuous selection theorem by E. Michael (Theorem 3.1''' of [9]), a theorem of alternative concerning the fixed points of lower semicontinuous multifunctions was established in [3]. Such a result, stated as Theorem 1.2 below, gives sufficient conditions for the existence of a fixed point which belongs to the relative boundary of the corresponding value. Here, by "relative boundary" of a set $A \subseteq \mathbf{R}^n$ (denoted by $\partial^* A$) we mean the boundary of A with respect to its affine hull $\mathrm{aff}(A)$.

Theorem 1.2 ([3], Theorem 3.1). *Let $X \subseteq \mathbf{R}^n$ be closed and convex, $\Gamma : X \to 2^X$ a lower semicontinuous multifunction with closed graph and convex values. Assume that there exists an increasing unbounded sequence $\{r_k\}$ of positive real numbers, with $X \cap B_{r_1} \neq \emptyset$, such that for each $k \in \mathbf{N}$ one has:*

$$\Gamma(x) \cap B_{r_k} \neq \emptyset \quad for\ all \quad x \in X \cap B_{r_k}.$$

Then, at least one of the following assertions holds: (a) the set $\{x \in X : x \in \Gamma(x)\}$ is unbounded; (b) there exists $\hat{x} \in X$ such that $\hat{x} \in \partial^* \Gamma(\hat{x})$; (c) for each $w \in \mathbf{R}^n$, there exists $x \in X$, with $x \in \Gamma(x)$, such that

$$\langle w, x - y \rangle = 0 \quad for\ all \quad y \in \Gamma(x).$$

As showed in [3], none of the assertions (a), (b), (c) can be removed from the conclusion of Theorem 1.2.

Henceforth, we denote by $T_{B_r}(x)$ the Bouligand's tangent cone to the closed ball B_r at the point $x \in B_r$ (see, for instance, [3, 10]), namely the set

$$T_{B_r}(x) = \begin{cases} \mathbf{R}^n & \text{if } \|x\| < r \\ \{v \in \mathbf{R}^n : \langle v, x \rangle \leq 0\} & \text{if } \|x\| = r. \end{cases}$$

As an application of Theorem 1.2, the following theorem of alternative concerning periodic solutions of the linear control system (1) was proved in [3].

Theorem 1.3 ([3], Theorem 4.2). *Let the matrix functions $A : [0, T] \to \mathbf{R}^{n \times n}$ and $B : [0, T] \to \mathbf{R}^{n \times m}$ have entries in $L^\infty([0, T])$. Assume that there exists an increasing unbounded sequence $\{r_k\}$ of positive real numbers such that for each $k \in \mathbf{N}$ one has*

$$\{u \in \operatorname{conv}(U) : A(t)\, x + B(t)\, u \in T_{B_{r_k}}(x)\} \neq \emptyset \qquad (4)$$

for all $(t, x) \in [0, T] \times B_{r_k}$.

Then, at least one of the following assertions holds:

(A) the set $\{v \in \mathbf{R}^n : v \in K_T(v)\}$ is unbounded; (B) there exists $x^* \in \mathbf{R}^n$ such that $x^* \in \partial^* K_T(x^*)$; (C) there exists $x^* \in \mathbf{R}^n$ such that $K_T(x^*) = \{x^*\}$.

It is important to point out (see Remark 4.1 of [3]) that Theorem 1.3 is significant in the case $\|J(T)\|^* \geq 1$, where $\|J(T)\|^*$ denotes the operator norm of $J(T)$. In fact, when $\|J(T)\|^* < 1$, by a different approach it can be proved (without assuming (4)) that at least one of conclusions (B) and (C) must hold, while (A) never holds. Conversely, when $\|J(T)\|^* \geq 1$, Theorem 1.3 is no longer true without assuming (4), and none of the assertions (A), (B), (C) can be removed from the conclusion.

At this point, note that if $\operatorname{int}K_T(0)$ (the interior of $K_T(0)$) is non-empty, then by (2) Theorem 1.3 gives the following alternative:

(A)′ the set $\{v \in \mathbf{R}^n : v \in K_T(v)\}$ is unbounded; (B)′ there exists $x^* \in \mathbf{R}^n$ such that $x^* \in \partial K_T(x^*)$.

The aim of this paper is to show that if the situation $\operatorname{int}(K_T(0))$ occurs (this happens, for instance, if the system (1) is *normal* and the set U contains more than one point [2]), then, by using a different approach, Theorem 1.3 can be meaningfully improved. That is, the tangential condition (4) can be weakened and simplified (the sequence $\{r_k\}$ being replaced by a unique $r > 0$) and a stronger conclusion can be obtained. That is, we want to prove the following result:

Theorem 1.4. *Let the matrix functions $A : [0, T] \to \mathbf{R}^{n \times n}$ and $B : [0, T] \to \mathbf{R}^{n \times m}$ have entries in $L^\infty([0, T])$. Assume that: (i) $\operatorname{int} K_T(0) \neq \emptyset$; (ii) there exists $r > 0$ such that*

$$E(t, x) := \{u \in \operatorname{conv}(U) : A(t)\, x + B(t)\, u \in T_{B_r}(x)\} \neq \emptyset$$

for all $(t, x) \in [0, T] \times B_r$.

Then, at least one of the following assertions holds:

(A)″ $x \in K_T(x)$ *for all* $x \in \mathbf{R}^n$; (B)″ *there exists* $x^* \in \mathbf{R}^n$ *such that* $x^* \in \partial K_T(x^*)$.

Moreover, if $J(T) \neq I$ *(the identity matrix), then* (A)″ *does not hold.*

In order to prove Theorem 1.4, we first establish a new theorem of alternative for the fixed points of lower semicontinuous multifunctions (Theorem 2.1 below). Then, we prove Theorem 1.4 by an argument similar to the one used in [3]. Examples of applications of Theorem 1.4 and also counterexamples to possible improvements are presented.

2. The proof of theorem 1.4.

Firstly, we fix some notations. If $S \subseteq V \subseteq \mathbf{R}^n$, we shall denote by $\partial_V S$ the boundary of S in V, by $\mathrm{int}_V S$ the interior of S in V, by $\mathrm{aff}(S)$ the affine hull of S, by ∂S the boundary of S in $\mathbf{R}^n$, by $\partial^* S$ the relative boundary of S (that is, the boundary of S in $\mathrm{aff}(S)$), and by $\mathrm{ri}(S)$ the relative interior of S (that is, the interior of S in $\mathrm{aff}(S)$). Finally, if $x \in \mathbf{R}^n$ and $r > 0$, we put $B(x, r) := \{v \in \mathbf{R}^n : \|x - v\| \leq r\}$ (as before, we put $B_r := B(0, r)$). For the basic definitions and results about multifunctions, we refer to [11, 12]. If $X \subseteq \mathbf{R}^n$ and $F : X \to 2^X$ is a multifunction, we denote by $\mathrm{Fix}(F)$ the set of all fixed points of F, that is

$$\mathrm{Fix}(F) := \{x \in X : x \in F(x)\}.$$

The following result will be a key tool in the proof of Theorem 1.4.

Theorem 2.1. *Let X be a non-empty closed connected subset of $\mathbf{R}^n$, and let $F : X \to 2^X$ be a lower semicontinuous multifunction with closed graph and convex values. Moreover, assume that $\mathrm{aff}(F(x)) = \mathrm{aff}(X)$ for all $x \in X$. Then, at least one of the following assertions holds:*

(A)‴ $\mathrm{Fix}(F) = \emptyset$; (B)‴ $\mathrm{Fix}(F) = X$; (C)‴ *there exists $\hat{x} \in X$ such that $\hat{x} \in \partial^* F(\hat{x})$.*

Proof. Assume that conclusions (A)‴ and (B)‴ do not hold. Therefore, $\mathrm{Fix}(F) \neq \emptyset$ and $\mathrm{Fix}(F) \neq X$. Thus, the connectedness of X implies that $\partial_X(\mathrm{Fix}(F)) \neq \emptyset$. Let $\hat{x}$ be any point in $\partial_X(\,fix)$. Since our assumptions imply that $\mathrm{Fix}(F)$ is closed, we get $\hat{x} \in \mathrm{Fix}(F)$. We claim that $\hat{x} \in \partial^* F(\hat{x})$. On the contrary, assume $\hat{x} \in \mathrm{ri}(F(\hat{x}))$. In this occurrence, all the assumptions of Proposition 3 of [4] are satisfied by taking $S = X$,

$V := \mathrm{aff}(X) = \mathrm{aff}(F(\hat{x}))$, and $s_0 = y_0 = \hat{x}$. Consequently, there exist a neighborhood H of $\hat{x}$ in X and $r > 0$ such that

$$B(\hat{x}, r) \cap V \subseteq F(x) \quad \text{for all} \quad x \in H. \tag{5}$$

Now, take $W = H \cap B(\hat{x}, r)$. By (5) we easily get $W \subseteq \mathrm{Fix}(F)$. Since W is a neighborhood of $\hat{x}$ in X, we obtain

$$\hat{x} \in \mathrm{int}_X(\mathrm{Fix}(F)),$$

and this is a contradiction since $\hat{x} \in \partial_X(\mathrm{Fix}(F))$. Such a contradiction implies that $\hat{x} \in \partial^* F(\hat{x})$, as desired. $\triangle$

Remark 2.2. It is worth pointing out that Theorem 2.1 is no longer true if the assumption $\mathrm{aff}(F(x)) = \mathrm{aff}(X)$ is deleted from the hypotheses. To see this, take $X = \mathbf{R}^2$, and

$$F(x, y) = \{-x\} \times [y - 1, y + 1] \quad \text{for each} \quad (x, y) \in \mathbf{R}^2.$$

It is a simple matter to check that F is lower semicontinuous with closed graph and non-empty convex values. Moreover, for each $(x, y) \in \mathbf{R}^2$ one has

$$\mathrm{aff}(F(x, y)) = \{-x\} \times \mathbf{R} \neq \mathrm{aff}(X) = \mathbf{R}^2.$$

Finally, one has

$$\mathrm{Fix}(F) = \{(0, y) : y \in \mathbf{R}\},$$

and also

$$(0, y) \in \mathrm{ri}(F(0, y)) \quad \text{for all} \quad y \in \mathbf{R},$$

hence conclusions (A)''', (B)''' and (C)''' do not hold.

Proof of Theorem 1.4. Of course, there is no loss of generality in assuming that A and B have bounded entries in $[0, T]$. Now we want to apply Theorem 2.1 with $X = \mathbf{R}^n$ and $F(x) = K_T^*(x)$. As we mentioned before, our assumptions imply that each set $K_T^*(x)$ is non-empty, compact and convex. Moreover, by (2) and Proposition 1.4.14 of [11] the multifunction K_T^* is both lower and upper semicontinuous (in particular, it has closed graph). Finally, by assumption (i) and by (2) we get $\mathrm{aff}(K_T^*(x)) = \mathbf{R}^n$ for all $x \in \mathbf{R}^n$. Thus, all the assumptions of Theorem 2.1 are satisfied. We now claim that $\mathrm{Fix}(K_T^*) \neq \emptyset$. To prove our claim, define the multifunction $\Phi : [0, T] \times \mathbf{R}^n \to 2^{\mathbf{R}^n}$ by setting

$$\Phi(t, x) = \bigcup_{u \in \mathrm{conv}(\Omega)} \{A(t) x + B(t) u\}.$$

Let $M > 0$ be such that U is contained in the closed ball of $\mathbf{R}^m$ with radius M centered at the origin. Observe what follows:

(a) each set $\Phi(t, x)$ is non-empty, compact and convex.

(b) the multifunction $\Phi(\cdot, x)$ is measurable for each $x \in \mathbf{R}^n$ (see Theorem 8.2.8 of [11]).

(c) for all $(t, x, y) \in [0, T] \times \mathbf{R}^n \times \mathbf{R}^n$, one has

$$d_H(\Phi(t, x), \Phi(t, y)) \le \|x - y\| \sup_{t \in [0,T]} \|A(t)\|^*,$$

where d_H is the Hausdorff distance and $\|A(t)\|^*$ is the operator norm of $A(t)$ as a linear function from $\mathbf{R}^n$ into itself.

(d) for each $(t, x) \in [0, T] \times \mathbf{R}^n$ one has

$$\Phi(t, x) \subseteq (1 + \|x\|) B(0, K),$$

with

$$K = \max \{ \sup_{t \in [0,T]} \|A(t)\|^*, \, M \cdot \sup_{t \in [0,T]} \|B(t)\|^* \}.$$

(e) $\Phi(t, x) \cap T_{B_r}(x) \ne \emptyset$ for each $(t, x) \in [0, T] \times B_r$ (by (ii)).

Consequently, by Theorem 7.1 of [13], for each fixed $v \in B_r$ there exists an absolutely continuous $x(\cdot) : [0, T] \to \mathbf{R}^n$ satisfying

$$\begin{cases} \dot{x}(t) \in \Phi(t, x(t)) & \text{a.e. in } [0, T] \\ x(0) = v & \\ x(t) \in B_r & \text{for all } t \in [0, T]. \end{cases} \tag{6}$$

By Filippov's theorem (Theorem 8.2.10 of [11]), the function $x(\cdot)$ is a solution of (3) in $[0, T]$, hence by (6) we get

$$K_T^*(v) \cap B_r \ne \emptyset.$$

Therefore, we have proved that

$$K_T^*(v) \cap B_r \ne \emptyset \quad \text{for all} \quad v \in B_r.$$

Thus, by the Fan-Kakutani fixed-point theorem (Theorem 1 of [14]) the multifunction $v \in B_r \to K_T^*(v) \cap B_r$ admits a fixed point, hence $\text{Fix}(K_T^*) \ne \emptyset$, as claimed. Consequently, by Theorem 2.1 and by (2) we get the first part of our conclusion (the alternative). To prove the remaining part, assume that $(A)''$ holds. By (2), we get

$$x - J(T) x \in K_T(0) \quad \text{for all} \quad x \in \mathbf{R}^n.$$

Since $K_T(0)$ is compact, this implies that the linear operator $x \mapsto x - J(T)\, x$ is identically equal to the origin of $\mathbf{R}^n$, hence $J(T) := I$. This completes the proof. $\triangle$

Remark 2.3. It is immediate to check that the alternative in the conclusion of Theorem 1.4 can be rewritten in the following equivalent way. *"One and only one of the following assertions holds : (A)* $x \in \mathrm{int}\, K_T(x)$ for all $x \in \mathbf{R}^n$; (B)* there exists $x^* \in \mathbf{R}^n$ such that $x^* \in \partial K_T(x^*)$."*

We now show a situation where Theorem 1.4 applies, while it is not possible to find a sequence $\{r_k\}$ satisfying the assumption of Theorem 1.3 (from now on, we use subscripts to denote component of vectors).

Example 2.4. Consider the linear control system (with $T > 0$)

$$\begin{cases} \dot{x}_1 = \tfrac{1}{2} x_1 + u_1 \\[2mm] \dot{x}_2 = \tfrac{1}{2} x_2 + u_2, \qquad u \in [-1, 1]^2. \end{cases}$$

By standard arguments ([1, 2]) it is easy to check that $J(T) = e^{T/2}\, I$ and

$$K_T(x) = e^{T/2}\, x + [-2\,(e^{T/2} - 1),\ 2\,(e^{T/2} - 1)]^2 \quad \text{for all} \quad x \in \mathbf{R}^2.$$

If $x \in \mathbf{R}^2$, with $\|x\| = 1$, then one has $-\tfrac{1}{2} x \in E(t, x)$ for each $t \in [0, T]$. Therefore, the assumptions of Theorem 1.4 are satisfied with $r = 1$. Conversely, Theorem 1.3 can not be applied. To see this, observe that if $r > 4$ and

$$x = \left(\frac{r}{\sqrt{2}},\ \frac{r}{\sqrt{2}} \right)$$

(hence, $\|x\| = r$), then for all $t \in [0, T]$ one has

$$\Phi(t, x) := \bigcup_{u \in U} \{ A(t)\, x + B(t)\, u \} = \frac{1}{2} x + [-1, 1]^2 = \left[\frac{r}{2\sqrt{2}} - 1,\ \frac{r}{2\sqrt{2}} + 1 \right]^2.$$

Since

$$\frac{r}{2\sqrt{2}} - 1 > 0,$$

then $\Phi(t, x)$ is contained in the (open) positive orthant, hence it can not intersect $T_{B_r}(x)$.

Remark 2.5. Taking into account Remark 4.1 of [3], it is worth pointing out that in Example 2.4 is $\|J(T)\|^* = e^{T/2} > 1$.

Remark 2.6. Example 4.2 of [3] shows that Theorem 1.4 is no longer true if condition (ii) is deleted from the statement, while Example 4.3 of [3] provides a situation where the assumptions of Theorem 1.4 are satisfied and conclusion (B)'' does not hold. On the contrary, the next example provides a situation where the assumptions of Theorem 1.4 are satisfied and conclusion (A)'' does not hold.

Example 2.7. Consider the linear control system (with $T > 0$)

$$\begin{cases} \dot{x}_1 = -x_1 + u_1 \\ \dot{x}_2 = -x_2 + u_2, \end{cases} \quad u \in [-1,1]^2.$$

One has $0 \in E(t,x)$ for any $x \in \mathbf{R}^2$, with $\|x\| = 1$, and any $t \in [0,T]$. Therefore, assumption (ii) of Theorem 1.4 is satisfied with $r = 1$. Also, one has $J(T) = e^{-T} I$ and

$$K_T(x) = e^{-T} x + [-(1 - e^{-T}), (1 - e^{-T})]^2 \quad \text{for all} \quad x \in \mathbf{R}^2,$$

hence assumption (i) of Theorem 1.4 is satisfied. Moreover, since $J(T) \neq I$, conclusion (A)'' can not hold.

References

[1] H. Hermes, JP. Lasalle, *Functional Analysis and Time Optimal Control*, Academic Press, New York, 1969

[2] E.B. Lee, L. Markus, *Foundations of Optimal Control Theory*, John Wiley and Sons, New York, 1967.

[3] P. Cubiotti, *Application of quasi-variational inequalities to linear control systems*, J. Optim. Theory Appl. 89 (1996), 101–113.

[4] P. Cubiotti *An existence theorem for generalized quasi-variational inequalities*, Set-Valued Anal. 1 (1993), 81–87.

[5] P. Cubiotti, N.D. Yen, *A result related to Ricceri's conjecture on generalized quasi-variational inequalities*, Arch. Math. 69 (1997), 507–514.

[6] M. De Luca, *Generalized Quasi-Variational Inequalities and Traffic Equilibrium Problem*. In Variational Inequalities and Network Equilibrium Problems, F. Giannessi and A. Maugeri eds., Plenum Press, New York, 1995.

[7] M. De Luca M, A. Maugeri, *Discontinuous Quasi-Variational Inequalities and Applications to Equilibrium Problems*, In Nonsmooth Optimization. Methods and Applications, F. Giannessi ed., Gordon and Breach, 1992.

[8] B. Ricceri, *Basic Existence Theorems for Generalized Variational and Quasi-Variational Inequalities*, In Variational Inequalities and Network Equilibrium Problems, F. Giannessi and A. Maugeri eds., Plenum Press, New York, 1995.

[9] E. Michael *Continuous selections I*, Ann. Math. 63 (1956), 361–382.

[10] JP. Aubin, A. Cellina, *Differential Inclusions*, Springer-Verlag, Berlin, 1984.

[11] JP. Aubin, H. Frankowska, *Set-Valued Analysis*, Birkhäuser, Boston, 1990.

[12] E. Klein, A.C. Thompson, *Theory of Correspondences*, John Wiley and Sons, New York, 1984.

[13] C. Castaing, M. Moussaoui, A. Syam, *Multivalued differential equations on closed convex sets in Banach spaces*, Set-Valued Anal. 1 (1994), 329–3531.

[14] K. Fan *Fixed point and minimax theorems in locally convex topological linear spaces*, Proc. Nat. Acad. Sc. USA 38 (1952), 121–126.

VARIATIONAL INEQUALITIES FOR STATIC EQUILIBRIUM MARKET. LAGRANGEAN FUNCTION AND DUALITY

Patrizia Daniele

Dipartimento di Matematica

Università di Catania

Viale A. Doria, 6 - 95125 Catania

e-mail: daniele@dipmat.unict.it

Abstract   The spatially distributed economic market is considered in the case of excess of the supplies and of the demands. The equilibrium conditions that describe this "disequilibrium" model are expressed in terms of Variational Inequalities for which the existence of solutions is provided by recent existence results. Mainly, the Lagrangean theory for the model is studied and as an interesting consequence, we obtain that Lagrangean variables provide the excesses of supply and of demand. Hence, the Lagrangean theory allows us to obtain the most important data of the economic problem, exactly the excesses. Also a computational procedure is presented based on the direct method (see [6], [7]) that in this case reveals itself to be effective.

Keywords: supply excess, demand excess, Lagrangean function, direct method.

1. Introduction.

In this paper we consider the spatial price equilibrium problem in the case of excess of the supplies and of the demands. The equilibrium conditions that describe this "disequilibrium" model in the case of price formulation are expressed in terms of Variational Inequalities, for which the existence of solution is provided by recent existence results.

F. Giannessi et al (eds.),

Equilibrium Problems: Nonsmooth Optimization and Variational Inequality Models, 43–58.

© *2001 Kluwer Academic Publishers.*

This more general and realistic model, which generalizes the classical spatial price equilibrium problems formulated by A. A. Cournot, A. C. Pigou, S. Enke, P. A. Samuelson, T. Takamaya and G. G. Judge, A. Nagurney, adopts, unchanged, the concept of equilibrium, namely that the demand price is equal to the supply price plus the cost of transportation, if there is trade between the pair of supply and demand markets.

It is worth remarking that this classical equilibrium principle follows the user–optimized approach instead of the system–optimized one and that these different approaches have been noted first by A. C. Pigou.

Moreover S. Enke established the connection between spatial price equilibrium problems and electronic circuit networks and T. Takayama and G. G. Judge presented a lot of potential applications, for example to study problems in agriculture, in finance, in mineral economics, etc.

In this paper, we deal with the Lagrangean theory of the model and, as an interesting consequence, we obtain that Lagrangean variables provide the excesses of supply and of demand, which represent important features of the economic problem. Moreover we provide a dual formulation of the equilibrium conditions where among the dual variables there are the excesses of supply and of demand. These results have a certain importance since the price model can never be symmetric (see [1]) and hence can never be cast into an equivalent convex minimization problem in the usual sense, namely without an appeal to some kind of Gap function or to the Lagrangean theory.

Also a computational procedure is presented, based on the direct method (see [6], [7] also for an evaluation of the method) that in this case reveals itself to be effective.

Some computational examples conclude the paper.

Now let us present in details notations, assumptions and results.

We consider a single commodity that is produced at n supply markets and consumed at m demand markets. There is a total supply g_i in each supply market i, where $i = 1, 2, \ldots, n$ and a total demand f_j in each demand market j, where $j = 1, 2, \ldots, m$. Since the markets are spatially separated, x_{ij} units of commodity are transportated from i to j.

If we consider the excess supply s_i and the excess demand t_j, we must have

$$g_i = \sum_{j=1}^{m} x_{ij} + s_i \qquad i = 1, \ldots, n \qquad (1.1)$$

$$f_j = \sum_{i=1}^{n} x_{ij} + t_j \qquad j = 1, \ldots, m. \qquad (1.2)$$

We associate with each supply market i a supply price p_i and with each demand market j a demand price q_j. A fixed minimal supply price $\underline{p}_i \geq 0$ (price floor) for each supply market i and a fixed maximum demand price $\bar{q}_j > 0$ (price ceiling) for each demand market j are given. Moreover the transportation from i to j gives rise to unit costs π_{ij} and upper bounds $\bar{x}_{ij} > 0$ for the transportation fluxes are included.

Grouping the introduced quantities in vectors, we have the total supply vector $g \in \mathbb{R}^n$, the total supply price vector $p \in \mathbb{R}^n$, the total demand vector $f \in \mathbb{R}^m$, the total demand price vector $q \in \mathbb{R}^m$, the flux vector $x \in \mathbb{R}^{nm}$, the unit flux cost vector $\pi \in \mathbb{R}^{nm}$. Then the feasible set for the vectors $u = (p, q, x)$ is given by the product set

$$\mathbb{K} = \prod_{i=1}^{n} [\underline{p}_i, \infty[\times \prod_{j=1}^{m} [0, \bar{q}_j] \times \prod_{i=1}^{n} \prod_{j=1}^{m} [0, \bar{x}_{ij}] = \mathbb{K}_1 \times \mathbb{K}_2 \times \mathbb{K}_3.$$

As in unconstrained market equilibria (see [1], [2]), we assume that we are given the functions

$$g : \mathbb{K}_1 \to \mathbb{R}^n, \quad f : \mathbb{K}_2 \to \mathbb{R}^m, \quad \pi : \mathbb{K}_3 \to \mathbb{R}^{nm}$$

which express:

the dependence of the total supply g on the price p;

the dependence of the total demand f on the price q;

the dependence of trasportation unit cost π on the flux vector x.

According to perfect equilibrium, the economic market conditions governing the "disequilibrium" model take the following form (see [1], [8], [9]):

Definition 1.1 *A vector* $u = (p, q, x) \in \mathbb{K}$ *is a market equilibrium if*

$$s_i > 0 \Rightarrow p_i = \underline{p}_i, \qquad p_i > \underline{p}_i \Rightarrow s_i = 0 \qquad i = 1, \ldots, n; \qquad (1.3)$$

$$t_j > 0 \Rightarrow q_j = \bar{q}_j, \qquad q_j < \bar{q}_j \Rightarrow t_j = 0 \qquad j = 1, \ldots, m; \qquad (1.4)$$

$$p_i + \pi_{ij} \begin{cases} \geq q_j & se & x_{ij} = 0 \\ = q_j & se & 0 < x_{ij} < \bar{x}_{ij} \\ & & i = 1, \ldots, n \quad j = 1, \ldots, m \\ \leq q_j & se & x_{ij} = \bar{x}_{ij}. \end{cases} \qquad (1.5)$$

As it is proved in [5] and in [3], such equilibrium conditions are equivalent to a Variational Inequality. In fact the following result holds (see [5], [3]):

Theorem 1.1 *Suppose that for each $i = 1, 2, \ldots, n$ and $j = 1, 2, \ldots, m$ there holds*

$$q_j = 0 \Rightarrow f_j(q) \geq 0 \qquad x_{ij} > 0 \Rightarrow \pi_{ij}(x) > 0. \qquad (1.6)$$

Then $u^* = (p^*, q^*, x^*) \in \mathbb{K}$ satisfies the market equilibrium conditions (1.3)–(1.5) if and only if u is a solution to

$$\ll v(u^*), u - u^* \gg \; = \; \sum_{i=1}^{n} \Big(g_i(p^*) - \sum_{j=1}^{m} x_{ij}^*\Big)(p_i - p_i^*) - \sum_{j=1}^{m} \Big(f_j(q^*) +$$

$$- \sum_{i=1}^{n} x_{ij}^*\Big)(q_j - q_j^*) \; + \; \sum_{i=1}^{n}\sum_{j=1}^{m} \Big(p_i^* + \pi_{ij}(x^*) - q_j^*\Big)(x_{ij} - x_{ij}^*) \geq 0$$

$$\forall u \; = \; (p, q, x) \in \mathbb{K}. \qquad (1.7)$$

In (1.6) $v(u)$ is the operator $v : \mathbb{K} \to \mathbb{R}^{n+m+nm}$ defined by setting

$$v(u) = \Big((g_i(p) - \sum_{j=1}^{m} x_{ij})_{i=1,\ldots,n}, \; -(f_j(q) - \sum_{i=1}^{n} x_{ij})_{j=1,\ldots,m},$$

$$(p_i + \pi_{ij}(x) - q_{ij})_{\substack{i=1,\ldots,n \\ j=1,\ldots,m}}\Big).$$

Also in [5] and [3] existence results for the Variational Inequality (1.6) can be find, under suitable assumptions of pseudomonotonicity and of hemicontuinity.

We are concerned with the Lagrangean theory associated to the Variational Inequality (1.6), using the following function

$$\mathcal{L}(u, \alpha, \beta, \gamma) = \ll v(u^*), u - u^* \gg - \sum_{i=1}^{n} \alpha_i(p_i - \underline{p}_i) - \sum_{j=1}^{m} \beta_j(\overline{q}_j - q_j) -$$

$$- \sum_{i=1}^{n}\sum_{j=1}^{m} \gamma_{ij}(\overline{x}_{ij} - x_{ij}),$$

where $u^* \in \mathbb{K}$.

In particular we prove the following theorem:

Theorem 1.2 $u^* = (p^*, q^*, x^*) \in \mathbb{K}$ *is a solution of (1.6) if and only if there exist* $\overline{\alpha} \in \mathbb{R}^n$, $\overline{\beta} \in \mathbb{R}^m$, $\overline{\gamma} \in \mathbb{R}^{nm}$ *such that:*

$$g_i(p^*) - \sum_{j=1}^{m} x_{ij}^* + \overline{\alpha}_i = 0, \quad -f_j(q^*) + \sum_{i=1}^{n} x_{ij}^* - \overline{\beta}_j = 0,$$

$$(1.8)$$

$$p_i^* + \pi_{ij}(x^*) - q_j^* - \overline{\gamma}_{ij} = 0$$

$$\overline{\alpha}_i(\underline{p}_i - p_i^*) = 0, \quad \overline{\beta}_j(q_j^* - \overline{q}_j) = 0, \quad \overline{\gamma}_{ij}(x_{ij}^* - \overline{x}_{ij}) = 0 \qquad (1.9)$$

$$\overline{\alpha}_i, \quad \overline{\beta}_j, \quad \overline{\gamma}_{ij} \geq 0 \quad \forall i = 1, 2, \ldots, n, \quad \forall j = 1, 2, \ldots, m \qquad (1.10)$$

$$\underline{p}_i - p_i^* \leq 0, \quad q_j^* - \overline{q}_j \leq 0, \quad x_{ij}^* - \overline{x}_{ij} \leq 0$$

$$(1.11)$$

$$\forall i = 1, 2, \ldots, n, \quad \forall j = 1, 2, \ldots, m.$$

Remark 1. The Lagrangean variables $\overline{\alpha}_i$, $\overline{\beta}_j$ play a very important role in the theory of the spatially distributed economic market. In fact the supply excess s_i and the demand excess t_j satisfy the same relations

$$s_i(\underline{p}_i - p_i) = 0 \quad \text{and} \quad t_j(q_j - \overline{q}_j) = 0.$$

Hence, by the relations (1.7), (1.3), (1.4), $\overline{\alpha}_i$, and $\overline{\beta}_j$ coincide with $-s_i$ and $-t_j$ respectively. Therefore the Lagrangean theory provides directly these important features of the market.

Moreover we characterize the solution of the Variational Inequality as a saddle point of $\mathcal{L}$ and we prove the following theorem:

Theorem 1.3 $u^* = (p^*, q^*, x^*) \in \mathbb{K}$ *is a solution of (1.6) if and only if* $(u^*, \overline{\alpha}, \overline{\beta}, \overline{\gamma}) \in \mathbb{R}^{n+m+nm} \times \mathbb{R}^n \times \mathbb{R}^m \times \mathbb{R}^{nm}$ *is a saddle point of the Lagrangean function*

$$\mathcal{L}(u, \alpha, \beta, \gamma) = \ll v(u^*), u - u^* \gg - \sum_{i=1}^{n} \alpha_i(p_i - \underline{p}_i) - \sum_{j=1}^{m} \beta_j(\overline{q}_j - q_j) +$$

$$- \sum_{i=1}^{n} \sum_{j=1}^{m} \gamma_{ij}(\overline{x}_{ij} - x_{ij}).$$

Finally, we present a computational procedure using the direct method of [6] which in this case reveals itself to be effective.

2. Proof of theorem 1.2.

In order to obtain such a result, we take into account Theorem 4.2 in [4], which can be applied to problem (1.6) written in the form

$$\min_{u \in \mathbb{K}} \ll v(u^*), u - u^* \gg = 0 \qquad (2.1)$$

and to the Lagrangean function

$$\mathcal{L}(u, \alpha, \beta, \gamma) = \ll v(u^*), u - u^* \gg - \sum_{i=1}^{n} \alpha_i(p_i - \underline{p}_i) - \sum_{j=1}^{m} \beta_j(\overline{q}_j - q_j) +$$

$$- \sum_{i=1}^{n} \sum_{j=1}^{m} \gamma_{ij}(\overline{x}_{ij} - x_{ij}).$$

Slater condition on the constraints is obviously satisfied and, then, by virtue of such a theorem $u^* = (p^*, q^*, x^*) \in \mathbb{K}$ is a solution of the problem (2.1) and hence of problem (1.6), if and only if

$$g_i(p^*) - \sum_{j=1}^{m} x_{ij}^* + \overline{\alpha}_i = 0, \quad -f_j(q^*) + \sum_{i=1}^{n} x_{ij}^* - \overline{\beta}_j = 0,$$

$$(2.2)$$

$$p_i^* + \pi_{ij}(x^*) - q_j^* - \overline{\gamma}_{ij} = 0$$

$$\sum_{i=1}^{n} \overline{\alpha}_i(p_i^* - \underline{p}_i) = 0, \quad \sum_{j=1}^{m} \overline{\beta}_j(\overline{q}_j - q_j^*) = 0, \quad \sum_{i=1}^{n} \sum_{j=1}^{m} \overline{\gamma}_{ij}(\overline{x}_{ij} - x_{ij}^*) = 0$$

$$(2.3)$$

$$\overline{\alpha}_i, \quad \overline{\beta}_j, \quad \overline{\gamma}_{ij} \geq 0 \quad \forall i = 1, 2, \ldots, n, \quad \forall j = 1, 2, \ldots, m \qquad (2.4)$$

$$\underline{p}_i - p_i^* \leq 0, \quad q_j^* - \overline{q}_j \leq 0, \quad x_{ij}^* - \overline{x}_{ij} \leq 0$$

$$(2.5)$$

$$\forall i = 1, 2, \ldots, n, \quad \forall j = 1, 2, \ldots, m$$

Since $\overline{\alpha}_i \geq 0$, $\underline{p}_i - p_i^* \leq 0$ and the first part of (2.3) holds, it must be

$$\overline{\alpha}_i(\underline{p}_i - p_i^*) = 0.$$

Analogously, since $\overline{\beta}_j \geq 0$, $q_j^* - \overline{q}_j \leq 0$ and the second part of (2.3) holds, it will be

$$\overline{\beta}_j(q_j^* - \overline{q}_j) = 0.$$

Finally, since $\overline{\gamma}_{ij} \geq 0$, $x^*_{ij} - \overline{x}_{ij} \leq 0$ and the third part of (2.3) holds, we have

$$\overline{\gamma}_{ij}(x^*_{ij} - \overline{x}_{ij}) = 0.$$

Then the assertion of Theorem 1.2 is achieved. ∎

3. Proof of theorem 1.3.

That $(u^*, \overline{\alpha}, \overline{\beta}, \overline{\gamma})$ is a saddle point of the Lagrangean function means:

$$\mathcal{L}(u^*, \alpha, \beta, \gamma) \leq \mathcal{L}(u^*, \overline{\alpha}, \overline{\beta}, \overline{\gamma}) \leq \mathcal{L}(u, \overline{\alpha}, \overline{\beta}, \overline{\gamma}) \tag{3.1}$$

$$\forall u \in \mathbb{R}^{n+m+nm}, \quad \forall \alpha \in \mathbb{R}^n_+, \quad \forall \beta \in \mathbb{R}^m_+, \quad \forall \gamma \in \mathbb{R}^{nm}_+,$$

that is:

$$-\sum_{i=1}^{n} \alpha_i(p^*_i - \underline{p}_i) - \sum_{j=1}^{m} \beta_j(\overline{q}_j - q^*_j) - \sum_{i=1}^{n}\sum_{j=1}^{m} \gamma_{ij}(\overline{x}_{ij} - x^*_{ij}) \leq$$

$$\leq \quad -\sum_{i=1}^{n} \overline{\alpha}_i(p^*_i - \underline{p}_i) - \sum_{j=1}^{m} \overline{\beta}_j(\overline{q}_j - q^*_j) - \sum_{i=1}^{n}\sum_{j=1}^{m} \overline{\gamma}_{ij}(\overline{x}_{ij} - x_{ij}) \leq$$

$$\tag{3.2}$$

$$\leq \quad \ll v(u^*), u - u^* \gg -\sum_{i=1}^{n} \overline{\alpha}_i(p_i - \underline{p}_i) +$$

$$-\sum_{j=1}^{m} \overline{\beta}_j(\overline{q}_j - q_j) - \sum_{i=1}^{n}\sum_{j=1}^{m} \overline{\gamma}_{ij}(\overline{x}_{ij} - x_{ij}).$$

Let us prove the sufficient condition and so let $(u^*, \overline{\alpha}, \overline{\beta}, \overline{\gamma}) \in \mathbb{R}^{n+m+nm}$ $\times \mathbb{R}^n \times \mathbb{R}^m \times \mathbb{R}^{nm}$ be saddle point of the Lagrangean function. In the left-hand side of (3.1) let us choose $\beta_j = \overline{\beta}_j$ and $\gamma_{ij} = \overline{\gamma}_{ij}$ and let α_i be running in $\mathbb{R}^n$. Then we obtain:

$$-\sum_{i=1}^{n} \alpha_i(p^*_i - \underline{p}_i) \leq -\sum_{i=1}^{n} \overline{\alpha}_i(p^*_i - \underline{p}_i)$$

that is

$$\sum_{i=1}^{n} \alpha_i(p^*_i - \underline{p}_i) \geq \sum_{i=1}^{n} \overline{\alpha}_i(\underline{p}_i - p^*_i) \tag{3.3}$$

and, if we choose $\alpha_i = 0$, we have:

$$\sum_{i=1}^{n} \overline{\alpha}_i(p_i^* - \underline{p}_i) \leq 0. \tag{3.4}$$

Since $\alpha_i \geq 0$, there cannot exist $\overline{i}$ such that $p_{\overline{i}}^* - \underline{p}_{\overline{i}} < 0$, because in this case the left-hand side of (3.3) would diverge to $-\infty$ when α_i diverges to ∞, and then it must be $p_i^* - \underline{p}_i \geq 0 \quad \forall i = 1, 2, \ldots, n$. As a consequence, (3.4) becomes:

$$\sum_{i=1}^{n} \overline{\alpha}_i(\underline{p}_i - p_i^*) = 0, \quad \overline{\alpha}_i \geq 0, \quad p_i^* - \underline{p}_i \geq 0 \tag{3.5}$$

and, therefore,

$$\overline{\alpha}_i(\underline{p}_i - p_i^*) = 0, \quad \forall i = 1, 2, \ldots, n.$$

If we now set $\alpha_i = \overline{\alpha}_i$ and $\beta_j = \overline{\beta}_j$ in the left-hand side of (3.1), we obtain:

$$-\sum_{i=1}^{n}\sum_{j=1}^{m} \gamma_{ij}(\overline{x}_{ij} - x_{ij}^*) \leq -\sum_{i=1}^{n}\sum_{j=1}^{m} \overline{\gamma}_{ij}(\overline{x}_{ij} - x_{ij}^*)$$

that is

$$\sum_{i=1}^{n}\sum_{j=1}^{m} \gamma_{ij}(\overline{x}_{ij} - x_{ij}^*) \geq \sum_{i=1}^{n}\sum_{j=1}^{m} \overline{\gamma}_{ij}(\overline{x}_{ij} - x_{ij}^*). \tag{3.6}$$

If we choose $\gamma_{ij} = 0$, we have:

$$\sum_{i=1}^{n}\sum_{j=1}^{m} \overline{\gamma}_{ij}(\overline{x}_{ij} - x_{ij}^*) \leq 0. \tag{3.7}$$

Let us remark that there cannot be any pair $\overline{i}$, $\overline{j}$ such that $\overline{x}_{\overline{i}\overline{j}} - x_{\overline{i}\overline{j}}^* < 0$ since in such a case the left-hand side of (3.6) would diverge to $-\infty$ and, then, it must be $\overline{x}_{ij} - x_{ij}^* \geq 0$. As a consequence (3.7) becomes

$$\sum_{i=1}^{n}\sum_{j=1}^{m} \overline{\gamma}_{ij}(\overline{x}_{ij} - x_{ij}^*) = 0 \quad \forall i = 1, 2, \ldots, n, \quad \forall j = 1, 2, \ldots, m \tag{3.8}$$

and, therefore,

$$\overline{\gamma}_{ij}(\overline{x}_{ij} - x_{ij}^*) = 0 \quad \forall i = 1, 2, \ldots, n, \quad \forall j = 1, 2, \ldots, m. \tag{3.9}$$

Finally, if we set $\alpha_i = \overline{\alpha}_i$ and $\gamma_{ij} = \overline{\gamma}_{ij}$ in the left-hand side of (3.1), we obtain:

$$-\sum_{j=1}^{m} \beta_j(\overline{q}_j - q_j^*) \leq -\sum_{j=1}^{m} \overline{\beta}_j(\overline{q}_j - q_j^*)$$

that is

$$\sum_{j=1}^{m} \beta_j(\bar{q}_j - q_j^*) \geq \sum_{j=1}^{m} \overline{\beta}_j(\bar{q}_j - q_j^*). \qquad (3.10)$$

If we choose $\beta_j = 0$ (3.10) becomes:

$$\sum_{j=1}^{m} \overline{\beta}_j(\bar{q}_j - q_j^*) \leq 0. \qquad (3.11)$$

Since $\beta \geq 0$, there cannot be $\bar{j}$ such that $\bar{q}_{\bar{j}} - q_{\bar{j}}^* < 0$ because in such a case the left-hand side of (3.10) would diverge to $-\infty$ and, then, it must be $\bar{q}_{\bar{j}} - q_{\bar{j}}^* \geq 0$ $\quad \forall j = 1, 2, \ldots, m$. As a consequence (3.11) becomes:

$$\sum_{j=1}^{m} \overline{\beta}_j(\bar{q}_j - q_j^*) = 0 \quad \forall j = 1, 2, \ldots, m \qquad (3.12)$$

and therefore

$$\overline{\beta}_j(\bar{q}_j - q_j^*) = 0.$$

From the right-hand side of (3.1), taking into account (3.5), (3.8) and (3.12), we obtain:

$$\mathcal{L}(u, \overline{\alpha}, \overline{\beta}, \overline{\gamma}) \geq 0 = \mathcal{L}(u^*, \overline{\alpha}, \overline{\beta}, \overline{\gamma}) \quad \forall u \in \mathbb{K}.$$

Since the convex and differentiable function $\mathcal{L}(u, \overline{\alpha}, \overline{\beta}, \overline{\gamma})$ assumes its minimum value at u^*, then it must be:

$$\frac{\partial \mathcal{L}}{\partial p_i} = g_i(p^*) - \sum_{j=1}^{m} x_{ij}^* + \overline{\alpha}_i = 0, \qquad \frac{\partial \mathcal{L}}{\partial q_j} = -f_j(q^*) + \sum_{i=1}^{n} x_{ij}^* - \overline{\beta}_j = 0,$$

$$\frac{\partial \mathcal{L}}{\partial x_{ij}} = p_i^* + \pi_{ij}(x^*) - q_j^* - \overline{\gamma}_{ij} = 0.$$

Hence, all the conditions of theorem 1.2 are satisfied, and then $u^* = (p^*, q^*, x^*) \in \mathbb{K}$ is a solution of (1.6).

Viceversa, let us assume that $u^* = (p^*, q^*, x^*) \in \mathbb{K}$ is solution of the Variational Inequality (1.6) and then $(u^*, \overline{\alpha}, \overline{\beta}, \overline{\gamma})$ satisfies (1.7)–(1.10). Let us prove that $(u^*, \overline{\alpha}, \overline{\beta}, \overline{\gamma})$ is a saddle point of the Lagrangean function, that is

$$\mathcal{L}(u^*, \alpha, \beta, \gamma) \leq \mathcal{L}(u^*, \overline{\alpha}, \overline{\beta}, \overline{\gamma}) \leq \mathcal{L}(u, \overline{\alpha}, \overline{\beta}, \overline{\gamma}).$$

Taking into account (1.7)–(1.10), we get:

$$\mathcal{L}(u^*, \overline{\alpha}, \overline{\beta}, \overline{\gamma}) = 0.$$

So, we have to prove that $\mathcal{L}(u^*, \alpha, \beta, \gamma) \leq 0$ and that $\mathcal{L}(u, \overline{\alpha}, \overline{\beta}, \overline{\gamma}) \geq 0$.

Let us prove that $\mathcal{L}(u^*, \alpha, \beta, \gamma) \leq 0$. Since

$$\mathcal{L}(u^*, \alpha, \beta, \gamma) = -\sum_{i=1}^{n} \alpha_i(\underline{p}_i - p_i^*) - \sum_{j=1}^{m} \beta_j(q_j^* - \overline{q}_j) - \sum_{i=1}^{n}\sum_{j=1}^{m} \gamma_{ij}(x_{ij}^* - \overline{x}_{ij}),$$

our request is verified by assumption.

Let us now prove that $\mathcal{L}(u, \overline{\alpha}, \overline{\beta}, \overline{\gamma}) \geq \mathcal{L}(u^*, \overline{\alpha}, \overline{\beta}, \overline{\gamma}) = 0$, where

$$\mathcal{L}(u, \overline{\alpha}, \overline{\beta}, \overline{\gamma}) = \ll v(u^*), u - u^* \gg - \sum_{i=1}^{n} \overline{\alpha}_i(p_i - \underline{p}_i) - \sum_{j=1}^{m} \overline{\beta}_j(\overline{q}_j - q_j) +$$

$$- \sum_{i=1}^{n}\sum_{j=1}^{m} \overline{\gamma}_{ij}(\overline{x}_{ij} - x_{ij}).$$

The function $\mathcal{L}(u, \overline{\alpha}, \overline{\beta}, \overline{\gamma})$ is convex with respect to u and, then, the conditions (1.7) are sufficient to ensure that $\mathcal{L}(u, \overline{\alpha}, \overline{\beta}, \overline{\gamma})$ assumes its minimum value at $u = u^*$.   ∎

The results above allow us to obtain a dual formulation of the Variational Inequality (1.6). In fact

$$\mathbb{K}^* = \{(u, \alpha, \beta, \gamma) \in \mathbb{R}^{n+m+nm} \times \mathbb{R}^n \times \mathbb{R}^m \times \mathbb{R}^{nm} :$$

$$\alpha_i = \sum_{i=1}^{n} x_{ij} - g_i(p), \quad \beta_j = \sum_{i=1}^{n} x_{ij} - f_j(q), \quad \gamma_{ij} = q_j - p_i - \pi_{ij}(x),$$

$$\underline{p}_i - p_i \leq 0, \quad q_j - \overline{q}_j \leq 0, \quad x_{ij} - \overline{x}_{ij} \leq 0\}.$$

We will call Dual Variational Inequality the problem:

"find $(u^*, \overline{\alpha}, \overline{\beta}, \overline{\gamma}) \in \mathbb{K}^* :$

$$\sum_{i=1}^{n}(\alpha_i - \overline{\alpha}_i)(\underline{p}_i - p_i) + \sum_{j=1}^{m}(\beta_j - \overline{\beta}_j)(q_j - \overline{q}_j) + \sum_{i=1}^{n}\sum_{j=1}^{m}(\gamma_{ij} - \overline{\gamma}_{ij})(x_{ij} - \overline{x}_{ij}) \geq 0$$

$$\forall(u, \alpha, \beta, \gamma) \in \mathbb{K}^*."$$

It seems to be interesting to solve the Dual Variational Inequality in view of a market evaluation, since it provides the excesses of the supply and of the demand and allows one to organize a better management and control of the market.

4. Calculation of the equilibrium.

We suppose, for sake of simplicity, that the monotonicity condition holds in order to ensure the uniqueness of the solution. We may use the direct method of [6]. This method consists of the following steps.

1 Check if the system

$$\begin{cases} v(u) = 0 \\ u \in \mathbb{K}. \end{cases}$$

has solutions. If this system has no solutions, the solution of the Variational Inequality lies on the boundary of the polyedron $\mathbb{K}$, whose boundary is formed by faces.

2 Let us fix an index $i \in \{1, 2, \dots, n\}$ and consider the face $\mathbb{K} \cap \{p_i = \underline{p}_i\} = \mathbb{K}_i$ and the system

$$\begin{cases} g_l(p) - \sum_{j=1}^m x_{lj} = 0 & l = 1, 2, \dots, n \quad l \neq i \\ f_j(q) - \sum_{l=1}^n x_{lj} = 0 & j = 1, 2, \dots, m \\ p_l + \pi_{lj}(x) - q_j = 0 & l = 1, 2, \dots, n \\ & j = 1, 2, \dots, m \\ u \in \mathbb{K}_i. \end{cases} \tag{4.1}$$

If this system admits a solution $(p, q, x) \in \mathbb{K}_i$, it is the solution of the Variational Inequality if and only if

$$g_i(p) - \sum_{j=1}^m x_{ij} > 0. \tag{4.2}$$

Analogously, if we fix an index $j \in \{1, 2, \dots, m\}$ and consider the face $\mathbb{K} \cap \{q_j = 0\} = \mathbb{K}^j$, we must consider the system

$$\begin{cases} g_i(p) - \sum_{k=1}^m x_{ik} = 0 & i = 1, 2, \dots, n \\ f_k(q) - \sum_{i=1}^n x_{ik} = 0 & k = 1, 2, \dots, m \quad k \neq j \\ p_i + \pi_{ik}(x) - q_k = 0 & i = 1, 2, \dots, n \\ & k = 1, 2, \dots, m \\ u \in \mathbb{K}^j \end{cases} \tag{4.3}$$

and the eventual solution of this system is the solution of the Variational Inequality if and only if

$$f_j(q) - \sum_{i=1}^n x_{ij} > 0. \tag{4.4}$$

If we consider the face $\mathbb{K} \cap \{q_j = \bar{q}_j\} = \check{\mathbb{K}}^j$ and check if the system

$$\begin{cases} g_i(p) - \sum_{k=1}^{m} x_{ik} = 0 & i = 1, 2, \ldots, n \\ f_k(q) - \sum_{i=1}^{n} x_{ik} = 0 & k = 1, 2, \ldots, m, \quad k \neq j \\ p_i + \pi_{ik}(x) - q_k = 0 & i = 1, 2, \ldots, n \\ & k = 1, 2, \ldots, m \\ u \in \tilde{\mathbb{K}}^j \end{cases} \tag{4.5}$$

admits solution, this eventual solution is the one of the Variational Inequality if and only if

$$f_j(q) - \sum_{i=1}^{n} x_{ij} < 0. \tag{4.6}$$

Finally, if we fix two indexes $i \in \{1, 2, \ldots, n\}$ and $j \in \{1, 2, \ldots, m\}$ and consider the face $\mathbb{K} \cap \{x_{ij} = 0\} = \mathbb{K}_{ij}$ and the system

$$\begin{cases} g_i(p) - \sum_{j=1}^{m} x_{ij} = 0 & i = 1, 2, \ldots, n \\ f_j(q) - \sum_{i=1}^{n} x_{ij} = 0 & j = 1, 2, \ldots, m \\ p_l + \pi_{lk}(x) - q_k = 0 & l = 1, 2, \ldots, n, \quad l \neq i, \\ & k = 1, 2, \ldots, m \quad k \neq j \\ u \in \mathbb{K}_{ij}, \end{cases} \tag{4.7}$$

the eventual solution of this system is the one of the Variational Inequality if and only if

$$p_i + \pi_{ij}(x) - q_j > 0. \tag{4.8}$$

On the contrary, if we consider the face $\mathbb{K} \cap \{x_{ij} = \overline{x}_{ij}\} = \tilde{\mathbb{K}}_{ij}$ and the system

$$\begin{cases} g_i(p) - \sum_{j=1}^{m} x_{ij} = 0 & i = 1, 2, \ldots, n \\ f_j(q) - \sum_{i=1}^{n} x_{ij} = 0 & j = 1, 2, \ldots, m \\ p_l + \pi_{lk}(x) - q_k = 0 & l = 1, 2, \ldots, n, \quad l \neq i, \\ & k = 1, 2, \ldots, m \quad k \neq j \\ u \in \tilde{\mathbb{K}}_{ij}, \end{cases} \tag{4.9}$$

the eventual solution is the one of the Variational Inequality if and only if

$$p_i + \pi_{ij}(x) - q_j < 0. \tag{4.10}$$

3 If the conditions (4.2), (4.4), (4.6), (4.8), (4.10) are not fulfilled, the Variational Inequality cannot have solution in the interior of $\mathbb{K}_i$, $\mathbb{K}^j$, $\tilde{\mathbb{K}}^j$, $\mathbb{K}_{ij}$, $\tilde{\mathbb{K}}_{ij}$, whereas if the systems (4.1), (4.3), (4.5),

(4.7), (4.9) admit no solutions, the solution of the Variational Inequality has to lie on the boundary of someone of these faces. In any case, if (4.2), (4.4), (4.6), (4.8), (4.10) are not verified or the systems admit no solutions, the solution of the Variational Inequality must be searched on the boundary of the faces, that is on faces whose dimension is reduced of 1 unit, which we obtain by fixing two indexes $i_1, i_2 \in \{1, 2, \ldots, n\}$ or $j_1, j_2 \in \{1, 2, \ldots, m\}$ or $(i_1, j_1), (i_2, j_2) \in \{1, 2, \ldots, n\} \times \{1, 2, \ldots, m\}$ or a mixture of them. For these new faces we can repeat the procedure 2) with the same criteria on the sign and we may obtain the solution or iterate the method.

An evaluation of the "effectiveness" of this direct method is given in [6] where also a comparison with other methods is provided.

5. Example.

Let us consider an example consisting of two supply markets and two demand markets.

The supply functions and the demand functions depend on two non negative parameters h and k, by means of which it is possible to control the production and the demand.

The supply functions are:

$$g_1(p_1, p_2) = p_1 + h, \quad g_2(p_1, p_2) = -p_1 + p_2 + k;$$

the demand functions are:

$$f_1(q_1, q_2) = q_2, \quad f_2(q_1, q_2) = q_1 - q_2 + h;$$

the transportation cost functions are:

$$\pi_{11}(x) = x_{11} + 1, \quad \pi_{12}(x) = x_{12} + 1,$$

$$\pi_{21}(x) = x_{21} + x_{12}, \quad \pi_{22}(x) = x_{22} + 1,$$

where $x = (x_{11}, x_{12}, x_{21}, x_{22})$. As the feasible set $\mathbb{K}$ of the vectors $u = (p_1, p_2, q_1, q_2, x_{11}, x_{12}, x_{21}, x_{22})$, we assume the positive orthant $\mathbb{K} = \mathbb{R}^8_+$ and the components of the operator $v = v(u)$ are:

$$\begin{cases} p_1 - x_{11} - x_{12} + h, \\ -p_1 - x_{21} - x_{22} + p_2 + k, \\ q_2 - x_{11} - x_{21}, \\ q_1 - q_2 - x_{12} - x_{22} + h, \\ p_1 - q_1 + x_{11} + 1, \\ p_1 - q_2 + x_{12} + 1, \\ p_2 - q_1 + x_{21} + x_{12}, \\ p_2 - q_2 + x_{22} + 1. \end{cases}$$

The equilibrium supply prices, demand prices and shipments are given as a solution of the system

$$\begin{cases} v(u) = 0 \\ u \in \mathbb{K} \end{cases}$$

if

$$-6h + k - 1 \geq 0, \quad -17h + k - 1 \geq 0, \quad h + k - 1 \geq 0, \quad 4h + k - 1 \geq 0,$$

$$-h + 2k - 2 \geq 0, \quad -2h + 7k - 4 \geq 0, \quad -h + 5k - 2 \geq 0,$$

$$19h + 4k - 4 \geq 0, \quad -6h + k - 2 \geq 0, \quad -h + 5k - 2 \geq 0,$$

and we obtain as equilibrium solution:

$$\begin{cases} p_1 = \frac{16h+4k-4}{3}, \\ p_2 = \frac{-2h+4k-4}{3}, \\ q_1 = \frac{-2h+7k-4}{3}, \\ q_2 = \frac{-h+5k-2}{3}, \\ x_{11} = \frac{-18h+3k-3}{3} = -6h + k - 1, \\ x_{12} = \frac{-17h+k-1}{3}, \\ x_{21} = \frac{17h+2k+1}{3}, \\ x_{22} = \frac{h+k-1}{3}. \end{cases}$$

For such values of h and k no supply and demand excesses are present. But, for example, if $\frac{h}{2} + 1 < k < 1 + 6h$, the system

$$\begin{cases} v(u) = 0 \\ u \in \mathbb{K} \end{cases}$$

does not admit any solution because $x_{11} < 0$. Following the procedure of the direct method, we find that the solution, in this case, lies on the face $\mathbb{K} \cap \{p_2 = 0\}$ and is given by

$$
\begin{cases}
p_1 = 0, \\
q_1 = 1 + \frac{h}{2}, \\
q_2 = 1 + \frac{h}{2}, \\
x_{11} = \frac{h}{2}, \\
x_{12} = \frac{h}{2}, \\
x_{21} = 1, \\
x_{22} = \frac{h}{2}
\end{cases}
$$

because

$$
s_2 = -p_1 - x_{21} - x_{22} + p_2 + k = -1 - \frac{h}{2} + k > 0
$$

(recall that s_2 is the excess of production). Moreover:

$$
\begin{cases}
s_1 = 0 \\
t_1 = 0 \\
t_2 = 0
\end{cases}
$$

and

$$
\begin{cases}
g_1 = p_1 + h = h, \\
g_2 = -p_1 + p_2 + k = k, \\
f_1 = q_2 = 1 + \frac{h}{2}, \\
f_2 = q_1 - q_2 + h = h.
\end{cases}
$$

In this case, we have excess of production in the market 2 and the production in the market 1 is equal to the minimum. Then it is possible to organize the production in order to avoid the situation $\frac{h}{2} + 1 < k < 1 + 6h$ which leads to the disequilibrium. Analogous remarks can be obtained by studing the other cases in which the solution is not given by the system.

References

[1] S. C. Dafermos, *Exchange price equilibrium and Variational Inequalities*, Math. Programming 46 (1990), 391–402.

[2] P. Daniele, *Duality Theory for Variational Inequalities*, Communications in Applied Analysis 1, no. 2, (1997), 257–267.

[3] P. Daniele, A. Maugeri, *On Dynamical Equilibrium Problems and Variational Inequalities*, in this book.

[4] F. Giannessi, *Metodi Matematici della Programmazione. Problemi lineari e non lineari*, Pitagora Editrice, Bologna, 1982.

[5] J. Gwinner, *Stability of Monotone Variational Inequalities with Various Applications*, Variational Inequalities and Network Equilibrium Problems, F. Giannessi - A. Maugeri Eds., Plenum Press, New York, 1995, 123–142.

[6] A. Maugeri, *Convex Programming, Variational Inequalities and applications to the traffic equilibrium problem*, Appl. Math. Optim., 16 (1987), 169–185.

[7] A. Maugeri, *Disequazioni Variazionali e Quasi-variazionali e applicazioni a problemi di ottimizzazione su reti*, B.U.M.I. (7) 4-B (1990), 327–343.

[8] A. Nagurney, *Network economics. A Variational Inequality Approach*, Kluwer Academic Publishers, 1993.

[9] A. Nagurney, L. Zhao, *Disequilibrium and Variational Inequalities*, J. Comput. Appl. Math. 33 (1990), 181–198.

ON DYNAMICAL EQUILIBRIUM PROBLEMS AND VARIATIONAL INEQUALITIES

Patrizia Daniele
Dipartimento di Matematica
Università di Catania
Viale A. Doria, 6 - 95125 Catania
e-mail: daniele@dipmat.unict.it

Antonino Maugeri
Dipartimento di Matematica
Università di Catania
Viale A. Doria, 6 - 95125 Catania
e-mail: maugeri@dipmat.unict.it

Abstract   We consider a time–dependent economic market in order to show the existence of time–dependent market equilibrium (which we call dynamic equilibrium). The model we are concerned with is the spatial price equilibrium model in the presence of excesses of supplies and of demands.

   This kind of network problem is directly incorporated into the Variational Inequality model, which provides not only the existence, but also the computation, the stability and the sensitivity of the equilibrium patterns.

   The study of the time-dependent model seems to be important because it allows us to follow the evolution in time of prices and of commodities.

Keywords: Time-dependent model, spatial price equilibrium, Variational Inequalities, dynamical equilibrium.

F. Giannessi et al (eds.),
Equilibrium Problems: Nonsmooth Optimization and Variational Inequality Models, 59–69.
© 2001 *Kluwer Academic Publishers.*

1. Introduction.

The aim of this paper is to consider a dynamic economic market model and to provide the existence of a dynamic market equilibrium. The model we are concerned with is the spatial price equilibrium model which is also connected with electronic circuit networks. It is surprising that this kind of network problems can be incorporated directly into the Variational Inequality models (see [6], [5], [1], [2]), not only in the static case, but also in the time-dependent case (see section 3).

The interest in the dynamic case is quite recent and only a few papers (see [8], [3], [4]) are devoted to this topic regarding general kinds of traffic equilibrium problem. In this paper we present the equilibrium formulation for the spatial price equilibrium models when the data depend on the time and we provide a Variational Inequality formulation from which it is possible to derive not only the existence of the dynamic equilibrium, but also the computation, the stability and the sensitivity of the equilibrium patterns.

In our opinion it is important to study the dynamic model since, in such a way, we are able to describe concrete situations in which the price, the commodities are not stable with respect to the time.

For what concerns references and a survey of the results in the static case, we refer to [7].

2. A static market model.

We present a finite dimensional inequality which describes a constrained equilibrium of spatially distributed economic market with given bounds on prices and transportation fluxes. We follow the model formulated by [6] and extended by [5].

Let us consider a single commodity that is produced at n supply markets and consumed at m demand markets. There is a total supply g_i in each supply market i, where $i = 1, 2, \ldots, n$ and a total demand f_j in each demand market j, where $j = 1, 2, \ldots, m$. Since the markets are spatially separated, x_{ij} units of commodity are transported from i to j.

If we consider the excess supply s_i and the excess demand t_j, we must have

$$g_i = \sum_{j=1}^{m} x_{ij} + s_i \qquad i = 1, \ldots, n \tag{2.1}$$

$$f_j = \sum_{i=1}^{n} x_{ij} + t_j \qquad j = 1, \ldots, m. \tag{2.2}$$

We associate with each supply market i a supply price p_i and with each demand market j a demand price q_j. A fixed minimal supply price $\underline{p}_i \geq 0$ (price floor) for each supply market i and a fixed maximum demand price $\bar{q}_j > 0$ (price ceiling) for each demand market j are given. Moreover, the transportation from i to j gives rise to unit costs π_{ij} and upper bounds $\bar{x}_{ij} > 0$ for the transportation fluxes are included.

Grouping the introduced quantities in vectors, we have the total supply vector $g \in \mathbb{R}^n$, the total supply price vector $p \in \mathbb{R}^n$, the total demand vector $f \in \mathbb{R}^m$, the total demand price vector $q \in \mathbb{R}^m$, the flux vector $x \in \mathbb{R}^{nm}$, the unit flux cost vector $\pi \in \mathbb{R}^{nm}$. Then the feasible set for the vectors $u = (p, q, x)$ is given by the product set

$$\mathcal{M} = \prod_{i=1}^{n} [\underline{p}_i, \infty[\times \prod_{j=1}^{m} [0, \bar{q}_j] \times \prod_{i=1}^{n} \prod_{j=1}^{m} [0, \bar{x}_{ij}] = \mathcal{M}_1 \times \mathcal{M}_2 \times \mathcal{M}_3.$$

As in unconstrained market equilibria (see [1], [2]), we assume that we are given the functions

$$g : \mathcal{M}_1 \to \mathbb{R}^m, \quad f : \mathcal{M}_2 \to \mathbb{R}^m, \quad \pi : \mathcal{M}_3 \to \mathbb{R}^{nm}$$

which express:
the dependence of the total supply g on the price p;
the dependence of the total demand f on the price q;
the dependence of transportation unit cost π on the flux vector x.

According to perfect equilibrium, the economic market conditions take the following form:

Definition 2.1 *A vector* $u = (p, q, x) \in \mathcal{M}$ *is a market equilibrium if*

$$s_i > 0 \Rightarrow p_i = \underline{p}_i, \qquad p_i > \underline{p}_i \Rightarrow s_i = 0 \qquad i = 1, \dots, n; \qquad (2.3)$$

$$t_j > 0 \Rightarrow q_j = \bar{q}_j, \qquad q_j < \bar{q}_j \Rightarrow t_j = 0 \qquad j = 1, \dots, m; \qquad (2.4)$$

$$p_i + \pi_{ij} \begin{cases} \geq q_j & if \quad x_{ij} = 0 \\ = q_j & if \quad 0 < x_{ij} < \bar{x}_{ij} \quad i = 1, \dots, n \\ & \qquad\qquad\qquad\quad j = 1, \dots, m \\ \leq q_j & if \quad x_{ij} = \bar{x}_{ij}. \end{cases} \qquad (2.5)$$

We can characterize a market equilibrium as a solution to a Variational Inequality

Suppose that for each $i = 1, 2, \ldots, n$ and $j = 1, 2, \ldots, m$ there holds

$$q_j = 0 \Rightarrow f_j(q) \geq 0 \qquad x_{ij} > 0 \Rightarrow \pi_{ij}(x) > 0. \tag{2.6}$$

Then $u = (p, q, x) \in \mathcal{M}$ satisfies the market equilibrium conditions (2.1)-(2.5) if and only if u is a solution to

$$\ll v(u^*), u - u^* \gg \; = \; \sum_{i=1}^{n} \left(g_i(p^*) - \sum_{j=1}^{m} x_{ij}^* \right)(p_i - p_i^*) - \sum_{j=1}^{m} \left(f_j(q^*) + \right.$$

$$- \sum_{i=1}^{n} x_{ij}^* \right)(q_j - q_j^*) \; + \; \sum_{i=1}^{n} \sum_{j=1}^{m} \left(p_i^* + \pi_{ij}(x^*) - q_j^* \right)(x_{ij} - x_{ij}^*) \geq 0$$

$$\forall u \; = \; (p, q, x) \in \mathbb{K}. \tag{2.7}$$

Proof. For what follows, we need a sketch of the proof.

Let $u = (p, q, x) \in \mathcal{M}$ satisfy the market conditions (2.1)-(2.5) with

$$s_i = g_i(p) - \sum_{j=1}^{m} x_{ij} \geq 0$$

$$t_j = f_j(q) - \sum_{i=1}^{n} x_{ij} \geq 0.$$

If $p_i = \underline{p}_i$, then $\tilde{p}_i - p_i \geq 0$ and the product in the first sum in (2.6) is nonnegative, otherwise by (2.3) (if $p_i > \underline{p}_i$) $s_i = 0$ and the product vanishes. By similar case distinctions, one obtains that each product in the second sum is nonpositive and each product in the third sum is nonnegative. This proves (2.6).

Conversely, let (2.6) hold. By the choices

$$\tilde{p}_k = p_k, \qquad \tilde{q}_l = q_l, \qquad \tilde{x}_{kl} = x_{kl}$$

for all indexes k except some fixed index i and for all indexes l except some fixed index j, one obtains that in (2.6) all products in the first and in the third sum are nonnegative, whereas all products in the second sum are nonpositive.

Now let us suppose that (2.1) is not verified, that is there exists i^* such that

$$s_{i^*} = g_{i^*}(p) - \sum_{j=1}^{n} x_{i^*j} < 0$$

and consider the first sum

$$\sum_{i=1}^{n} \left(g_i(p) - \sum_{j=1}^{m} x_{ij} \right)(\tilde{p}_i - p_i) \geq 0 \quad \forall \tilde{p} \in \mathcal{M}_1.$$

Then chosen $\tilde{p}_i = p_i$ for $i \neq i^*$ and $\tilde{p}_{i^*} > p_{i^*}$, we obtain a contradiction.

Let us suppose that (2.3) are not verified. Then there exists i^* such that either $s_{i^*} > 0$ and $p_{i^*} > \underline{p}_{i^*}$ or $p_{i^*} > \underline{p}_{i^*}$ and $s_{i^*} > 0$. In both the cases, the choice

$$\tilde{p}_i = p_i \quad \text{for} \quad i \neq i^*$$

$$\tilde{p}_{i^*} = \underline{p}_{i^*}$$

leads to a contradiction.

Now, let us turn to the third sum

$$\sum_{i=1}^{n}\sum_{j=1}^{m} \left(p_i + \pi_{ij}(x) - q_j \right)(\tilde{x}_{ij} - x_{ij}) \geq 0 \quad \forall \tilde{x} \in \mathcal{M}_3$$

and suppose that conditions (2.5) are not verified, that is there exist i^* and j^* such that

$$p_{i^*} + \pi_{i^*j^*}(x) > q_{j^*} \quad \text{and} \quad x_{i^*j^*} > 0.$$

Then the choice $x_{ij} = \tilde{x}_{ij}$ for $i \neq i^*$, $j \neq j^*$ and $\tilde{x}_{i^*j^*} = 0$ leads to a contradiction. Analogously if there exist i^* and j^* such that

$$p_{i^*} + \pi_{i^*j^*}(x) < q_{j^*} \quad \text{and} \quad x_{i^*j^*} < \overline{x}_{ij}.$$

Let us consider the second sum

$$\sum_{j=1}^{m} \left(f_j(q) - \sum_{i=1}^{n} x_{ij} \right)(\tilde{q}_j - q_j) \leq 0 \quad \forall q \in \mathcal{M}_2.$$

If $q_j \in (0, \bar{q}_j]$, let us choose $\tilde{q}_j = 0$. Then, obviously, $t_j = f_j(q) - \sum_{i=1}^{n} x_{ij}$ is nonnegative. Thus the case that there exists i^* such that $q_{i^*} = 0$ and $t_{j^*} < 0$ remains. It is an absurdity.

In fact from

$$t_{j^*} = f_{j^*}(q) - \sum_{i=1}^{n} x_{ij^*} < 0,$$

it follows

$$\sum_{i=1}^{n} x_{ij^*} > f_{j^*} \geq 0$$

and therefore there exists some index i^* such that

$$x_{i^*j^*} > 0.$$

The already proved condition (2.5) provides us

$$p_i^* + \pi_{i^*j^*}(x) \leq q_j^* = 0$$

from which

$$\pi_{i^* j^*}(x) \leq 0$$

what contradicts (2.6) with $x_{i^* j^*} > 0$. To verify (2.4), let us apply an analogous as in the proof of (2.3): $t_j > 0$ and $q_j < \bar{q}_j$ cannot hold simultaneously.

3. The time-dependent market model.

Now we consider the dynamic case. The markets, whose geometry remains fixed, are considered at all time $t \in T = [0, T]$. For each time $t \in T$ we have a total supply vector $g(t) \in \mathbb{R}^n$, a supply price vector $p(t) \in \mathbb{R}^n$, the total demand vector $f(t) \in \mathbb{R}^m$, the demand price vector $q(t) \in \mathbb{R}^m$, the flux vector $x(t) \in \mathbb{R}^{nm}$ and the unit cost vector $\pi(t) \in \mathbb{R}^{nm}$.

The feasible vectors $u(t) = (p(t), q(t), x(t))$ have to satisfy the time-dependent bounds on prices and transportation fluxes, namely that, almost everywhere on T

$$u(t) \in \prod_{i=1}^{n} [\underline{p}_i(t), \infty[\times \prod_{j=1}^{m} [0, \bar{q}_j(t)] \times \prod_{i=1}^{n} \prod_{j=1}^{m} [0, \bar{x}_{ij}(t)]$$

where $\underline{p}_i(t)$, $\bar{q}_j(t)$, $\bar{x}_{ij}(t)$ are given.

For technical reasons, the functional setting for the trajectories u is the reflexive Banach space: L

$$L^2(T, \mathbb{R}^n) \times L^2(T, \mathbb{R}^m) \times L^2(T, \mathbb{R}^{nm})$$

which we abbreviate by

$$L = L_1 \times L_2 \times L_3.$$

The set of feasible vectors $u = (p, q, x)$ is given by

$$K = K_1 \times K_2 \times K_3 =$$
$$= \{p \in L_1 : \underline{p}(t) \leq p(t) \quad \text{a. e. on} \quad T\} \times$$
$$\times \{q \in L_2 : 0 \leq q(t) \leq \bar{q}(t) \quad \text{a. e. on} \quad T\} \times$$
$$\times \{x \in L_3 : 0 \leq x(t) \leq \bar{x}(t) \quad \text{a. e. on} \quad T\}.$$

It is easily seen that $\mathbb{K}$ is a convex, closed but not bounded set because $\mathbb{K}_1$ is unbounded. Furthermore we are giving the mappings:

$$g : \mathbb{K}_1 \to L_1, \quad f : \mathbb{K}_2 \to L_2, \quad \pi : \mathbb{K}_3 \to L_3$$

which assign to each price trajectories $p \in \mathbb{K}_1$ and $q \in \mathbb{K}_2$ the supply $g \in L_1$ and the demand $f \in L_2$ respectively and to the flow trajectory $x \in \mathbb{K}_3$ the cost $\pi \in L_3$.

Introducing the excess supply $s_i(t)$ and the excess demand $t_j(t)$ we must have:

$$g_i(p(t)) = \sum_{j=1}^{m} x_{ij}(t) + s_i(t) \quad i = 1, 2, \ldots, n \tag{3.1}$$

$$f_j(q(t)) = \sum_{i=1}^{n} x_{ij}(t) + t_j(t) \quad j = 1, 2, \ldots, m. \tag{3.2}$$

Obviously $s \in L_1$ and $t \in L_2$. Dynamic market equilibrium takes the following form:

Definition 3.2 $u = (p, q, x) \in L$ *is a dynamic market equilibrium if and only if for each* $i = 1, 2, \ldots, n$ *and* $j = 1, 2, \ldots, m$ *and a. e. on* T *there hold:*

$$s_i(t) > 0 \Rightarrow p_i(t) = \underline{p}_i(t) \ , \quad p_i(t) > \underline{p}_i(t) \Rightarrow s_i(t) = 0 \atop i = 1, \ldots, n; \tag{3.3}$$

$$t_j(t) > 0 \Rightarrow q_j(t) = \overline{q}_j(t) \ , \quad q_j(t) < \overline{q}_j(t) \Rightarrow t_j(t) = 0 \atop j = 1, \ldots, m; \tag{3.4}$$

$$p_i(t) + \pi_{ij}(t) \begin{cases} \geq q_j(t) & if \quad x_{ij}(t) = 0 \\ = q_j(t) & if \quad 0 < x_{ij}(t) < \overline{x}_{ij}(t) \\ \leq q_j(t) & if \quad x_{ij}(t) = \overline{x}_{ij}(t). \end{cases} \tag{3.5}$$

Let $v : \mathbb{K} \to L$ be the operator defined setting

$$v = v(p(t), q(t), x(t)) =$$

$$\left((g_i - \sum_{j=1}^{m} x_{ij})_{i=1,\ldots,n}, (f_j - \sum_{i=1}^{n} x_{ij})_{j=1,\ldots,m}, (p_i + \pi_{ij}(x) - q_j)_{\substack{i=1,\ldots,n \\ j=1,\ldots,m}} \right).$$

The following characterization holds.

Suppose that for each $i = 1, 2, \ldots, n$ and $j = 1, 2, \ldots, m$ there hold

1 $q_j(t) = 0$ on a set $E \subseteq T$ having positive measure

$$\Rightarrow f_j(q(t)) \geq 0 \ in \ E; \tag{3.6}$$

2 $x_{ij}(t) > 0$ on a set $E \subseteq T$ having positive measure $\Rightarrow \pi_{ij}(x(t)) > 0$ in E.

Then $u = (p, q, x) \in K$ is a dynamic market equilibrium if and only if u is a solution to

$$\ll v, \tilde{u} - u \gg = \int_0^T \left\{ \sum_{i=1}^n \left(g_i(p(t)) - \sum_{j=1}^m x_{ij}(t) \right) (\tilde{p}_i(t) - p_i(t)) + \right.$$

$$- \sum_{j=1}^m \left(f_j(q(t)) - \sum_{i=1}^n x_{ij}(t) \right) (\tilde{q}_j(t) - q_j(t)) + \qquad (3.7)$$

$$\left. + \sum_{i=1}^n \sum_{j=1}^m \left(p_i(t) + \pi_{ij}(x(t)) - q_j(t) \right) (\tilde{x}_{ij}(t) - x_{ij}(t)) \right\} dt \geq 0$$

$\forall \tilde{u} = (\tilde{p}, \tilde{q}, \tilde{x}) \in K$.

Proof. Assume that (3.1)-(3.5) hold. Let $\tilde{u} \in K$. Since the union of finitely nullsets is a nullset, it follows from theorem 2

$$\sum_{i=1}^n \left(g_i(p(t)) - \sum_{j=1}^m x_{ij}(t) \right) (\tilde{p}_i(t) - p_i(t)) - \sum_{j=1}^m \left(f_j(q(t)) - \sum_{i=1}^n x_{ij}(t) \right) (\tilde{q}_j(t) -$$

$$q_j(t)) + \sum_{i=1}^n \sum_{j=1}^m \left(p_i(t) + \pi_{ij}(x(t)) - q_j(t) \right) (\tilde{x}_{ij}(t) - x_{ij}(t)) \geq 0 \quad \text{a. e. on} \quad T.$$

Hence (3.6) follows.

Viceversa, let us assume that (3.6) holds. As a consequence we obtain, assuming in turns $\tilde{q} = q$ and $\tilde{x} = x$, $\tilde{p} = p$ and $\tilde{x} = x$, $\tilde{p} = p$ and $\tilde{q} = q$:

$$\int_0^T \sum_{i=1}^n \left(g_i(p(t)) - \sum_{j=1}^m x_{ij}(t) \right) (\tilde{p}_i(t) - p_i(t)) \, dt \geq 0 \quad \forall p \in K_1,$$

$$\int_0^T \sum_{j=1}^m \left(f_j(q(t)) - \sum_{i=1}^n x_{ij}(t) \right) (\tilde{q}_j(t) - q_j(t)) \, dt \geq 0 \quad \forall q \in K_2,$$

$$\int_0^T \sum_{i=1}^n \sum_{j=1}^m \left(p_i(t) + \pi_{ij}(x(t)) - q_j(t) \right) (\tilde{x}_{ij}(t) - x_{ij}(t)) \right\} dt \geq 0 \quad \forall x \in K_3$$

from which (3.1) and (3.3), (3.2) and (3.4), (3.5) follow respectively.

In fact, let us assume that (3.1) does not hold. Then exists an index i^* together with a set $E \subseteq T$ having positive measure such that

$$s_{i^*}(t) = g_{i^*}(t) - \sum_{j=1}^m x_{i^*j}(t) < 0 \quad \text{on} \quad E.$$

Then the choice

$$\tilde{p}_i(t) = p_i(t) \quad \text{for} \quad i \neq i^*$$

$$\tilde{p}_i^*(t) \begin{cases} = p_i(t) & \text{on} \quad \mathcal{T} - E \\ > p_i(t) & \text{on} \quad E \end{cases}$$

leads to a contradiction:

$$\int_0^T \sum_{i=1}^n \left(g_i(p(t)) - \sum_{j=1}^m x_{ij}(t) \right) (\tilde{p}_i(t) - p_i(t)) \, dt =$$

$$\int_E \left(g_i^*(p(t)) - \sum_{j=1}^m x_{i^*j}(t) \right) (\tilde{p}_i(t) - p_i(t)) \, dt < 0.$$

Suppose now that (3.3) is not verified. Then there exists an index i^* together with a set $E \subset \mathcal{T}$ having positive measure such that either $s_{i^*}(t) > 0$ and $p_{i^*}(t) > \underline{p}_{i^*}(t)$ or $p_{i^*}(t) > \underline{p}_{i^*}(t)$ and $s_{i^*}(t) > 0$ on E.
In both the cases the choice

$$\tilde{p}_i(t) = p_i(t) \quad \text{for} \quad i \neq i^*$$

$$\tilde{p}_i^*(t) = \begin{cases} p_i^*(t) & \text{on} \quad \mathcal{T} - E \\ \underline{p}_i^*(t) & \text{on} \quad E \end{cases}$$

leads to a contradiction. Similarly one can proceed in other cases and the equivalence is achieved.

4. Existence of equilibria.

Let us recall some concepts that will be useful in the following. Let E be a real topological vector space, $K \subseteq E$ convex. Then $v : K \to E^*$ is said to be:

1 *pseudomonotone* if and only if

$$\forall u_1, u_2 \in K \quad \langle v(u_1), u_2 - u_1 \rangle \geq 0 \Rightarrow \langle v(u_2), u_1 - u_2 \rangle \leq 0;$$

2 *hemicontinuous* if and only if

$$\forall u \in K \quad \text{the function} \quad z \to \langle v(z), u - z \rangle$$

is upper semicontinuous on K;

3 *hemicontinuous along line segments* if and only if

$$\forall u_1, u_2 \in K \quad \text{the function} \quad z \to \langle v(z), u_2 - u_1 \rangle$$

is upper semicontinuous on the line segment $[u_1, u_2]$.

From the results by [3], we can derive the following abstract existence theorem:

Let E be a real topological vector space and $K \subseteq E$ be convex and nonempty. Let $v : K \to E^*$ be given such that

    1 there exist $A \in K$ nonempty, compact and $B \in K$ compact, convex such that, for every $u_1 \in K \setminus A$, there exists $u_2 \in B$ with $\langle v(u_1), u_2 - u_1 \rangle < 0$;

and either

    2 *v is hemicontinuous*

or

    3 *v is pseudomonotone and hemicontinuous along line segments.*

Then exists $u \in A$ such that

$$\langle v(u), \tilde{u} - u \rangle \geq 0 \quad \forall \tilde{u} \in K.$$

We apply this result to derive the existence theorem for the Variational Inequality (3.6).

Each of the following conditions is sufficient to ensure the existence of the solution of (3.6):

    1

$$v = v(p(t), q(t), x(t)) =$$

$$\left((g_i - \sum_{j=1}^{m} x_{ij})_{i=1,\dots,n}, \ (f_j - \sum_{i=1}^{n} x_{ij})_{j=1,\dots,m}, \ (p_i + \pi_{ij}(x) - q_j)_{\substack{i=1,\dots,n \\ j=1,\dots,m}} \right)$$

is hemicontinuous with respect to the strong topology and there exist $A_1 \subseteq K_1$, $A_2 \subseteq K_2$, $A_3 \subseteq K_3$ compact and there exist $B_1 \subseteq K_1$, $B_2 \subseteq K_2$, $B_3 \subseteq K_3$ compact, convex with respect to the strong topology such that

$$\forall u_1 = (p_1, q_1, x_1) \in (K_1 \setminus A_1) \times (K_2 \setminus A_2) \times (K_3 \setminus A_3)$$
$$\exists u_2 = (p_2, q_2, x_2) \in B_1 \times B_2 \times B_3$$

such that

$$\emptyset \ll v(u_1), u_2 - u_1 \gg < 0;$$

2 v is pseudomonotone, v is hemicontinuous along line segments and there exist $A_1 \subseteq K_1$ compact and $B_1 \subseteq K_1$ compact, convex with respect to the weak topology such that

$$\forall p \in K_1 \setminus A_1 \quad \exists \tilde{p} \in B_1 :$$

$$\ll v(p), \tilde{p} - p \gg = \int_0^T \sum_{i=1}^n g_i(p(t))(\tilde{p}_i(t) - p_i(t)) \, dt < 0;$$

3 v is hemicontinuous on $\mathbb{K}$ with respect to the weak topology, $\exists A_1 \subseteq K_1$ compact, $\exists B_1 \subseteq K_1$ compact, convex with respect to the weak topology such that

$$\forall p \in K_1 \setminus A_1 \quad \exists \tilde{p} \in B_1 :$$

$$\ll v(p), \tilde{p} - p \gg = \int_0^T \sum_{i=1}^n g_i(p(t))(\tilde{p}_i(t) - p_i(t)) \, dt < 0.$$

References

[1] S. Dafermos, *Traffic Equilibrium and Variational Inequalities*, Transportation Sc. 14 (1980), 42–54.

[2] S. Dafermos, *Exchange price equilibrium and Variational Inequalities*, Math. Programming 46 (1990), 391–402.

[3] P. Daniele, A. Maugeri, W. Oettli, *Variational Inequalities and time-dependent traffic equilibria*, C. R. Acad. Sci. Paris t. 326, serie I, (1998)

[4] P. Daniele, A. Maugeri, W. Oettli, *Time Dependent Traffic Equilibria*, J. Optim. Th. Appl. 103, No. 3, 543–555.

[5] J. Gwinner, *Stability and Monotone Variational Inequalities with Various Applications*, Variational Inequalities and Network Equilibrium Problems, (F. Giannessi - A. Maugeri, Eds). Plenum Press, (1995), 123–142.

[6] A. Nagurney, L. Zhao, *Disequilibrium and Variational Inequalities*, I Comput. Appl. Math. 33 (1990), 181–198.

[7] A. Nagurney, *Network Economics - A Variational Inequality Approach*, Kluwer Academic Publishers, 1993.

[8] M. J. Smith, *A New Dynamic Traffic Model and the Existence and calculation of Dynamic user Equilibria on Congested Capacity-constrained Road Network*, Transp. Res. B 27B (1993), 49-63.

NONLINEAR PROGRAMMING
METHODS FOR SOLVING
OPTIMAL CONTROL PROBLEMS*

Carla Durazzi

Dipartimento di Matematica
Università di Bologna
Piazza di Porta San Donato 5
40126   Bologna, Italia
e-mail: dzc@dns.unife.it

Emanuele Galligani

Dipartimento di Matematica
Università di Modena e Reggio Emilia
Via Campi 213/b, 41100   Modena, Italia
e-mail: galligani@unimo.it

Abstract   This paper concerns with the problem of solving optimal control problems by means of nonlinear programming methods. The technique employed to obtain a mathematical programming problem from an optimal control problem is explained and the Newton interior–point method, chosen for its solution, is presented with special regard to the choice of the involved parameters. An analysis of the behaviour of the method is reported on four optimal control problems, related to improving water quality in an aeration process and to the study of diffusion convection processes.

Keywords: Optimal Control, Nonlinear Programming, Newton Interior–Point Method, Finite Difference Approximations.

*Work carried out by INdAM grant at the Department of Mathematics, University of Bologna, Italy

71

F. Giannessi et al (eds.),
Equilibrium Problems: Nonsmooth Optimization and Variational Inequality Models, 71–99.
© *2001 Kluwer Academic Publishers.*

1. Introduction.

Let us consider an optimal control problem as follows: given a set of ν_1 ordinary differential equations (state equations)

$$\dot{y}(t) = \phi(y(t), u(t)) \quad t_0 \leq t \leq t_f \tag{1.1}$$

(the ν_1–vector $y(t)$ is the state vector and the ν_2–vector $u(t)$ is the control or input vector), an initial state at a given time t_0

$$y(t_0) = y_0 \tag{1.2}$$

an objective cost functional

$$J(u(t)) = \psi(y(t_f)) + \int_{t_0}^{t_f} \varphi(y(\tau), u(\tau))d\tau \tag{1.3}$$

and a set of U of allowable piecewise continuous control vectors $u(t)$, find a control $u^*(t) \in U$ and a corresponding trajectory $y^*(t)$ which transfer the system from $y(t_0)$ to $y(t_f)$ in such a way that the functional J is minimized (or maximized). There is no reason to suppose that such a control exists. Pontryagin maximum principle (e.g. see [20, p. 80]) gives a necessary condition to distinguish an optimal control to the other controls that transfer the system from $y(t_0)$ to $y(t_f)$.

Problems in which the dynamic system is linear and the objective function is quadratic, represent an extremely important special family of optimal control problems (Linear-Quadratic (LQ) optimal control problems). In this case the optimal control can be expressed in linear feedback form solving the Riccati matrix differential equation.

Many computational methods for solving this LQ optimal control problem have been investigated. The approaches based on the solution of the corresponding Riccati equation turn out to be the most efficient and reliable ones (e.g. see [3], [6]).

Let us consider the approach based on the possibility to transcribe an optimal control problem as a nonlinear programming problem with a special structure (e.g. see [5]).

The optimal control problem (1.1)-(1.3) can be transformed into a finite dimensional mathematical programming problem.

In details, we consider $K + 1$ points in $[t_0, t_f]$ such that $t_0 < t_1 < ... < t_K = t_f$ and $u(t)$ piecewise constant, that is $u(i) = u(t)$, $t \in [t_i, t_{i+1})$, $i = 0, ..., K - 1$. Let $y(t; y_0, u)$ be the solution of (1.1)-(1.2) where the $K \cdot \nu_2$ vector u is $u = (u(0)^T, u(1)^T, ..., u(K - 1)^T)^T$.

If we set $y(i) = y(t_i; y_0, u)$ and $y_i(t)$ is the solution of (1.1) in $[t_i, t_{i+1})$ corresponding to $u(i)$ with the initial condition $y_i(t_i) = y(i)$, then from (1.1)

$$y(t_{i+1}) = y(t_i) + \int_{t_i}^{t_{i+1}} \phi(y(\tau), u(\tau))d\tau \tag{1.4}$$

so that

$$y(i+1) = y(i) + \int_{t_i}^{t_{i+1}} \phi(y_i(\tau), u(i))d\tau$$

it follows

$$y(i+1) = y(i) + \tilde{\phi}(y(i), u(i))$$

Analogously (1.3) becomes

$$J(u) = \sum_{i=0}^{K-1} \tilde{\varphi}(y(i), u(i)) + \psi(y(k))$$

obtaining the mathematical programming problem

$$\min J(u) \qquad (1.5)$$

$$\begin{aligned} y(i+1) &= y(i) + \tilde{\phi}(y(i), u(i)) \\ y(0) &= y_0 \end{aligned} \qquad (1.6)$$

Problem (1.5)–(1.6) can also be considered as a discrete–time optimal control problem.

When ϕ is linear, that is $\phi(y(t), u(t)) = Ay(t) + Bu(t)$, the solution of system (1.1) has the expression

$$y(t) = e^{A(t-t_0)}y(t_0) + \int_{t_0}^{t} e^{A(t-\tau)}Bu(\tau)d\tau$$

thus the solution $y_i(t)$ of (1.1) in $[t_i, t_{i+1}]$ has the form (set $h_i = t_{i+1} - t_i$, $\tau' = \tau - t_i$ and $0 \le t \le h_i$)

$$\begin{aligned} y_i(t) &= e^{At}y(i) + \int_0^t e^{A(t-\tau')}Bu(t_i + \tau')d\tau' \\ &= e^{At}y(i) + A^{-1}(e^{At} - I)Bu(i) \end{aligned}$$

therefore, formula (1.4) has the expression

$$\begin{aligned} y(i+1) &= y(i) + \int_0^{h_i} \phi(y(t_i + \tau'), u(t_i + \tau'))d\tau' \\ &= y(i) + \int_0^{h_i} \phi(y_i(\tau'), u(i))d\tau' \\ &= y(i) + \int_0^{h_i} (A(e^{A\tau'}y(i) + A^{-1}(e^{A\tau'} - I)Bu(i)) + Bu(i))d\tau' \\ &= e^{Ah_i}y(i) + A^{-1}(e^{Ah_i} - I)Bu(i) \end{aligned}$$

The discretization of the continuous optimal control problem (1.1)–(1.3) can also be performed in many other ways. The problem (1.1)–(1.3) may generate a finite dimensional mathematical programming problem without simultaneously giving rise to a discrete optimal control problem. For instance, this would be the case if we found it necessary to restrict the control $u(t)$ to a sum of Chebyshev polynomials for $t \in [t_0, t_f]$ (e.g. see [23], [13]).

Since it is often required, in practice, that the state and the control vectors have physical meaning, then equality and/or inequality constraints have to be added to the system (1.6) so that problem (1.5)–(1.6) can be formulated as

$$
\begin{aligned}
\min_{\substack{h(z) = 0 \\ g(z) \le 0}} \quad f(z)
\end{aligned}
\tag{1.7}
$$

where $z = (y^T, u^T)^T$. Well known approaches to problem (1.7) are based on sequential quadratic programming (SQP) method (see ([4], [9] for recent references); in this paper, we describe the Newton interior–point method for solving problem (1.7) and we analyse the behaviour of the method on four well known different optimal control problems related to improving water quality in an aeration process and to the study of diffusion convection processes.

2. Framework of the Method.

Consider the nonlinear programming problem

$$
\begin{aligned}
\min_{\substack{h(z) = 0 \\ g(z) \le 0}} \quad f(z)
\end{aligned}
\tag{2.1}
$$

with $z \in \mathbb{R}^n$, $h(z) = (h_1(z), ..., h_m(z))^T$ and $g(z) = (g_1(z), ..., g_p(z))^T$. Introducing the slack variables $s = (s_1, ..., s_p)$ problem (2.1) is equivalent to

$$
\begin{aligned}
\min_{\substack{h(z) = 0 \\ g(z) + s = 0 \\ s \ge 0}} \quad f(z)
\end{aligned}
\tag{2.2}
$$

The Karush–Kuhn–Tucker (KKT) conditions for problem (2.2) are:

$$
\begin{cases}
\nabla f(z) + \nabla h(z)\lambda + \nabla g(z)w = 0 \\
\quad\quad h(z) = 0 \\
\quad\quad g(z) + s = 0 \\
SWe = 0; \quad s \ge 0; \quad w \ge 0
\end{cases}
$$

with

$$
\lambda = (\lambda_1, ..., \lambda_m)^T, \quad w = (w_1, ..., w_p)^T, \quad S = \text{diag}(s_1, ... s_p), \quad W = \text{diag}(w_1, ..., w_p)
$$

$e = (1, ..., 1)^T$ (p–vector), $\nabla f(z)$ indicates the gradient of $f(z)$ and the $n \times m$ and $n \times p$ matrices $\nabla h(z)$ and $\nabla g(z)$ indicate the transpose of the jacobian matrices of $h(z)$ and $g(z)$.

If we set,

$$v = \begin{bmatrix} z \\ \lambda \\ w \\ s \end{bmatrix} \quad H(v) = \begin{bmatrix} H_1(v) \\ SWe \end{bmatrix} = \begin{bmatrix} \nabla f(z) + \nabla h(z)\lambda + \nabla g(z)w \\ h(z) \\ g(z) + s \\ SWe \end{bmatrix}$$

then the KKT conditions become

$$H(v) = 0 \tag{2.3}$$
$$s \geq 0; \quad w \geq 0$$

When we solve the system (2.3) by Newton method, we set $v^{(0)}$ such that $s^{(0)} > 0$ and $w^{(0)} > 0$ and we compute

$$v^{(k+1)} = v^{(k)} + \alpha_k \Delta v^{(k)}$$

where $\Delta v^{(k)}$ is the solution of the linear system

$$H'(v^{(k)})\Delta v = -H(v^{(k)}) \tag{2.4}$$

Here $\Delta v = (\Delta z^T, \Delta \lambda^T, \Delta w^T, \Delta s^T)^T$ and

$$H'(v) = \begin{bmatrix} \nabla^2 \mathcal{L}(z, \lambda, w) & \nabla h(z) & \nabla g(z) & 0 \\ \nabla h(z)^T & 0 & 0 & 0 \\ \nabla g(z)^T & 0 & 0 & I \\ 0 & 0 & S & W \end{bmatrix}$$

$\mathcal{L}(z, \lambda, w) = f(z) + \lambda^T h(z) + w^T g(z)$ denotes the lagrangean associated with (2.1), $\nabla^2 \mathcal{L}(z, \lambda, w) = \nabla^2 f(z) + \sum_{i=1}^{m} \lambda_i \nabla^2 h_i(z) + \sum_{j=1}^{p} w_i \nabla^2 g_j(z)$ its hessian matrix, $\nabla^2 f(z)$, $\nabla^2 h_i(z)$ and $\nabla^2 g_j(z)$ denote the hessian matrices of $f(z)$, $h_i(z)$ $(i = 1, ..., m)$ and $g_j(z)$ $(j = 1, ..., p)$ respectively, and I is the identity matrix of order p.

We notice that the fourth equation of (2.4) is

$$S^{(k)}\Delta w + W^{(k)}\Delta s = -S^{(k)}W^{(k)}e$$

It implies that if $s_i^{(\bar{k})} = 0$ (or $w_i^{(\bar{k})} = 0$) for an iteration $\bar{k}$, then $s_i^{(k)} = 0$ (or $w_i^{(k)} = 0$) for $k > \bar{k}$. This means that when the iterates reach the boundary of the feasible region, they are forced to stick to the boundary.

Let us consider the barrier method for problem (2.2):

$$\min_{\substack{h(z) = 0 \\ g(z) + s = 0 \\ s > 0}} f(z) - \rho_k \sum_{i=1}^{p} \ln s_i \tag{2.5}$$

with $\rho_k > 0$. The KKT conditions for (2.5) are

$$\begin{cases} \nabla f(z) + \nabla h(z)\lambda + \nabla g(z)w = 0 \\ h(z) = 0 \\ g(z) + s = 0 \\ SWe = \rho_k e; \quad s > 0; \quad w > 0 \end{cases}$$

If we set $\tilde{e} = (0, ..., 0, e^T)^T$ then the KKT conditions for (2.5) become

$$H(v) = \rho_k \tilde{e} \qquad (2.6)$$
$$s > 0; \quad w > 0$$

and they can be considered as a perturbation of (2.3), where the perturbation appears only on the complementarity equation $SWe = 0$.

The perturbation on the complementary equation forces the iterates to be sufficiently far from the boundary of the feasible region.

Let us solve the system (2.3) (i.e. the KKT conditions for (2.1)) by inexact Newton method ([7], [17, p. 95]). If we assume that $\sigma_k \in (0,1)$ is the forcing term, that $\sigma_k \mu_k \tilde{e}$ is the residual such that

$$\|\mu_k \tilde{e}\|_2 \le \|H(v^{(k)})\|_2 \qquad (2.7)$$

and that $v^{(0)}$ is interior to the feasible region, then the inexact Newton method is expressed by

$$v^{(k+1)} = v^{(k)} + \alpha_k \Delta v^{(k)} \qquad (2.8)$$

where $\Delta v^{(k)}$ is the solution of

$$H'(v^{(k)})\Delta v = -H(v^{(k)}) + \sigma_k \mu_k \tilde{e} \qquad (2.9)$$

On the other hand, when we solve system (2.6) (i.e. the KKT conditions for the k-th barrier problem (2.5)) by means of Newton interior–point method ([12]), we "solve inexactly" the k-th barrier problem by one iteration of Newton method and we choose the barrier parameter ρ_k in order to minimize the merit function

$$\Phi(v) = H(v)^T H(v) = \|H(v)\|_2^2 \qquad (2.10)$$

(with ρ_0 prespecified and $v^{(0)}$ feasible). If we set the barrier parameter $\rho_k = \sigma_k \mu_k$, then we state the Newton interior–point method by formulae (2.8) and (2.9) and it can be seen as an inexact Newton method (see [2], [10]).

Now the problem is how to choose the parameter μ_k in order to respect the condition (2.7) on the residual of inexact Newton method and to guarantee that $\Delta v^{(k)}$ is a descent direction for $\Phi(v)$. Furthermore, we should choose the parameter α_k in order to obtain feasible points $v^{(k)}$ and a global convergence to a solution. Before analysing the different choices for the parameters μ_k and α_k, we recall standard Newton assumptions for problem (2.1) or (2.3) (see [12]).

Consider the following conditions

A1 *Existence:* let $(z^{*T}, \lambda^{*T}, w^{*T})^T$ be a solution of (2.3).

A2 *Smoothness:* suppose $\nabla^2 f(z)$, $\nabla^2 h_i(z)$ $(i = 1, ..., m)$ and $\nabla^2 g_j(z)$ $(j = 1, ..., p)$ be locally Lipschitz continuous at z^*.

A3 *Regularity:* let the gradients of the active constraints in the solution be linearly independent. That is $\nabla h_1(z^*), ..., \nabla h_m(z^*), \nabla g_r(z^*)$ $r \in \mathcal{I}_A$, are linearly independent. Here $\mathcal{I}_A$ is the set of the indices of the inequality constraints which are active in z^*.

A4 *Second-Order Sufficient Conditions:* suppose the hessian of the lagrangian $\mathcal{L}(z, \lambda, w)$ associated with (2.1) be positive definite on the subspace $\mathcal{M} = \{\tilde{z} \mid \tilde{z} \neq 0, \nabla h_i(z^*)^T \tilde{z} = 0, i = 1, ..., m; \nabla g_r(z^*)^T \tilde{z} = 0, r \in \mathcal{I}_A\}$.

A5 *Strict Complementarity:* suppose the positiveness of the multipliers associated to the active constraints in z^*. That is $w_j^* - g_j(z^*) > 0$, $j = 1, ..., p$.

Suppose the conditions A1 and A2 hold: then the following statements are equivalent ([12]):

- Conditions A3–A5 also hold.

- The jacobian matrix $H'(v^*)$ is nonsingular.

Here $v^* = (z^{*T}, \lambda^{*T}, w^{*T}, s^{*T})^T$ and $s^* = -g(z^*)$

3. Choice of the Parameters.

In the first we consider the choice of the parameter μ_k at the iteration k of the method described by equations (2.9) and (2.8). As seen before, this parameter has to satisfy the criterium of inexact Newton method (2.7) and has to minimize the merit function $\Phi(v)$ defined in (2.10), that is μ_k has to satisfy the descent criterium for $\Phi(v)$

$$\nabla \Phi(v^{(k)})^T \Delta v^{(k)} \leq 0$$

where $\Delta v^{(k)}$ is the solution of (2.8).
From the definition of $\Phi(v)$ (2.10), from the formula (2.9) for $\Delta v = \Delta v^{(k)}$ and since $H(v^{(k)})^T \tilde{e} = s^{(k)T} w^{(k)}$, we have that

$$\nabla \Phi(v^{(k)})^T \Delta v^{(k)} = -2\|H(v^{(k)})\|_2^2 + 2\sigma_k \mu_k (s^{(k)T} w^{(k)})$$

In fact,

$$\begin{aligned}
\nabla \Phi(v^{(k)})^T \Delta v^{(k)} &= [2\nabla H(v^{(k)}) H(v^{(k)})]^T \Delta v^{(k)} = 2H(v^{(k)})^T H'(v^{(k)}) \Delta v^{(k)} \\
&= 2H(v^{(k)})^T [-H(v^{(k)}) + \sigma_k \mu_k \tilde{e}] \\
&= -2\|H(v^{(k)})\|_2^2 + 2\sigma_k \mu_k H(v^{(k)})^T \tilde{e}
\end{aligned}$$

We notice that, if

$$\mu_k \leq \frac{\|H(v^{(k)})\|_2^2}{s^{(k)T}w^{(k)}} \leq \frac{\|H(v^{(k)})\|_2^2}{\sigma_k(s^{(k)T}w^{(k)})} \tag{3.1}$$

then $\nabla\Phi(v^{(k)})^T\Delta v^{(k)} \leq 0$. Two different choices of μ_k that satisfies (3.1) are the one in [12]

$$\mu_k^{(1)} = \frac{s^{(k)T}w^{(k)}}{p} \tag{3.2}$$

and from formula (2.7) with the equality (see [10])

$$\mu_k^{(2)} = \frac{\|H(v^{(k)})\|_2}{\sqrt{p}} \tag{3.3}$$

We have also that

$$\mu_k^{(1)} \leq \mu_k^{(2)} \tag{3.4}$$

In order to prove that the values (3.2) and (3.3) of μ_k satisfy (3.1) and that inequality (3.4) holds, we use the following relations:

$$\frac{s^{(k)T}w^{(k)}}{p} = \frac{\|S^{(k)}W^{(k)}e\|_1}{p} \leq \frac{\|S^{(k)}W^{(k)}e\|_2}{\sqrt{p}} \leq \frac{\|S^{(k)}W^{(k)}e\|_1}{\sqrt{p}} = \frac{s^{(k)T}w^{(k)}}{\sqrt{p}}$$

and

$$\|S^{(k)}W^{(k)}e\|_2 \leq \|H(v^{(k)})\|_2$$

In fact,

$$\mu_k^{(1)} = \frac{s^{(k)T}w^{(k)}}{p} \leq \frac{(\sqrt{p}\|S^{(k)}W^{(k)}e\|_2)^2}{p(s^{(k)T}w^{(k)})} \leq \frac{\|H(v^{(k)})\|_2^2}{(s^{(k)T}w^{(k)})}$$

and

$$\mu_k^{(2)} = \frac{\|H(v^{(k)})\|_2}{\sqrt{p}} = \frac{\|H(v^{(k)})\|_2^2}{\sqrt{p}\|H(v^{(k)})\|_2} \leq \frac{\|H(v^{(k)})\|_2^2}{\sqrt{p}\|S^{(k)}W^{(k)}e\|_2}$$

$$\leq \frac{\|H(v^{(k)})\|_2^2}{(s^{(k)T}w^{(k)})}$$

finally, we have

$$\frac{(s^{(k)T}w^{(k)})}{p} \leq \frac{\|S^{(k)}W^{(k)}e\|_2}{\sqrt{p}} \leq \frac{\|H(v^{(k)})\|_2}{\sqrt{p}}$$

The choice $\mu_k^{(1)}$ is strictly connected with the idea of the adherence to the central path which is the basis of interior–point methods. Sometimes it happens that the value of $\mu_k^{(1)}$ is "too small" when we are far away from a solution. It implies that the system (2.6) is "too close" to the system (2.3) and it can produce stagnation of the current solution on the boundary of the feasible region as seen in the previous section.

In this case we propose to use other values for μ_k, for instance the value $\mu_k^{(2)}$ or any value between $\mu_k^{(1)}$ and $\mu_k^{(2)}$, to solve "more inexactly" system (2.3) and to obtain a different descent direction; we call "safeguards" these values, that are intended to prevent the forcing terms (if the method is seen as inexact Newton method) or the barrier terms (if the method is seen as Newton interior–point method) from becoming too small too quickly.

Moreover, we observe that, if μ_k belongs to the range $[\mu_k^{(1)}, \mu_k^{(2)}]$, formula (2.7) of inexact Newton method is satisfied ([10]).

The choice of the parameter α_k has to guarantee the feasibility and the global convergence of the method.

In the case of feasibility, we should choice α_k in such a way that $v^{(k)}$ belongs to the feasible region (i.e. $s^{(k)} > 0$, $w^{(k)} > 0$), the point $v^{(k+1)} = v^{(k)} + \alpha_k \Delta v^{(k)}$ belongs to the feasible region $s^{(k+1)} > 0$, $w^{(k+1)} > 0$.

In order to satisfy this condition, following [12], let us define two functions of α

$$\varphi_1^{(k)}(\alpha) = \min_{i=1,\dots,p} (S^{(k)}(\alpha)W^{(k)}(\alpha)e) - \gamma_k \tau_1 \left(\frac{s^{(k)}(\alpha)^T w^{(k)}(\alpha)}{p} \right)$$

$$\varphi_2^{(k)}(\alpha) = s^{(k)}(\alpha)^T w^{(k)}(\alpha) - \gamma_k \tau_2 \| H_1(v^{(k)}(\alpha)) \|_2$$

where

$$\tau_1 = \frac{\min_{i=1,\dots,p} (S^{(0)}W^{(0)}e)}{\left(\frac{s^{(0)^T} w^{(0)}}{p} \right)} \qquad \tau_2 = \frac{s^{(0)^T} w^{(0)}}{\| H_1(v^{(0)}) \|_2} \qquad \gamma_k \in [\tfrac{1}{2}, 1)$$

and

$$v^{(k)}(\alpha) = \begin{bmatrix} x^{(k)}(\alpha) \\ \lambda^{(k)}(\alpha) \\ w^{(k)}(\alpha) \\ s^{(k)}(\alpha) \end{bmatrix} = \begin{bmatrix} x^{(k)} \\ \lambda^{(k)} \\ w^{(k)} \\ s^{(k)} \end{bmatrix} + \alpha \begin{bmatrix} \Delta x^{(k)} \\ \Delta \lambda^{(k)} \\ \Delta w^{(k)} \\ \Delta s^{(k)} \end{bmatrix} = v^{(k)} + \alpha \Delta v^{(k)}$$

with $s^{(0)} > 0$, $w^{(0)} > 0$ that imply $\tau_1 > 0$ and $\tau_2 > 0$.

Furthermore, let us consider the domain

$$\Omega(\varepsilon) = \left\{ v \mid 0 < \varepsilon \le \Phi(v) \le \Phi(v^{(0)}); \quad \begin{cases} \min_{i=1,\dots,p}(SWe) \ge \frac{\tau_1}{2} \left(\frac{s^T w}{p} \right) \\ s^T w \ge \frac{\tau_1}{2} \| H_1(v) \|_2 \end{cases} \right\}$$

Let $\alpha_k^{(1)}$ and $\alpha_k^{(2)}$ be in $(0, 1]$ such that

$$\varphi_1^{(k)}(\alpha) \ge 0 \ \ \forall \alpha \in (0, \alpha_k^{(1)}] \qquad \text{and} \qquad \varphi_2^{(k)}(\alpha) \ge 0 \ \ \forall \alpha \in (0, \alpha_k^{(2)}]$$

These conditions imply

$$\min_{i=1,\dots,p} (S^{(k)}(\alpha)W^{(k)}(\alpha)e) \ge \gamma_k \tau_1 \left(\frac{s^{(k)}(\alpha)^T w^{(k)}(\alpha)}{p} \right) \qquad (3.5)$$

and

$$s^{(k)}(\alpha)^T w^{(k)}(\alpha) \geq \gamma_k \tau_2 \| H_1(v^{(k)}(\alpha)) \|_2 \tag{3.6}$$

This means that condition (3.5) keeps the iterates $v^{(k)}$ sufficiently far from the boundary of the feasible region and condition (3.6) obliges the sequence $\{s^{(k)}(\alpha)^T w^{(k)}(\alpha)\}$ to converge to zero slower than $\{\| H_1(v^{(k)}(\alpha)) \|_2\}$. Regarding to (3.5), in details, let $v^{(k)} \in \Omega(\varepsilon)$, since

$$(s^{(k)^T} w^{(k)})^2 \geq \| S^{(k)} W^{(k)} e \|_2^2$$

and for (3.6)

$$s^{(k)^T} w^{(k)} \geq \gamma_k \tau_2 \| H_1(v^{(k)}) \|_2 \quad \Rightarrow \quad (s^{(k)^T} w^{(k)})^2 \geq [\gamma_k \tau_2 \| H_1(v^{(k)}) \|_2]^2$$

then we have

$$
\begin{aligned}
2(s^{(k)^T} w^{(k)})^2 &\geq \| S^{(k)} W^{(k)} e \|_2^2 + [\gamma_k \tau_2 \| H_1(v^{(k)}) \|_2]^2 \\
s^{(k)^T} w^{(k)} &\geq \frac{1}{\sqrt{2}} \sqrt{\| S^{(k)} W^{(k)} e \|_2^2 + [\gamma_k \tau_2 \| H_1(v^{(k)}) \|_2]^2} \\
&\geq \min(1, \tfrac{1}{2} \tau_2) \sqrt{\| S^{(k)} W^{(k)} e \|_2^2 + \| H_1(v^{(k)}) \|_2^2} \\
&= \min(1, \tfrac{1}{2} \tau_2) \| H(v^{(k)}) \|_2
\end{aligned}
$$

If $H(v^{(k)}) \neq 0$ (i.e. $\| H(v^{(k)}) \|_2 \neq 0$, or $\Phi(v^{(k)}) \geq \varepsilon$) then $s^{(k)^T} w^{(k)} > 0$. From (3.5)

$$s_i^{(k)} w_i^{(k)} \geq \frac{\gamma_k \tau_1}{p} s^{(k)^T} w^{(k)} > 0 \quad i = 1, ..., p$$

thus, if $\{s^{(k)}\}$ and $\{w^{(k)}\}$ are bounded, we have $\{s_i^{(k)}, w_i^{(k)}\} > 0$.
In order to satisfy both conditions (3.5) and (3.6), the parameter α_k is chosen as the minimum between $\alpha_k^{(1)}$ and $\alpha_k^{(2)}$.
About the practical computation of $\alpha_k^{(1)}$ and $\alpha_k^{(2)}$ we can proceed as follows (see [10, Theor. 3.1, §4]):

$$\alpha_k^{(1)} = \sigma_k \mu_k (1 - \tau_1 \gamma_k) / \bar{L}$$

(here $\bar{L} = \max_{i=1,...,p} |\Delta s_i^{(k)} \Delta w_i^{(k)} - (\tau_1 \gamma_k / p) \Delta s^{(k)^T} \Delta w^{(k)}|$) and

$$\alpha_k^{(2)} = \sigma_k \mu_k p / |\Delta w^{(k)^T} \Delta s^{(k)}|$$

when the objective is quadratic, otherwise we choose $\alpha_k^{(2)} = \alpha_k^{(1)}$, check if this choice satisfies (3.6), else we moltiply $\alpha_k^{(2)}$, at most three times, by a factor $(0.8)^i$, $i = 1, 2, 3$, to satisfy (3.6). If it does not happen, we set $\alpha_k^{(2)} = \alpha_k^{(1)}$. Thus we call

$$\tilde{\alpha}_k = \min(\alpha_k^{(1)}, \alpha_k^{(2)}, 1)$$

and we determine the value α_k from $\tilde{\alpha}_k$ by using backtracking line search strategies as that of Armijo and Goldstein (e.g. see [8], [18]) or that of Eisenstat and Walker ([11]).

We recall here, both the backtracking procedures in pseudocode.

In both the cases, we set $\alpha_k = \tilde{\alpha}_k$, $\hat{v}^{(k)} = \alpha_k \Delta v^{(k)}$, $\theta \in (0,1)$, $\beta \in (0, \frac{1}{2}]$ and $\sigma_k \in (0,1)$.

For Armijo and Goldstein technique in pseudocode we have

$$
\begin{aligned}
&\textbf{while} \quad (\Phi(v^{(k)} + \hat{v}^{(k)}) > \Phi(v^{(k)}) + \alpha_k \beta \Phi'(v^{(k)})) \quad \textbf{do} \\
&\quad \left|
\begin{array}{ll}
\alpha_k &\leftarrow \quad \theta \alpha_k \\
\hat{v}^{(k)} &\leftarrow \quad \alpha_k \Delta v^{(k)}
\end{array}
\right. \\
&\textbf{end} \\
&v^{(k+1)} \leftarrow v^{(k)} + \hat{v}^{(k)}
\end{aligned}
\tag{3.7}
$$

with

$$
\Phi'(v^{(k)}) = \frac{\mathrm{d}}{\mathrm{d}\alpha} \Phi(v^{(k)}(\alpha))|_{\alpha=0}
$$

The "while" condition can be stated as to find $\alpha_k = \theta^l \tilde{\alpha}_k$, where l is the smallest nonnegative integer such that

$$
\Phi(v^{(k)}(\alpha_k)) \leq \Phi(v^{(k)}) + \alpha_k \beta \Phi'(v^{(k)})
\tag{3.8}
$$

then $v^{(k+1)} = v^{(k)}(\alpha_k)$.

In details, we have:

$$
\begin{aligned}
\frac{\mathrm{d}}{\mathrm{d}\alpha} \Phi(v^{(k)}(\alpha)) &= \frac{\mathrm{d}}{\mathrm{d}\alpha} \|H(v^{(k)} + \alpha \Delta v^{(k)})\|_2^2 \\
&= 2H(v^{(k)} + \alpha \Delta v^{(k)})^T H'(v^{(k)} + \alpha \Delta v^{(k)})\Delta v^{(k)} \\
&= 2H(v^{(k)}(\alpha))^T H'(v^{(k)}(\alpha))\Delta v^{(k)}
\end{aligned}
$$

thus

$$
\begin{aligned}
\Phi'(v^{(k)}) &= 2H(v^{(k)})^T H'(v^{(k)})\Delta v^{(k)} \\
&= 2H(v^{(k)})^T [-H(v^{(k)}) + \sigma_k \mu_k \tilde{e}] \\
&= -2\Phi(v^{(k)}) + 2\sigma_k \mu_k \left[(H_1(v^{(k)}))^T \quad (S^{(k)}W^{(k)}e)^T \right]^T \begin{bmatrix} 0 \\ e \end{bmatrix} \\
&= -2\Phi(v^{(k)}) + 2\sigma_k \mu_k s^{(k)T} w^{(k)}
\end{aligned}
$$

since from (3.1)

$$
\mu_k \leq \frac{\|H(v^{(k)})\|_2^2}{s^{(k)T} w^{(k)}} = \frac{\Phi(v^{(k)})}{s^{(k)T} w^{(k)}}
$$

we have

$$
\Phi'(v^{(k)}) \leq -2(1 - \sigma_k)\Phi(v^{(k)}) < 0
$$

and the condition (3.8) with the bound for $\Phi'(v^{(k)})$ computed above, implies

$$\Phi(v^{(k+1)}) \le [1 - 2\alpha_k\beta(1 - \sigma_k)]\Phi(v^{(k)}) \tag{3.9}$$

or

$$\|H(v^{(k+1)})\|_2 \le \sqrt{1 - 2\alpha_k\beta(1 - \sigma_k)}\|H(v^{(k)})\|_2$$

Because of the inequality

$$[1 - 2\alpha_k\beta(1 - \sigma_k)] < [1 - \alpha_k\beta(1 - \sigma_k)]^2 \tag{3.10}$$

we can also obtain

$$\Phi(v^{(k+1)}) \le [1 - \alpha_k\beta(1 - \sigma_k)]^2\Phi(v^{(k)}) \tag{3.11}$$

that suggests to formulate the Eisenstat and Walker backtracking technique as follows

$$\begin{aligned}
&\text{while} \quad (\|H(v^{(k)} + \hat{v}^{(k)})\|_2 > (1 - \alpha_k\beta(1 - \sigma_k))\|H(v^{(k)})\|_2) \;\; \text{do} \\
&\left|\begin{array}{l} \alpha_k \;\; \leftarrow \;\; \theta\alpha_k \\ \hat{v}^{(k)} \;\; \leftarrow \;\; \alpha_k\Delta v^{(k)} \end{array}\right. \\
&\text{end} \\
&v^{(k+1)} \;\; \leftarrow \;\; v^{(k)} + \hat{v}^{(k)}
\end{aligned} \tag{3.12}$$

We note that if α_k is computed by the Armijo and Goldstein technique and condition (3.1) holds, then condition (3.9) is satisfied and then (from (3.10)) α_k also satisfies Eisenstat and Walker backtracking condition (3.11) (see [11, Prop. 2.1]).

4. A Global Algorithm.

A global interior-point (GIP) algorithm for problem (2.1) can be stated as follows:

Set:

$$v^{(0)} = \begin{bmatrix} z^{(0)} \\ \lambda^{(0)} \\ w^{(0)} \\ s^{(0)} \end{bmatrix} \quad \text{with } s^{(0)} > 0, \; w^{(0)} > 0$$

$\theta \in (0, 1); \quad \beta \in (0, \frac{1}{2}]$ (backtracking parameters)

$\gamma_{-1} \in [\frac{1}{2}, 1)$ (parameter for α_k)

$\Phi(v^{(0)}) = \|H(v^{(0)})\|_2^2$

For $k = 0, 1, ...$ until $\Phi(v^{(k)}) \leq \varepsilon_{\text{exit}}$

- for some $\mu_k \in [\frac{s^{(k)T} w^{(k)}}{p}, \frac{\|H(v^{(k)})\|_2}{\sqrt{p}}]$

  $\sigma_k \in (0, 1)$

- $\Delta v^{(k)}$ solution of $H'(v^{(k)})\Delta v = -H(v^{(k)}) + \sigma_k \mu_k \tilde{e}$

- $\gamma_k \in [\frac{1}{2}, \gamma_{k-1}]; \quad \tilde{\alpha}_k = \min(\alpha_k^{(1)}, \alpha_k^{(2)}, 1)$

- $\alpha_k = \theta^l \tilde{\alpha}_k$ computed by a backtracking strategy ((3.7) or (3.12))

- $v^{(k+1)} = v^{(k)} + \alpha_k \Delta v^{(k)}$

end

The global convergence of the algorithm is obtained under the following assumptions which are also stated in [12].

(a) In the set $\Omega(0)$, the functions $f(z)$, $h_i(z)$, $(i = 1, ..., m)$ and $g_j(z)$, $(j = 1, ..., p)$ are twice continuously differentiable, $H_1'(v)$ is Lipschitz continuous and $h_i(x)$, $(i = 1, ..., m)$ are linearly independent.

(b) The sequence $\{z^{(k)}\}$ is bounded.

(c) The matrix $\nabla^2 \mathcal{L} + \nabla g(z) S^{-1} W \nabla g(z)^T$ is invertible for v in any compact set of $\Omega(0)$, where $s > 0$. Here $\nabla^2 \mathcal{L} = \nabla^2 f(z) + \sum_{i=1}^{m} \lambda_i \nabla^2 h_i(z) + \sum_{j=1}^{p} w_j \nabla^2 g_j(z)$.

(d) Let $\mathcal{I}_g^0 = \{i \mid 1 \leq i \leq p, \liminf_{k \to \infty} s_i^{(k)} = 0\}$, then the gradients $\nabla h_1(z^{(k)}), ..., \nabla h_m(z^{(k)}), \nabla g_r(z^{(k)})$ $r \in \mathcal{I}_g^0$, are linearly independent for k "sufficiently" large (i.e., if $g(z^{(k)}) - s^{(k)} \to 0$, for $k \to \infty$, then assumption (d) is equivalent to the regularity assumption A3).

It is possible to show the following theorems.
Theorem 1. Under the assumptions (a)–(d) and if $v^{(k)} \in \Omega(\varepsilon)$, $\varepsilon > 0$, then

- $\{v^{(k)}\}$ bounded above (Lemma 6.1 in [12])

- $\{(s^{(k)}, w^{(k)})\}$ is componentwise bounded away from zero

  (i.e. $\liminf_{k \to \infty}(s_i^{(k)}, w_i^{(k)}) > 0$), (Lemma 6.1 in [12]).

- $\{H'(v^{(k)})^{(-1)}\}$ is bounded (Lemma 6.2 in [12]).

- $\{\Delta v^{(k)}\}$ is bounded (Corollary 6.1 in [12]).

Theorem 2. Under the assumptions (a)–(d) and if $v^{(k)} \in \Omega(\varepsilon)$, $\varepsilon > 0$, $\{\sigma_k\}$ bounded away from zero (i.e. $\liminf_{k\to\infty} \sigma_k > 0$) and $\mu_k^{(1)} \leq \mu_k \leq \mu_k^{(2)}$ then

- $\{\tilde{\alpha}_k\}$ is bounded away from zero (i.e. $\liminf_{k\to\infty} \tilde{\alpha}_k > 0$), (Theorem 3.1 in [10]).

Corollary Under the assumptions (a)–(d) and if $v^{(k)} \in \Omega(\varepsilon)$, $\varepsilon > 0$, σ_k bounded away from zero and one and α_k bounded away from zero ($\alpha_k \leq 1$ for definition of $\tilde{\alpha}_k$), then from the "while" condition in (3.7) or in (3.12), we have that $\{\Phi(v^{(k)})\}$ (i.e. $\{\|H(v^{(k)})\|_2\}$) is a bounded, monotone nonincreasing sequence then it converges in $\mathbb{R}$.

Theorem 3. ([10, Theor. 3.2]) Let $v^{(k)}$ be generated by the global algorithm with $\varepsilon_{\text{exit}} = 0$ and backtracking technique (3.7); assume that σ_k is bounded away from zero and one and that $\nabla\Phi(v)$ is Lipschitz continuous. Then the sequence $\{H(v^{(k)})\}$ converges to zero (for $k \to \infty$). Furthermore any limit point v^* satisfies the KKT conditions for problem (2.1).

The same theorem holds when the backtracking strategy is (3.12); The hypothesis on the Lipschitz continuity of $\nabla\Phi(v)$ has to be dropped ([10, Theor. 3.6]).

Theorem 4. ([10, Theor. 3.3, Theor. 3.7]) If hypotheses in Theorem 3 are satisfied and v^* is a limit point of $v^{(k)}$ such that $H'(v^*)$ is invertible, then the sequence $\{v^{(k)}\}$ converges to v^* (for $k \to \infty$) and $H(v^*) = 0$.

5. Computational Experience.

In this section, four example problems are included to demonstrate the effectiveness of the method presented in this paper. The first example is a well known optimal control problem related to improving water quality by means of in-stream aeration process. The other examples are optimal control problems with linear and weakly nonlinear state equations related to the study of diffusion–convection processes.

These classical problems are sufficient to show how much the efficiency of the method can be increased when we make different choices of the descent direction in the Newton equation, safeguarding the method by a breakdown state and retaining fast local convergence in the final phase of the iteration.

5.1. Optimal In–Stream Aeration.

The concentration of dissolved oxygen (DO) in a river has come to be accepted as a criterion of water quality. In–water tests have shown that artificial aeration by means of in–stream mechanical or diffuser type aerators can be an economically attractive supplent or alternative to advanced wastewater treatment as a means of improving water quality. (Certain sources of pollution such as agricultural and storm runoff cannot reasonably be handled by treatment plants, and an in-stream process – such as chemical treatment or supplemental artificial aeration – is the only possible alternative). Once in–stream aeration has been accepted as a means of improving water quality, there is then the question of when, where and how much aeration is necessary to achieve the maximum increase in the DO level at minimum cost. This is a typical problem of optimal control of distributed parameter systems, in which the system to be controlled is described by a diffusion equation (Streeter–Phelps equation [21]), see for example [16]:

$$\frac{\partial C}{\partial t} = \frac{1}{A}\frac{\partial}{\partial x}\left(AE\frac{\partial C}{\partial x}\right) - V\frac{\partial C}{\partial x} + K_3(C_S - C) + U - K_1 L$$

with the initial condition

$$C(x,0) = C_0(x)$$

and boundary conditions

$$C(0,t) = \varphi_1(t) \qquad C(x_f,t) = \varphi_2(t)$$

where
x_f= estuary length (ft.).
$C(x,t)$= Dissolved Oxygen (DO) concentration (mg/l).
$C_S(x,t)$= DO saturation value (mg/l).
$A(x,t)$= estuary cross section area (sq ft).
$E(x,t)$= tidal dispersion (sq ft /day).
$V(x)$= time averaged velocity of the river (ft/day).
K_1= Biochemical Oxygen Demand (BOD) oxidation rate (1/day).
K_3: atmospheric aeration rate (1/day).
$L(x,t)$: BOD concentration (mg/l).
$U(x,t)$: aeration rate ((mg/l)/day).

The Biochemical Oxygen Demand concentration, as a function of effluent input, obeys a similar equation

$$\frac{\partial L}{\partial t} = \frac{1}{A}\frac{\partial}{\partial x}\left(AE\frac{\partial L}{\partial x}\right) - V\frac{\partial L}{\partial x} - K_1 L + S$$

The Dissolved Oxygen concentration $C(x,t)$ must satisfy a constraint of the form (*state–constraint*)

$$C(x,t) \le C_{max} \quad 0 \le x \le x_f, \ 0 \le t \le t_f \tag{5.1}$$

In the special case of a stream (no tidal action), assuming the cross sectional area constant, the model reduces to

$$\frac{\partial C}{\partial t} = -V\frac{\partial C}{\partial x} - K_3 C - K_1 L + K_3 C_S + U \tag{5.2}$$

$$C(x,0) = C_0(x) \tag{5.3}$$

$$C(0,t) = \varphi_1(t) \tag{5.4}$$

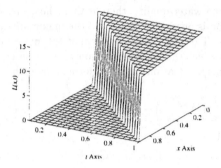

Figure 5.1.   Distribution of BOD.

and

$$\frac{\partial L}{\partial t} = -V\frac{\partial L}{\partial x} - K_1 L + S$$

where $S(x,t)$ = effluent discharge rate $((\text{mg/l})/\text{day})$.
In this paper we consider the following equation for the BOD concentration (J_1 is a constant)

$$L(x,t) = \begin{cases} \frac{J_1}{V}e^{-(K_1/V)x} & x < Vt \\ 0 & x \geq Vt \end{cases}$$

There are many relevant criterion functionals and the most popular one is the mean square. This functional has the advantage that the optimal control problem is a feedback control. The objective function to be minimized is therefore:

$$J(U) = \int_0^{t_f} \int_0^{x_f} D_1(x)\,(C - C_S)^2\,dxdt + \tilde{\rho}\int_0^{t_f} \int_{l_0}^{l_f} D_2(x)U^2 dxdt \qquad (5.5)$$

where the region of control is $[l_0, l_f] \subseteq [0, x_f]$ and t_f is the time duration of the control.

Minimizing this objective function over $U(x,t)$ will tend to drive the DO level towards the saturation value, while penalizing large control inputs. The non negative function $D_1(x)$ (dimensionless) and the positive function $D_2(x)$ (dimensionless) are appropriate weighting functions; $\tilde{\rho}$ is a scalar parameter (day^2).

The interval $0 \leq x \leq x_f$ is divided into N subintervals each of width h, so that $N \cdot h = x_f$. The space derivative in the differential equation (5.2) is replaced by the difference approximant

$$\frac{\partial C}{\partial x} = (C(x,t) - C(x - h,t))\,/h + \mathcal{O}(h) \qquad (5.6)$$

Equation (5.2) with (5.6), applied to the mesh points $x_i = i \cdot h$, $i = 1, \ldots, N$, produces a system of N first order ordinary differential equations of the form

$$\dot{y}(t) = Ay(t) + q(t) + u(t) \tag{5.7}$$

where
$y(t)$ is the approximation to the vector solution

$$(C(x_1, t), C(x_2, t), \ldots, C(x_N, t))^T \, ;$$

$u(t) = (U(x_1, t), \ U(x_2, t), \ \ldots, \ U(x_N, t))^T;$
$q(t) = ((\frac{V_1}{h}\varphi_1(t) + K_3 C_S(x_1, t) - K_1 L(x_1, t)), \ (K_3 C_S(x_2, t) - K_1 L(x_2, t)),$
   $\ldots, \ (K_3 C_S(x_N, t) - K_1 L(x_N, t)))^T;$

$$A = \begin{pmatrix} -\left(\frac{V_1}{h} + K_3\right) & & & \\ \frac{V_2}{h} & -\left(\frac{V_2}{h} + K_3\right) & & \\ & \ddots & \ddots & \\ & & \frac{V_N}{h} & -\left(\frac{V_N}{h} + K_3\right) \end{pmatrix}$$

with $V_i = V(x_i)$, $i = 1, 2, \ldots, N$.

The vector solution $y(t)$ of the matrix differential equation (5.7) is subject to the initial vector equation

$$y(0) = (C_0(x_1), \ C_0(x_2), \ \ldots, \ C_0(x_N))^T \tag{5.8}$$

The matrix A is a strictly diagonally dominant matrix, then it is non singular and the eigenvalues are $\lambda_i = -(\frac{V_i}{h} + K_3) < 0$, $i = 1, 2, \ldots N$. Thus, the continuous dynamic system (5.7) is *asymptotically stable* [19, p. 65].

Suppose that the fixed–time process (5.7)–(5.8) is of K steps duration. Then, the interval $0 \le t \le t_f$ is divided into K subintervals each of width Δt, so that $K \cdot \Delta t = t_f$; we assume that the controls are piecewise constant (i.e. $u(t) = u_j$ for $t \in [t_j, t_{j+1})$, $j = 0, 1, \ldots, K - 1$). Using the forward difference method for solving the matrix differential equation (5.7), we obtain at the time $t = (j+1)\Delta t$ for any time-level $j = 0, 1, \ldots, K - 1$ the following difference equations

$$\frac{y_{j+1} - y_j}{\Delta t} = Ay_j + q(t_j) + u_j$$

or

$$y_{j+1} = (I + \Delta t\, A)y_j + \Delta t\, q(t_j) + \Delta t\, u_j \tag{5.9}$$

for $j = 0, 1, \ldots, K - 1$, with $y_0 = y(0)$.

Here y_j is the approximation to $y(t)$ for

The eigenvalues of the matrix $\Omega = I + \Delta t\, A$ are $1 - \Delta t\,(V_i/h + K_3)$, $i = 1, \ldots, N$. Therefore, for $\Delta t < 2/(\bar{V}/h + K_3)$ with $\bar{V} = \max_{1 \le i \le N}\{V_i\}$, the discrete dynamic system (5.9) is *asymptotically stable* [19, p. 71].
Set

$$
z = \begin{pmatrix} y_1 \\ y_2 \\ \vdots \\ y_K \\ u_0 \\ u_1 \\ \vdots \\ u_{K-1} \end{pmatrix}, \qquad
\tilde{q} = \begin{pmatrix} \Delta t\, q(t_0) + \Omega y(0) \\ \Delta t\, q(t_1) \\ \vdots \\ \Delta t\, q(t_{K-1}) \end{pmatrix} \qquad \text{and}
$$

$$
R = \begin{pmatrix}
I & & & & \big| & -\Delta t I & & \\
-\Omega & I & & & \big| & & \ddots & \\
& \ddots & \ddots & & \big| & & & \ddots \\
& & -\Omega & I & \big| & & & -\Delta t I
\end{pmatrix}
$$

then equation (5.9) can be written in the form

$$
Rz = \tilde{q} \tag{5.10}
$$

An analogous discrete form for the state–constraints (5.1) is

$$
Gz \le C_{max} e \tag{5.11}
$$

where $G = [I\ 0]$ and $e = (1 \ldots 1)^T$.
For evaluating the double integral in (5.5) we use the trapezoidal rule to integrate with respect to the space variable, assuming that $[l_0, l_f] = [x_1, x_N]$, so we obtain

$$
\begin{aligned}
J(u(t)) &= \int_0^{t_f} \frac{1}{2} h[D_1(x_0)(\varphi_1(t) - C_S(x_0, t))^2 + \\
&\quad + 2 \sum_{i=1}^{N-1} D_1(x_i)(C(x_i, t) - C_S(x_i, t))^2 + D_1(x_N)(C(x_N, t) - C_S(x_N, t))^2] dt + \\
&\quad + \bar{\rho} \int_0^{t_f} \frac{1}{2} h[D_2(x_1) U^2(x_1, t) + \\
&\quad + 2 \sum_{i=2}^{N-1} D_2(x_i) U^2(x_i, t) + D_2(x_N) U^2(x_N, t)] dt
\end{aligned}
$$

then, the rectangular rule to integrate with respect to the time variable and we have

$$\tilde{J} = \frac{1}{2}h\Delta t \sum_{j=0}^{K-1} [D_1(x_0)(\varphi_1(t_j) - C_S(x_0,t_j))^2 +$$

$$+ \sum_{i=1}^{N-1} 2D_1(x_i)(C(x_i,t_j) - C_S(x_i,t_j))^2 + D_1(x_N)(C(x_N,t_j) - C_S(x_N,t_j))^2] +$$

$$+ \frac{1}{2}\tilde{\rho}h\Delta t \sum_{j=0}^{K-1} [D_2(x_1)U^2(x_1,t_j) + 2\sum_{i=2}^{N-1} D_2(x_i)U(x_i,t_j)^2 + D_2(x_N)U^2(x_N,t_j)]$$

Thus

$$\tilde{J} = \frac{1}{2}h\Delta t \sum_{j=1}^{K-1} [2D_1(x_1)(C^2(x_1,t_j) - 2C(x_1,t_j)C_S(x_1,t_j)) + \dots$$

$$\dots + 2D_1(x_{N-1})(C^2(x_{N-1},t_j) - 2C(x_{N-1},t_j)C_S(x_{N-1},t_j)) +$$

$$+ D_1(x_N)(C^2(x_N,t_j) - 2C(x_N,t_j)C_S(x_N,t_j))] +$$

$$+ \frac{1}{2}\tilde{\rho}h\Delta t \sum_{j=0}^{K-1} [D_2(x_1)U^2(x_1,t_j) + 2\sum_{i=2}^{N-1} D_2(x_i)U^2(x_i,t_j) + D_2(x_N)U^2(x_N,t_j)] + \hat{\gamma}$$

(here $\hat{\gamma}$ is a constant). If we define

$$D_1 = \begin{pmatrix} 2D_1(x_1) & & & \\ & \ddots & & \\ & & 2D_1(x_{N-1}) & \\ & & & D_1(x_N) \end{pmatrix} \qquad r = \begin{pmatrix} h\Delta t\, D_1 c_{S1} \\ h\Delta t\, D_1 c_{S2} \\ \vdots \\ h\Delta t\, D_1 c_{S(K-1)} \\ 0 \\ \vdots \\ 0 \end{pmatrix}$$

$$D_2 = \begin{pmatrix} D_2(x_1) & & & \\ & \ddots & & \\ & & 2D_2(x_{N-1}) & \\ & & & D_2(x_N) \end{pmatrix}$$

with $c_{Sj} = (C_S(x_1,t_j),\ C_S(x_2,t_j),\ \dots,\ C_S(x_N,t_j))^T$, and

$$Q = \begin{pmatrix} Q_y & 0 \\ 0 & \tilde{\rho}Q_u \end{pmatrix}$$

where

$$Q_y = \begin{pmatrix} h\Delta t D_1 & & & \\ & \ddots & & \\ & & h\Delta t D_1 & \\ & & & 0 \end{pmatrix}$$

$$Q_u = \begin{pmatrix} h\Delta t D_2 & & & \\ & \ddots & & \\ & & \ddots & \\ & & & h\Delta t D_2 \end{pmatrix}$$

we have that the optimal control problem described by formulae (5.1)–(5.4) and (5.5) has the form of the following quadratic programming problem

$$\text{minimize} \quad f(z) = \frac{1}{2} z^T Q z - r^T z \qquad (5.12)$$

$$\text{subject to} \quad Rz - \tilde{q} = 0$$

$$Gz - C_{max} e \leq 0$$

The matrix R has full row rank $K \cdot N$. The null spaces of Q and R have only the zero vector in common, i.e. the matrix $\begin{pmatrix} Q \\ R \end{pmatrix}$ has maximum rank $2K \cdot N$. Indeed:

$$\text{rank} \begin{pmatrix} Q \\ R \end{pmatrix} = \text{rank} \begin{pmatrix} Q_y & 0 \\ 0 & Q_u \\ R_1 & R_2 \end{pmatrix} = \text{rank} \begin{pmatrix} Q_y & 0 \\ R_1 & R_2 \\ 0 & Q_u \end{pmatrix}$$

$$= \text{rank} \begin{pmatrix} Q_y & R_1^T & 0 \\ 0 & R_2^T & Q_u \end{pmatrix}$$

Since Q_u is a symmetric positive definite matrix and rank $(Q_y \ R_1^T) = \text{rank } R_1 = K \cdot N$, the matrix $\begin{pmatrix} Q_y & R_1^T & 0 \\ 0 & R_2^T & Q_u \end{pmatrix}$ has maximum rank $2K \cdot N$, then the matrix $\begin{pmatrix} Q & R^T \\ R & 0 \end{pmatrix}$ is nonsingular.

Therefore, if there exists a feasible solution to the quadratic programming problem (5.12), then there exists a unique optimal solution [5, §§6.2, 6.3].

In test–problem # 1 the following data are given:

$x_f = 5$ miles	$t_f = 4$ days
$V(x) \equiv \bar{V} = 1$ mile/day	$J_1 = 15$
$K_1 = 0.16$ day^{-1}	$K_3 = 0.66$ day^{-1}
$C_0(x) = 6$ mg/l	$C_S(x,t) = 6$ mg/l
$\varphi_1(t) = 6$ mg/l	$C_{max}(x) = 7$ mg/l
$D_1(x) = 1$	$D_2(x) = 1$
$\tilde{\rho}$ is varying from 0.1 to 10	

We chose:

$N = 20, \ h = 5/20 \text{ mile} = 1/4 \text{ mile}$

$K = 20, \ \Delta t = 4/20 \text{ day} = 1/5 \text{ day}$

With these parameters the condition $\Delta t < 2/\left(\bar{V}/h + K_3\right)$ is satisfied.
At each iteration of the Newton method a system of $5 \cdot 20 \cdot 20 = 2000$ linear equations must be solved.

5.2. Diffusion Convection Processes.

A typical "diffusion–convection" process is obtained when a suspended material (effluent, sediment, etc.) is carried along ("convected") by a flow of fluid while at the same time its concentration is being attenuated ("diffused") within the flow. The equation describing this process, written for brevity in its one–dimensional form, is

$$\Psi_t = D\Psi_{xx} - V\Psi_x \tag{5.13}$$

where $\Psi(x,t)$ describes the concentration of a suspension convected with velocity V and diffusing according to the "diffusion coefficient" D. Of course equation (5.13) greatly oversimplifies the physical realities of turbulent flow and, in particular, practical difficulty attaches to the measurement of the number D. A useful measure of the degree of "convectiveness" in a problem is the dimensionless Peclet number $P_e = V \cdot x/D$, where x is some reference distance. In a numerical context a mesh step size h is associated with a "cell Peclet number" $P_e = V \cdot h/D$.

In general, we are concerned with a diffusion–convection process which is described by a differential equation of the form:

$$\Psi_t = a\Psi_{xx} + b\Psi_x - c\Psi + d + \tilde{\sigma}U \quad (x,t) \in [0,1] \times [0,1] \tag{5.14}$$

where $\Psi(x,t)$ is a function of the arguments x (space) and t (time) which characterizes the state of the controlled system and $U(x,t)$ is a function which characterizes the control actions of the system. The parameters a, b, c are known and constant with $a > 0$, $c \geq 0$; $\tilde{\sigma}$ is a constant or a given function of the argument Ψ. The function d is the source term; it may be a function of $\Psi(x,t)$, for example $d(x,t) = x\cos\left(\frac{\pi}{2}x\right)e^{-t} + 3e^{\Psi(x,t)}$.

The solution of (5.14) is usually discussed with respect to the initial condition

$$\Psi(x,0) = \Psi_0(x) \tag{5.15}$$

and the boundary conditions

$$\Psi(0,t) = g_0(t) \quad \Psi(1,t) = g_1(t) \tag{5.16}$$

where $\Psi_0(x)$, $g_0(x)$, $g_1(x)$ are given functions with $\Psi_0(0) = g_0(0)$, $\Psi_0(1) = g_1(0)$.

To obtain finite difference approximations of the diffusion–convection problem (5.14)-(5.16), we discretize first only the spatial variable x, leaving the time variable t continuous. The interval $0 \leq x \leq 1$ is divided into N subintervals each of width h, so that $N \cdot h = 1$. The space derivatives in the differential equation (5.14) are replaced by the central difference approximants

$$\Psi_{xx} = (\Psi(x - h, t) - 2\Psi(x, t) + \Psi(x + h, t))/h^2 + \mathcal{O}(h^2) \qquad (5.17)$$

$$\Psi_x = (\Psi(x + h, t) - \Psi(x - h, t))/(2h) + \mathcal{O}(h^2) \qquad (5.18)$$

Equation (5.14) with (5.17), (5.18), applied to the mesh points $x_i = i \cdot h$, $i = 1, \ldots N - 1$, produces a system of $N - 1$ first order ordinary differential equations of the form

$$\dot{y}(t) = Ay(t) + q(t) + Bu(t) \qquad (5.19)$$

where
$y(t)$ is the approximation to the vector solution

$$(\Psi(x_1, t), \Psi(x_2, t), \ldots, \Psi(x_{N-1}, t))^T;$$

$$A = \begin{pmatrix} -\left(\frac{2a}{h^2} + c\right) & \left(\frac{a}{h^2} + \frac{b}{2h}\right) & & & \\ \left(\frac{a}{h^2} - \frac{b}{2h}\right) & -\left(\frac{2a}{h^2} + c\right) & \left(\frac{a}{h^2} + \frac{b}{2h}\right) & & \\ & \ddots & \ddots & \ddots & \\ & & \ddots & \ddots & \left(\frac{a}{h^2} + \frac{b}{2h}\right) \\ & & & \left(\frac{a}{h^2} - \frac{b}{2h}\right) & -\left(\frac{2a}{h^2} + c\right) \end{pmatrix}$$

$u(t) = (U(x_1, t), U(x_2, t), \ldots, U(x_{N-1}, t))^T;$
$q(t) = \left(\left(d(x_1, t) + \left(\frac{a}{h^2} - \frac{b}{2h}\right) g_0(t)\right), d(x_2, t), \ldots, d(x_{N-2}, t),\right.$
$\left.\left(d(x_{N-1}, t) + \left(\frac{a}{h^2} + \frac{b}{2h}\right) g_1(t)\right)\right)^T$
and
$$B = \text{diag}\{\tilde{\sigma}(y_1(t)), \tilde{\sigma}(y_2(t)), \ldots, \tilde{\sigma}(y_{N-1}(t))\}$$

The vector solution $y(t)$ of the matrix differential equation (5.19) is subject to the initial vector condition

$$y(0) = (\Psi_0(x_1), \Psi_0(x_2), \ldots, \Psi_0(x_{N-1}))^T \qquad (5.20)$$

For each h such that

$$h < 2a/|b| \qquad (5.21)$$

the tridiagonal matrix A is irreducibly diagonally dominant [22, p. 23] and essentially positive [22, p. 257]. Thus, the matrix A is non singular and any eigenvalue λ_i ($i = 1, \ldots N - 1$) of A is different from zero. Besides, the matrix A has a real simple eigenvalue λ_1 such that the associated eigenvector z_1 is positive and $Re(\lambda_i) < \lambda_1$ for any other eigenvalue λ_i of A ($i = 2, \ldots N - 1$) [22, p. 258]. Now, the matrix $-A$ is an M–matrix with $-A^{-1} > 0$ [22, p. 85]. From $Az_1 = \lambda_1 z_1$ with $z_1 > 0$, it follows that $-\frac{1}{\lambda_1} z_1 = -A^{-1} z_1 > 0$. Thus, $\lambda_1 < 0$. Since $Re(\lambda_i) < \lambda_1 < 0$ ($i = 2, \ldots N - 1$), we have the result that, if the matrix B has constant entries, the continuous dynamic system (5.19) is *asymptotically stable* [19, p. 65].

Suppose that the fixed–time process (5.19)–(5.20) is of K steps duration. Then, the interval $0 \leq t \leq 1$ is divided into K subintervals each of width Δt, so that $K \cdot \Delta t = 1$; we assume that the controls are piecewise constant (i.e. $u(t) = u_j$ for $t \in (t_j, t_{j+1}]$, $j = 0, \ldots, K-1$). Using the backward difference implicit method for solving the matrix differential equation (5.19), we obtain at the time $t = (j+1) \cdot \Delta t$ for any time–level $j = 0, \ldots, K - 1$ the following difference equations

$$\frac{y_{j+1} - y_j}{\Delta t} = A y_{j+1} + q_{j+1} + B u_j$$

or

$$(I - \Delta t\, A)\, y_{j+1} = y_j + \Delta t\, q_{j+1} + \Delta t\, B u_j \qquad (5.22)$$

where B is the $(N - 1) \times (N - 1)$ diagonal matrix

$$B = \text{diag} \ \{\tilde{\sigma}(y_{1j}), \tilde{\sigma}(y_{2j}), \ldots, \tilde{\sigma}(y_{(N-1)j})\}$$

Here, y_j is the approximation to the vector $(y_1(t_j), y_2(t_j), \ldots, y_{N-1}(t_j))^T$ for $j = 0, \ldots, K - 1$, with $y_0 = y(0)$.

Since $-A$ is an M–matrix, the matrix $\Omega = I - \Delta t\, A$ is an M–matrix with $(I - \Delta t\, A)^{-1} \leq -(\Delta t\, A)^{-1}$ [19, p. 109]. The eigenvalues of the matrix $(I - \Delta t\, A)^{-1}$ are $(1 - \Delta t\, \lambda_i)^{-1}$. Since $Re(\lambda_i) < 0$, $i = 1, \ldots, N - 1$, the real part of $(1 - \Delta t\, \lambda_i)$ is greater than unity for all $\Delta t > 0$. Then the discrete dynamic system (5.22) is *asymptotically stable* [19, p. 71].

Set $z = \begin{pmatrix} y_1 \\ y_2 \\ \vdots \\ y_K \\ u_0 \\ u_1 \\ \vdots \\ u_{K-1} \end{pmatrix}$, $\quad \tilde{q} = \begin{pmatrix} \Delta t\, q_1 + y(0) \\ \Delta t\, q_2 \\ \vdots \\ \Delta t\, q_K \end{pmatrix}$ and

$$R = \begin{pmatrix} \Omega & & & & | & -\Delta t\, B \\ -I & \Omega & & & | & & \ddots \\ & \ddots & \ddots & & | & & & \ddots \\ & & -I & \Omega & | & & & & -\Delta t\, B \end{pmatrix}$$

then equations (5.22) can be written in the form

$$Rz = \tilde{q} \tag{5.23}$$

An analogous discrete form for the state–constraints (5.26) is

$$Gz \leq \Psi_{max} e \tag{5.24}$$

where $G = [I\ 0]$ and $e = (1 \ldots 1)^T$.

We now wish to control the system (5.14)–(5.16) in such a way that the objective function

$$J(u) = \int_0^1 \int_0^1 \left(\Psi^2(x,t) + \tilde{\rho} U^2(x,t) \right) dx\,dt + \int_0^1 \Psi^2(x,1)dx \tag{5.25}$$

is minimized ($\tilde{\rho} > 0$) with the state–constraint

$$\Psi(x,t) \leq \Psi_{max} \quad 0 \leq x \leq 1, 0 \leq t \leq 1 \tag{5.26}$$

For evaluating the double integral in (5.25), we use the rectangular rule to integrate respect to the space variable and, we assume $U(x_0, t) = 0$ ($t \in [0,1]$), so that we obtain

$$J(u(t)) = \int_0^1 h \sum_{i=0}^{N-1} [\Psi^2(x_i, t) + \tilde{\rho} U^2(x_i, t)]dt +$$

$$+h \sum_{i=0}^{N-1} \Psi^2(x_i, 1)$$

then, using again the rectangular rule to integrate respect to the time variable, we have

$$
\begin{aligned}
\tilde{J} &= h\Delta t \sum_{j=0}^{K-1} [\Psi^2(x_0, t_j) + \tilde{\rho}U^2(x_0, t_j) + \sum_{i=1}^{N-1}(\Psi^2(x_i, t_j) + \\
&\quad + \tilde{\rho}U^2(x_i, t_j))] + h\Psi^2(x_0, t_K) + h\sum_{i=1}^{N-1}\Psi^2(x_i, t_K) \\
&= h\Delta t[\sum_{j=1}^{K-1}\sum_{i=1}^{N-1}\Psi^2(x_i, t_j) + \sum_{j=0}^{K-1}\sum_{i=1}^{N-1}\tilde{\rho}U^2(x_i, t_j)] + \\
&\quad + h\sum_{i=1}^{N-1}\Psi^2(x_i, t_K) + \hat{\gamma}
\end{aligned}
$$

(here $\hat{\gamma}$ is a constant). If we define

$$
Q = \begin{pmatrix} Q_y & 0 \\ 0 & \tilde{\rho}Q_u \end{pmatrix}
$$

where

$$
Q_y = \begin{pmatrix} 2h\Delta t\,I & & & \\ & \ddots & & \\ & & 2h\Delta t\,I & \\ & & & 2h I \end{pmatrix}
$$

$$
Q_u = \begin{pmatrix} 2h\Delta t\,I & & & \\ & \ddots & & \\ & & \ddots & \\ & & & 2h\Delta t\,I \end{pmatrix}
$$

we have that the optimal control problem described by formulae (5.25), (5.14)–(5.16) and (5.26) has the form of the following nonlinear programming problem

$$
\begin{aligned}
\text{minimize} \quad & f(z) = \frac{1}{2}z^T Q z & (5.27) \\
\text{subject to} \quad & Rz - \bar{q} = 0 \\
& Gz - \Psi_{max}e \leq 0
\end{aligned}
$$

At each iteration of the Newton method a system of $= 5 \cdot K \cdot (N-1)$ linear equations must be solved.

In test–problems, the following data are given:

# 2	$K = 6$	$N = 41$	$a = 1$	$b = 70$	$c = 8$
	$K = 10$	$N = 41$	$a = 1$	$b = 70$	$c = 12$
	$\tilde{\sigma}$ constant	$d(x,t)$ independent on $\Psi(x,t)$			
# 3	$K = 6$	$N = 41$	$a = 1$	$b = 1$	$c = 12$
	$\tilde{\sigma}$ constant	$d(x,t) = \nu(x,t) - \delta \exp^{\Psi(x,t)}$			
# 4	$K = 10$	$N = 41$	$a = 1$	$b = 1$	$c = 12$
	$\tilde{\sigma}$ constant	$d(x,t) = \nu(x,t) - \delta \Psi^2(x,t)$			

With these choices, condition (5.21) is satisfied.

For test-problems $\# 2 - \# 4$, we always set $\Psi_{max} = 2$, $\tilde{\rho} = 0.5$ and $\Psi_0(x) = x\cos\left(\frac{\pi}{2}x\right)$.

5.3. Numerical Results.

In this section we report the numerical results obtained on test–problems $\#1$–$\#4$ by the (GIP) Algorithm implemented in Fortran code (double precision) and carried out on SUN Sparc Station Ultra 30.

The perturbed Newton equation is solved by the routines in [1] and we always use the Armijo and Goldstein condition for backtracking. In the following the vector of the unknowns is $v = (y^T, u^T, \lambda^T, w^T, s^T)^T$ and we denote with $\mu_{1:}^{(1)} = s^{(k)T} w^{(k)}/p$, $\mu_k^{(2)} = \|H(v^{(k)})\|_2/\sqrt{p}$ and $\mu_k^{(M)}$ a value between $\mu_k^{(1)}$ and $\mu_k^{(2)}$.

We choose the following values for the parameters: $\vartheta = 0.5$, $\beta = 10^{-4}$, $\sigma_k = \min\left(0.4, 100 \cdot s^{(k)T} w^{(k)}\right)$ and we use an approximation for the steplength α_k (see §3). Moreover in tables we adopt the following notations

dp: dimension of the problem.

$v^{(0)}$: starting point $= (y^{(0)T}, u^{(0)T}, \lambda^{(0)T}, w^{(0)T}, s^{(0)T})^T$. We consider two possibilities for $v^{(0)}$, $v_1 = (e^T, e^T, e^T, e^T, e^T)^T$ and $v_2 = (e^T, e^T, e^T, 0.1 \cdot e^T, 0.1 \cdot e^T)^T$, where $e = (1, \ldots, 1)^T$.

H_0: $\|H(v^{(0)})\|_2$.

it: number of iterations to obtain $\|H\|_2 < 10^{-8}$. The number in the square brackets represents the iteration $\tilde{k}$ at which the steplength $\alpha_{\tilde{k}}$ becomes one.

s.o.: order of the step in the first iterations.

In order to have a solution to compare with the one computed by (GIP), we solve Problem $\# 1$ and $\# 2$ without the inequality contraints, using the routines in [1] to obtain the solution $\tilde{z} = (\bar{y}^T, \bar{u}^T)^T$; this solution also enable us to choose a suitable value for C_{max} (or Ψ_{max}). Then, we solve Problem $\# 1$ and $\# 2$ (with the equality and inequality constraints) by (GIP), obtaining the solution $z = (y^T, u^T)^T$. The last columns in Tables 1 and 2 gives the absolute error for state and control ($\|z - \tilde{z}\|_\infty$).

Table 1: test–problem # 1

$\tilde{\rho}$	$v^{(0)}$	H_0	it $(\mu_k^{(1)})$	it $(\mu_k^{(2)})$	s.o. $(\mu_k^{(1)})$	s.o. $(\mu_k^{(2)})$	ECS $(\mu_k^{(1)})$	ECS $(\mu_k^{(2)})$
0.1	v_1	1.086D+2	17 [3]	19 [2]	1D–1	1D–1	1D–8	1D–10
	v_2	1.229D+2	42 [34]	22 [3]	1D–2	1D–1	1D–11	1D–9
1	v_1	1.086D+2	19 [5]	19 [2]	1D–1	1D–1	1D–8	1D–9
	v_2	1.229D+2	72 [63]	22 [3]	3D–3	1D–2	1D–9	1D–9
10	v_1	1.096D+2	22 [8]	19 [2]	1D–2	1D–1	1D–10	1D–10
	v_2	1.238D+2	106 [97]	22 [3]	1D–4	1D–2	1D–10	1D–13

Table 2

#	dp	$v^{(0)}$	H_0	it $(\mu_k^{(1)})$	it $(\mu_k^{(2)})$	s.o. $(\mu_k^{(1)})$	s.o. $(\mu_k^{(2)})$	ESC $(\mu_k^{(1)})$
2	1200	v_1	1.794D+3	17 [4]	24 [3]	1D–1	1D–1	1D–10
		v_2	1.792D+3	76 [68]	26 [5]	1D–3	1D–2	1D–12
2	2000	v_1	1.392D+3	18 [4]	24 [2]	1D–1	1D–1	1D–12
		v_2	1.389D+3	90 [82]	26 [5]	1D–4	1D–3	1D–11

Table 3

#	dp	δ	$v^{(0)}$	H_0	it $(\mu_k^{(1)})$	it $(\mu_k^{(2)})$	s.o. $(\mu_k^{(1)})$	s.o. $(\mu_k^{(2)})$
3	1200	0.1	v_1	1.3861D+3	16 [3]	24 [3]	1D–1	1D–2
			v_2	1.3834D+3	62 [57]	26 [6]	7D–4	9D–4
		1	v_1	1.388D+3	16 [4]	23 [3]	1D–1	1D–1
			v_2	1.3853D+3	63 [56]	26 [6]	7D–4	5D–4
		10	v_1	1.4091D+3	16 [3]	23 [2]	1D–1	2D–1
			v_2	1.4058D+3	63 [56]	25 [6]	8D–4	1D–3
		5000	v_1	4.09064D+4	15 [3]	25 [1]	4D–1	1D+0
			v_2	4.08943D+4	50 [43]	26 [2]	1D–3	6D–2
4	2000	1	v_1	1.076D+3	17 [5]	23 [2]	1D–1	2D–1
			v_2	1.0725D+3	66 [59]	26 [6]	7D–4	1D–3
		10	v_1	1.0871D+3	16 [5]	fail	1D–1	1D–4
			v_2	1.083D+3	65 [58]	fail	8D–4	5D–4
		100	v_1	1.2643D+3	17 [3]	29 [3]	2D–1	5D–1
			v_2	1.2556D+3	61 [53]	fail	1D–3	6D–4

Carla Durazzi – Emanuele Galligani

Table 4: Test–problem # 4

dp	δ	$v^{(0)}$	it $(\mu_k^{(1)})$	it $(\mu_k^{(M)})$	it $(\mu_k^{(2)})$	s.o. $(\mu_k^{(1)})$	s.o. $(\mu_k^{(M)})$	s.o. $(\mu_k^{(2)})$
1200	10	v_2	64 [57]	13 [9]	fail	8D–4	1D–2	1D–3
2000	10	v_2	65 [58]	12 [10]	fail	8D–4	1D–2	5D–4
	100	v_2	61 [53]	11 [10]	fail	1D–3	1D–2	6D–4

Tables 1–2 show the different behaviour of the (GIP) algorithm with different starting points and different value for μ: when $v^{(0)} = v_1$ the two choices for μ are almost equivalent even if we obtain less (few) iterations with $\mu = \mu_k^{(1)}$. On the other hand, when $v^{(0)} = v_2$ and $\mu = \mu_k^{(1)}$, the iterate $v^{(k)}$ gets too close to the boundary and consequently the steplength α_k is too small: this causes the algorithm to perform a large number of iterations before the stop condition. In the considered tests–problems, few iterations are instead required when $\mu = \mu_k^{(2)}$.

Moreover from Tables 3–4, it follows that a value $\mu_k^{(M)}$ (where $\mu_k^{(1)} < \mu_k^{(M)} < \mu_k^{(2)}$) is more suitable: in fact $\mu_k^{(2)}$ is too large ((GIP) fails) and $\mu_k^{(1)}$ is too small (too iterations for the convergence), while the number of iterations reduces by the choice $\mu_k^{(M)}$.

A relevant conclusion that can be drawn by these experiments is that the (GIP) algorithm leads to satisfactory results on the considered control problems, if we take care of choosing μ in such a way that the steplength α_k is sufficiently large.

References

[1] E. Anderson, et al., *LAPACK Users' Guide*. SIAM, Philadelphia, 1992.

[2] S. Bellavia, *Inexact interior–point method*, Journal of Optimization Theory and Applications 96 (1998), 109–121.

[3] P. Benner, *Computational methods for linear–quadratic optimization*, Proceedings of Numerical Methods in Optimization, June 1997, Cortona, Italy. Rendiconti del Circolo Matematico di Palermo Ser. II, Suppl. 58 (1999), 21–56.

[4] P.T. Boggs, A.J. Kearsley, J.W. Tolle, *A practical algorithm for general large scale nonlinear optimization problems*, SIAM Journal on Optimization 9 (1999), 755–778.

[5] M.D. Canon, C.D. Cullum, E. Polak, *Theory of Optimal Control and Mathematical Programming*, Mc Graw–Hill, New York, 1970.

[6] C.H. Choi, A.J. Laub, *Efficient matrix–valued algorithms for solving stiff Riccati differential equations*, Proceedings of the 28th Conference on Decision and Control, December 1989, Tampa, Florida, 885–887.

[7] R.S. Dembo, S.C. Eisenstat, T. Steihaug, *Inexact Newton methods*, SIAM Journal on Numerical Analysis 19 (1982), 400–408.

[8] J.E. Dennis, R.B. Schnabel, *Numerical Methods for Unconstrained Optimization and Nonlinear Equations*, Prentice–Hall, Englewood Cliffs, 1983.

[9] C.R. Dohrmann, R.D. Robinett, *Dynamic programming method for constrained discrete–time optimal control*, Journal of Optimization Theory and Applications 101 (1999), 259–283.

[10] C. Durazzi, *On the Newton interior–point method for nonlinear programming problems*, Journal of Optimization Theory and Applications 104 (2000), 73–90.

[11] S.C. Eisenstat, H.F. Walker, *Globally convergent inexact Newton methods*, SIAM Journal on Optimization 4 (1994), 393–422.

[12] A.S. El-Bakry, R.A. Tapia, T. Tsuchiya, Y. Zhang, *On the formulation and theory of the Newton interior–point method for nonlinear programming*, Journal of Optimization Theory and Applications 89 (1996), 507–541.

[13] T.M. El-Gindy, H.M. El-Hawary, M.S. Salim, M. El-Kady, *A Chebyshev approximation for solving optimal control problems*, Computers & Mathematics with Applications 29 (1995), 35–45.

[14] C.A. Floudas, *Deterministic Global Optimization in Design, Control and Computational Chemistry*, IMA Proceedings: Large Scale Optimization with Applications. Part II: Optimal Design and Control, Biegler LT, Conn A, Coleman L, Santosa F, (editors) 93 (1997), 129–187.

[15] W. Hager, P.M. Pardalos (editors), *Optimal Control: Theory, Algorithms and Applications*, Kluwer Academic Publishers, 1998.

[16] W. Hullet, *Optimal estuary aeration: an application of distributed parameter control theory*, Applied Mathematics and Optimization 1 (1974), 20–63.

[17] C.T. Kelley, *Iterative Methods for Linear and Nonlinear Equations*, SIAM, Philadelphia, 1995.

[18] D.G. Luenberger, *Linear and Nonlinear Programming*, Addison–Wesley, Reading, 1984.

[19] J.M. Ortega, *Numerical Analysis: A Second Course*, Academic Press, New York, 1972.

[20] E.R. Pinch, *Optimal Control and the Calculus of Variations*, Oxford University Press, Oxford, 1993.

[21] H.W. Streeter, E.B. Phelps, *A study of the pollution and natural purification of the Ohio river*, U.S. Public Health Bulletin 1925; n. 146.

[22] R.S. Varga, *Matrix Iterative Analysis*. Prentice–Hall, Englewood Cliffs, 1962.

[23] J. Vlassenbroeck, R. Van Dooren, *A Chebyshev technique for solving nonlinear optimal control problems*. IEEE Transactions on Automatic Control 33 (1988), 333–340.

OPTIMAL FLOW PATTERN IN ROAD NETWORKS

Paolo Ferrari
University School of Engineering of Pisa

Abstract   The traditional theory of road networks defines as system optimal a feasible flow vector which minimises the network total transport cost, and shows that this optimal vector can be transformed into an equilibrium one if additional costs, equal to the difference between marginal costs and private costs, are imposed on each link of the network. However some authors have shown that, could tolls be imposed on all links of a network, there would be an infinity of such optimal vectors. But in real life tolls can be charged only on some links of road networks, and in this case very often the set of optimal toll vectors is empty. Moreover, the real reason for which tolls are imposed on some network links is not to minimise total transportation costs, but to recover road maintenance expenses and, at least in part, construction costs. Therefore, the optimal toll vector and corresponding flow pattern are those which produce a partition of road costs between road users and society as a whole in such a way as to maximise social welfare. This paper presents a theory of optimal flow pattern founded on this principle. A method of computing optimal tolls is proposed and is applied to a real network.

Keywords: system optimal vector, optimal road pricing, marginal cost of public funds, downhill simplex.

1. Introduction.

The theory of road networks defines the *system optimal*, a vector of link flows belonging to the set of feasible flow vectors that minimises the total transport cost of the network. Although the system optimal vector is not an equilibrium vector, it can be transformed into one if additional costs, in the form of tolls paid by drivers, are imposed on the

F. Giannessi et al (eds.),
Equilibrium Problems: Nonsmooth Optimization and Variational Inequality Models, 101–117.
© 2001 *Kluwer Academic Publishers.*

network links. Beckmann et al. (1956) have shown that, if the link cost functions are symmetric, the desired result can be obtained by imposing a toll equal to the difference between marginal costs and private costs on each link of the network.

Recently some authors (Bergendorff et al., 1997) have shown that an infinite number of toll vectors exists that can transform the system optimal vector into an equilibrium vector. And, for the sake of completeness, it can be seen that an infinite number of toll vectors exists that can transform any feasible flow vector into an equilibrium vector. The additional costs which transform the system optimal vector into an equilibrium vector may be positive or negative, i.e., they may represent money transfer from road users to society as a whole, or viceversa. As the theory does not consider the social costs of such transfers, the additional costs vector remains indeterminate. However, in attempting to make this vector definite, some authors (Hearn and Ramana, 1998) have introduced conditions which implicitly consider such transfers to have social costs, though these latter are not defined: for instance they dictate that additional costs cannot be negative, i.e., they must be tolls paid out by drivers, and that the total additional cost borne by road users must be minimised.

Moreover, in real life tolls can be charged only on some links of road networks, so that many components of the additional cost vector are necessarily zero. This constraint, along with the condition that non-zero components must be positive, defines the subset of the feasible flow vectors which can be transformed into equilibrium vectors; very often this subset does not include the system optimal vector.

As a matter of fact, the real reason for which tolls are imposed on some network links is not to minimise total transportation costs, but to recover road maintenance expenses and, at least in part, construction costs. In some cases, it is also used as a means to avoid exceeding the networks capacity. Therefore, the optimal toll vector and corresponding flow pattern are those which produce a partitioning of road costs between road users and society as a whole in such a way as to maximise social welfare.

The purpose of this paper is to present a theory which defines the conditions determining an optimal division of road costs between road users and society as a whole. Section 2 presents a critical review of the traditional theory of system optimization. Section 3 is dedicated to a new theory of optimal toll pricing in road networks. Its starting point is definition of a social welfare function, which accounts for both the willingness of road users to pay and the damage caused to the economy if road costs are subsidised through public financing by imposing taxes on society as a whole. Section 4 shows a method of computing the optimal vector of additional costs and optimal flow pattern. An application of the optimal toll pricing theory to a real case is the subject of Section 5. Lastly, some concluding remarks are presented in Section 6.

2. The traditional theory of system optimization.

Consider the graph $G(N.L)$ of a car road network in which N is the set of nodes and L is the set of links, n and l being the number of nodes and links, respectively. Some nodes are the origins and/or destinations of transportation demand. $w = (i, j)$, $i \neq j$, represents an ordered pair of centroids; W is the set of pairs w, and m is their number. Let:

d_w=the transportation demand between $w \in W$

d=the transportation demand vector between all $w \in W$

P_w=the set of paths p connecting $w \in W$

$P = \bigcup_{w \in W}$

M=the number of paths $p \in P$

B=the incidence matrix $(m \cdot M)$ between pairs $w \in W$ and paths $p \in P$

h_p=the flow on path $p \in P$

h=the flow vector on all $p \in P$

f_i=the flow on link $i \in L$

f=the vector of flows on all links $i \in L$

A=the incidence matrix $(l \cdot M)$ between links $i \in L$ and paths $p \in P$.

We have $Bh = d$ and $Ah = f$. We assume that the network capacity constraints are not binding, so that the set of the feasible flows on paths $p \in P$ is:

$$\Omega = \{h : h \in \mathbb{R}^M, \ h \geq 0, \ Bh - d = 0\} \tag{1}$$

and the set of the feasible link flows is:

$$\Theta = \{f : f \in \mathbb{R}^l, \ f = Ah, \ h \in \Omega\} . \tag{2}$$

Let:
$c(f)$=the vector of link costs
$C(h) = A^t c(f)$=the vector of path costs.

Since capacity constraints are not binding, $\tilde{f} \in \Theta$ is an equilibrium vector iff it is a solution to the following variational inequality (VI):

$$\langle c(\tilde{f}), f - \tilde{f} \rangle \geq 0 \quad \forall f \in \Theta . \tag{3}$$

Let: $f \in \Theta \ T(\hat{f}) = \{u : u \in \mathbb{R}^l, \langle u, f - \hat{f} \rangle \geq 0 \ \forall f \in \Theta\} =$the cone normal to Θ at $\hat{f}$.

Consider the following variational inequality:

$$\langle F(\hat{f}), f - \hat{f} \rangle \geq 0 \ \forall f \in \Theta \} . \tag{4}$$

We have that $\hat{f}$ is a solution to VI (4), iff $F(\hat{f}) \in T(\hat{f})$. This condition is certainly satisfied if we put:

$$F(\hat{f}) = c(\hat{f}) + [u - c(\hat{f})] \tag{5}$$

so that, if we add any vector $x \in \tilde{T}(\hat{f}) = T(\hat{f}) - c(\hat{f})$, where $\tilde{T}(\hat{f})$ is the cone obtained by shifting the vertex of $\tilde{T}(\hat{f})$ from the origin to point $-c(\hat{f})$, $\hat{f}$ becomes an equilibrium vector. If $c(f)$ is strictly monotone, given any $x \in \tilde{T}(\hat{f})$, the solution of VI(4) is unique, and is given by $\hat{f}$, so that the cones $\tilde{T}(\hat{f})$ relative to the different $\hat{f} \in \Theta$ are disjoint.

If we set $\bar{c}(\hat{f}) = c(\hat{f}) + x$, $x \in \tilde{T}(\hat{f})$, the equilibrium variational inequality can be written as:

$$\langle \bar{c}(\hat{f}), f - \hat{f} \rangle \geq 0 \quad \forall f \in \Theta \tag{6}$$

VI (6) can be expressed in terms of path flows as follows:

$$\langle \bar{C}(\hat{h}), h - \hat{h} \rangle \geq 0 \quad \forall h \in \Omega \tag{7}$$

where $A\hat{h} = \hat{f}$ and $\bar{C}(\hat{h}) = A^t \bar{c}(\hat{f})$.

Let $s(h) = -h \leq 0$ and $G(h) = d - Bh = 0$ be the constraints which define the set Ω of path feasible flows. Let $\nabla s(h)$ and $\nabla G(h)$ be the Jacobians of $s(h)$ and $G(h)$. A vector $\hat{h} \in \Omega$ is a solution to VI (7) iff there are two sets of multipliers $v \in \mathbb{R}^M$ and $\rho \in \mathbb{R}^m$, such that:

$$\bar{C}(\hat{h}) + \nabla s(\hat{h})^t v + \nabla G(\hat{h})^t \rho = 0 \tag{8}$$

with $v \geq 0$ and $\langle \hat{h}, v \rangle = 0$ (Ferrari, 1997).

We have $\nabla s(\hat{h}) = -I$, where I is the identity matrix, and $\nabla G(\hat{h}) = -B$. By taking into account that $v \geq 0$ and $\langle \hat{h}, v \rangle = 0$, from Eq. (8) we obtain:

$$\bar{C}(\hat{h}) - B^t \rho \geq 0$$
$$\hat{h}^t \bar{C}(\hat{h}) - d^t \rho = 0 . \tag{9}$$

If we put $\bar{C}(\hat{h}) = A^t \bar{c}(\hat{f})$ and $\bar{c}(\hat{f}) = c(\hat{f}) + x$, system (9) can be written as follows:

$$A^t[c(\hat{f}) + x] - B^r ho \geq 0$$

$$\hat{f}[c(\hat{f}) + x] - d^t \rho = 0 . \tag{10}$$

System (10) defines a polyhedron $\bar{T}(\hat{f})$ in $\mathbb{R}^{l+m}$. The set $\tilde{T}(\hat{f})$ of the additional costs x, which transform $\hat{f}$ into an equilibrium flow vector, is the projection of $\bar{T}(\hat{f})$ onto $\mathbb{R}^l$.

In general, the components of x could have any sign. If any one of these components is positive, it can interpreted as a toll paid by drivers; if it is negative, it represents a subsidy to road users. In general the possibility of some component of x being negative is excluded, because the social cost of subsidising road users is considered unacceptable.

The foregoing theory was developed (Bergendorff et al. 1997) for the particular case in which $\hat{F}$ is the system optimal vector. Hearn and Ramana (1998) have suggested some methods for choosing x in the set $\tilde{T}(\hat{f})$. For instance, if we wish to minimise the damage to users consequent to their paying tolls, the total toll $\langle \hat{f}, x \rangle$ must be minimised. Thus, the following problem of linear programming must be solved:

$$\min_{x,\rho} \langle \hat{f}, x \rangle \quad \text{s.t.} \quad x \geq 0, \quad (x, \rho) \in \tilde{T}(\hat{f}) . \tag{11}$$

However, in practice, such results are of little value, because they are based on the assumption that tolls can be levied on all links of a road network. This is simply not true of real road networks, where tolls can be collected only on a limited number of links, such as, for example, links representing motorway carriageways or stretches connecting such carriageways to adjacent roads, or links crossing cordons around urban-areas subjected to road pricing, or links directed to centroids of urban zones where parking fees are imposed.

Therefore, in general the set I of links on which tolls can be collected is a small subset of L, so that the set of *feasible tolls* is defined as follows:

$$\Phi = \{x : x \in \mathbb{R}^l, x_i \geq 0 \quad \forall i \in I, x_i = 0 \quad \forall i \notin I\} . \tag{12}$$

As the number of elements of I is generally very small in road networks, set $\tilde{T}(\hat{f}) \cap \Phi$ is often empty for many $\hat{f} \in \Theta$, in particular when $\hat{f}$ is the system optimal.

Consider, for instance, the small network illustrated in figure 1, made up of two links connecting two centroids, between which a transport demand of 2000 vph exists. Link 1 represents a motorway carriageway on which a toll can be imposed, whereas link 2 is a local road where tolls are impracticable. Thus, the set of *feasible tolls* is $\{x : x_1 \geq 0, x_2 = 0\}$, and is represented in figure 1b by the non-negative axis $x_1 \geq 0$. Segment AB in figure 1a measures the total transport demand between the centroids. Each point f in AB represents a flow vector on the two links, with the origin of flow f_1 on link 1 in A and that of flow f_2 on link 2 in B. The private-costs functions $c_i(f_i)$ are linear and indicated in the figure along with the marginal-costs functions $C_i(f_i)$. The abscissa Z of the intersection N of the private-costs functions represents the equilibrium flow vector $\bar{f}$ in the network, whose components are $\bar{f}_1=1167$ vph, $\bar{f}_2=833$ vph. The abscissa W of the intersection T of the marginal-costs functions identifies the system optimal vector $\hat{f}$, whose components are $\hat{f}_1=1383$ vph, $\hat{f}_2=617$ vph.

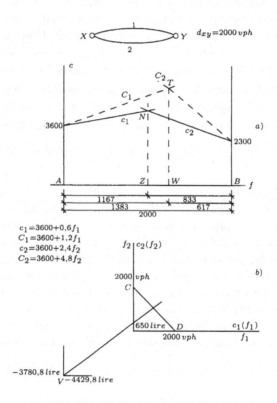

Figure 1 – The set of toll vectors that transform the system optimal vector of a road network into an equilibrium vector.

Any vector (x_1, x_2) which translates the private-costs functions, such that they intersect on the vertical at W, is a point in set $\tilde{T}(\hat{f})$. The set of feasible flows is represented in figure 1b by segment CD, and the cone $\tilde{T}(\hat{f})$ is the straight-line normal to CD passing through point V whose co-ordinates are $-c_1(\hat{f}_1)$=4429.8 and $-c_2(\hat{f}_2)$=3780.8. It does not intersect the non-negative axis x_1; thus, no feasible toll exists that can transform $\hat{f}$ into an equilibrium vector.

On the other hand, we can see from figure 1b that, if tolls could be imposed on both links of the network, the minimum additional cost to users would be obtained by imposing a 650 lire toll on link 2 alone: the total tolls paid by all users would be 650·617=401,050 lire. From expressions $c_1(f_1)$ and $c_2 = (f_2)$, we thereby deduce that the total costs to drivers, exclusive of toll, before and after instituting the toll, are 8,600,000 and 8,459,617 lire, respectively. Thus, optimising the system would provide a benefit of 140,833 lire to users, much less than the total tolls they are paying out.

3. A new theory of optimal flow pattern.

From the foregoing it appears evident that the traditional theory of system optimization can hardly be considered realistic. Often, the system optimal vector cannot be obtained, because tolls can be collected on only few links of the network, and even when it is determinable, the overall benefit to drivers turns out to be less than the total additional costs incumbent upon them.

A new theory of optimal flow pattern in road networks is presented in this section. It considers that the actual reason for which tolls are imposed on road networks is to recover maintenance costs and at least a part of the road construction costs. The optimal division of road costs between drivers and society as a whole is sought by accounting for both the willingness of motorists to pay and the social burden consequent to public financing of road costs. Public financing requires increasing the taxes levied on society as a whole. This, in turn engenders a loss of economy efficiency, which is measured by the *marginal cost of public funds* (Laffont, 1994), an index which quantifies, in monetary terms, the damage to the economy consequent to a unit increase in taxes.

Consider a road network represented as usual by a graph $G(N, L)$, where N is the set of nodes and L the set of links. Each year of the network's lifetime is divided into a sequence of time intervals, collected in s sets j. Each set is made up of m_j intervals, each of which is formed by the same number of hours, x_j, and characterised by the same vector of transportation demand. Transportation demand is constant over each interval, which is large enough for the traffic flow during it to be considered stationary.

A road A, belonging to the network, is formed by a subset Z of the network links and is divided into z sections. Each section has a toll-barrier where drivers are automatically charged a toll as they pass through without stopping. The road under examination may be a new road yet to be built, or an old road needing modifications. The road may be built (or modified) and operated by a publi c agency or franchised to a private company.

Let:

W = the set of pairs w of origin/destination centroids

d_{wj} = the hourly transportation demand between the pair w during a time interval $\tau \in j$

I = the set of links on which the toll-barriers are installed

π_{ij} = the toll imposed on drivers travelling link $i \in I$ during a time interval $\tau \in j$ π_j = the vector of tolls π_{ij} $\forall i \in L$

f_{ij} = the hourly traffic flow on link $i \in L$ during a time interval $\tau \in j$

f_j = the vector of flows f_{ij} $\forall i \in L$

Θ_j = the set of feasible flow vectors f_j.

We assume that no network capacity constraints are active at equilibrium. If the network is studied through a deterministic approach, and the cost functions $c_i(f_{ij})$ on links $i \in L$ are presumed to be separable, then the equilibrium relationship for the road network during interval $\tau \in j$ is given by the following minimisation problem:

$$\min f_j \in \Theta_j F_{1j}(f_j) = \sum_{i \in I} \int_0^{f_{ij}} c_i(x)dx + \sum_{i \in I} \int_0^{f_{ij}} (c_i(x) + \pi_{ij})dx . \quad (13)$$

Adopting a stochastic approach, let:
$C_{wj}(f_j)$ = the mean of the minimum travel costs borne by drivers between $w \in W$ during interval $\tau \in j$, when the vector of link flows is f_j.

$c_i(f_{ij})$ = the mean costs charged to drivers travelling link $i \in L$, during $\tau \in j$, which is a function of flow f_{ij}.
The equilibrium relationship for the road network during time interval $\tau \in j$ is given by the following minimisation problem (Sheffi, 1985):

$$\min_{f_j \in \Theta_j} F_{2j}(f_j) = - \sum_{w \in W} C_{wj}(f_j)d_{wj} + \sum_{i \in I} \left[f_{ij}c_i(f_{ij} - \int_0^{f_{ij}} c_i(x)dx \right] +$$

$$+ \sum_{i \in I} \left[f_{ij}(c_i(f_{ij}) + \pi_{ij}) - \int_0^{f_{ij}} (c_i(x) + \pi_{ij}) dx \right] . \qquad (14)$$

As problems (13) and (14) each has a unique solution, both implicitly define an equilibrium vector f_j as a function of π_j.

Let:

$\hat{c}_j$ =the transportation cost borne by all drivers on the road network during one hour of time interval $\tau \in j$, before the construction or modification of road A, thus in the absence of any toll.

$c_j(f_j(\pi_j))$ =the transportation cost borne by all drivers on the network, exclusive of toll, during one hour of time interval $\tau \in j$, after the construction or modification of road A and institution of a toll on it.

$S = \sum_{j=1}^{s} m_j x_j [\hat{c}_j - c_j(f_j(\pi_j))]$ =the difference in the yearly costs borne by all drivers on the network before and after the construction or modification of road A and imposition of tolls on it.

$f_i \sum_{j=1}^{s} m_j x_j f_{ij}$ =the number of vehicles travelling link $i \in L$ in one year.

$K = \sum_{i \in Z} (a_i + b_i f_i)$ =the annual cost borne by the road operator to maintain road A and amortise the cost of construction, where a_i is a coefficient representing the fixed cost of link i, while b_i represents the marginal maintenance cost.

$R = \sum_{i \in I} \sum_{j=1}^{s} m_j x_j \pi_{ij} f_{ij}$ =the amount of tolls paid by all drivers during a one-year period.

T =the annual road costs incumbent on public financing; in the event that road A is built and operated on a franchise contract, these are to be paid out to the private company.

λ =marginal social cost of public funds.

$S - R$ =the difference in yearly utility of the network users consequent to the construction or modification of road A, and imposition of tolls on it.

$T(1 + \lambda)$ =total cost borne by society as a whole consequent to levying the taxes needed to obtain the sum T.

$U = T + R - K$ =annual utility of the road operator.

$B = S - R - T(1 + \lambda) + U$ =the difference in yearly social welfare consequent to the construction or modification of road A.

Under the hypothesis that the road operator's profit is zero (Laffont J.J., 1994), we have $T + R - K = 0$, so that $T = K - R$ and:

$$B = S + \lambda R - (1 + \lambda)K \ . \tag{15}$$

The optimal toll vector maximises function (15) under the constraints that $\pi_j \geq 0 \ \forall j \in (1, s)$ and the network is in equilibrium. Thus, the problem to be solved is:

$$\max_{\pi_j \ \forall j \in (1,s)} B = \min_{\pi_j \ \forall j \in (1,s)} m_j x_j \Phi_j(f_j(\pi_j), \pi_j)$$

$$\text{s.t.} \ \pi_j \geq 0 \ \ \forall j \in (1, s) \tag{16}$$

where

$$\Phi_j(f_j(\pi_j), \pi_j) = c_j(f_j(\pi_j)) + \sum_{i \in Z} b_i(\pi_j) -$$

$$- \lambda \left(\sum_{i \in I} \pi_{ij} f_{ij}(\pi_j) - \sum_{i \in Z} b_i f_{ij}(\pi_j) \right) \tag{17}$$

and $f_{ij}(\pi_j) \ \forall i \in L, \ \forall j \in (1, s)$ are implicitly defined by minimum problem (13) or (14), depending upon whether the network is studied through a deterministic or a stochastic approach.

It is worthwhile noting that the road operator's fixed costs do not appear in Eq.(17): this signifies that construction costs have no influence on the optimum tolls.

As the domains of functions, Φ_j, F_{1j}, F_{2j} relative to the different sets j of time intervals are disjoint, we have:

$$\min_{\pi_j \ \forall j \in (1,s)} \sum_{j=1}^{s} m_j x_j \Phi_j(f_j(\pi_j), \pi_j) = \sum_{j=1}^{s} m_j x_j \min_{\pi_j} \Phi_j(f_j(\pi_j), \pi_j) \tag{18}$$

so that the optimal vector π_j relative to each $j \in (1, s)$ can be computed by solving the following problem:

$$\min_{\pi_j} \Phi_j(f_j(\pi_j), \pi_j) \ \text{s.t.} \ \pi_j \geq 0 \tag{19}$$

where $f_j(\pi_j)$ is defined as solution to the problem

$$\min_{f_j \in \Theta_j} F_{1j}(f_j, \pi_j) \ \text{ or } \ \min_{f_j \in \Theta_j} F_{2j}(f_j, \pi_j) \tag{20}$$

depending on whether the method adopted to study the network is deterministic or stochastic.

4. Calculation of the optimal toll vector.

The approach we propose to solving problem (19) is a variant of the *downhill simplex* algorithm set forth by Nelder and Mead, 1965 as a modification to the simplex method by Spendley et al., 1962. Herein we propose some variations to Nelder and Mead's original algorithm which enable it to account for the non-negativity constraint on π_j.

Let $\pi \in \mathbb{R}^m_+$ (the suffix j is omitted for brevity's sake) be a vector of the tolls imposed on links $i \in I$ of the road network. Consider $n + 1$ points $\pi_0 \ldots \pi_n$, which form the vertices of a simplex in $\mathbb{R}^m_+$, where n is the number of links $i \in I$. For instance, having assumed π_0 as the starting point, we can define the other n points as

$$\pi_r = \pi_0 + \delta e_r \tag{21}$$

where e_r is a vector whose r–component is 1, while the others are zero, and δ is a constant whose value depends on the characteristic length scale of the problem.

Function $\Phi(f(\pi_r), \pi_r)$ (the suffix j is omitted as before) is evaluated at the vertices of the simplex, where $f(\pi_r)$ is the equilibrium flow vector corresponding to toll vector π_r, i.e., $f(\pi_r)$ is the solution to one of problems (20). Therefore, an equilibrium flow vector $f(\pi_r)$ is computed for each π_r by assigning the transportation demand to the network by a deterministic or stochastic procedure, and then calculating $\Phi(f(\pi_r), \pi_r)$ through Eq. (17).

We indicate Φ as the function's value at π_r, and define h as the suffix such that $\Phi_h = \max_r \Phi_r$, and l as the suffix such that $\Phi_l = \min_r \Phi_r$. Furthermore, we define $\bar{\pi}$ as the centroid of points π_r with $r \neq h$: $\bar{\pi} = \frac{1}{n} \sum_{r \neq h} \pi_r$.

Computation of the optimum π follows an iterative process. At each stage of the process, π_h is replaced by a new point by applying one of these three operations: *reflection, contraction and expansion*.

Each step of the process begins with a reflection of π_h, which gives rise to a point π^*, whose co-ordinates $\pi_i^* \ \forall i \in I$ are defined by:

$$\pi_i^* = (1 + \alpha_i)\bar{\pi}_i - \alpha_i \pi_{ih} \tag{22}$$

where α_i are positive constants whose values must be defined while taking the non-negativity constraint on π^* into account. If $\bar{\pi}_i > \pi_{ih}, \alpha_i$, may have any value α^*; in the application to a real case presented in the next section, we have taken $\alpha^* = 1$. On the other hand, if $\bar{\pi}_i > \pi_{ih}$, we compute

$$\beta_i = \frac{\bar{\pi}_i}{\pi_{ih} - \bar{\pi}_i} \ . \tag{23}$$

If $\beta_i \geq \alpha^*$, we take $\alpha_i = \alpha^*$; otherwise, $\alpha_i = \beta_i$.

If $\Phi(\pi^*) < \Phi_l$, i.e. the reflection has produced a new minimum, we expand π^* to a point π^{**}, whose co-ordinates are

$$\pi_i^{**} = (1 - \lambda_i)\bar{\pi}_i + \gamma_i \pi_i^* \tag{24}$$

and where γ_i is a constant greater than unity. If $\pi_i^* > \bar{\pi}_i, \gamma_i$, may have any value $\gamma^* > 1$; in the application to a real case presented in the next section we have taken $\gamma^* = 2$. On the other hand, if $\pi_i^* < \bar{\pi}_i$, we compute:

$$\mu_i = \frac{\bar{\pi}_i}{\bar{\pi}_i - \pi_i^*} . \tag{25}$$

If $\mu_i \geq \gamma^*$, we take $\gamma_i = \gamma^*$; otherwise, $\gamma_i = \mu_i$. If $\Phi(\pi^{**}) < \Phi(\pi^*)$, π_h is replaced by π^{**}, while if $\Phi(\pi^{**}) > \Phi(\pi^*)$, π_h is replaced by π^*. In both cases a new simplex is obtained, function Φ is evaluated at the vertex replacing π_h, and the process starts over again with this new simplex.

If on reflecting π_h to π^*, we find that $\Phi(\pi^*) > \Phi_l$, then π_h is replaced by π^*, and the process starts over with the new simplex; unless $\Phi(\pi^*) > \Phi(\pi_r) \,\forall r \neq h$, in which case a new point π^{**} is defined by a one-dimensional contraction, whose co-ordinates are

$$\pi_i^{**} = \rho_i \pi_i^* + (1 - \rho_i)\bar{\pi}_i \tag{26}$$

and where $\rho_i = \frac{\alpha_i}{2}$. Then π_h is replaced by π^{**}, yielding a new simplex with which to start the process over again, unless $\Phi(\pi^{**}) > \Phi(\pi_h)$. When this occurs, all attempts to obtain a point better than π_h have failed, so that we replace each π_r, $r \neq l$, by a point π_r^* whose co-ordinates are

$$\pi_{ri}^* = \frac{\pi_{li} + \pi_{ri}}{2} . \tag{27}$$

In other words, we perform a contraction of the simplex around its best vertex, and the process starts over again with this new simplex.

The process terminates when the difference between the extreme values of Φ at the vertices of the simplex is fractionally less than a tolerance value η, i.e.:

$$\frac{2|\Phi_h - \Phi_l|}{|\Phi_h| + |\Phi_l|} < \eta . \tag{28}$$

In the application to a real case presented in the next section we have assumed $\eta = 10^{-6}$.

5. An application to the real case.

The method presented in the foregoing has been applied to the double-carriageway road Firenze-Pisa-Livorno (FI-PI-LI), which is part of the Arno Valley road network in Tuscany (Italy). The graph of the network has 292 links and 96 nodes, 48 of which are origin/destination centroids. The FI-PI-LI road is in need of improvements, the cost of which has been estimated at 250 billion Italian lire. The maintenance costs of one kilometre of road amount to nearly 100 million lire per year. The road-wor k and maintenance costs could be recovered, at least in part, by imposing a toll on motorists travelling the road. In the following, we apply the method presented in the foregoing to evaluate the optimal toll pricing, and then calculate the proportion of road costs which could be offset by the tolls collected.

We have presumed that tolls are charged automatically to drivers as they pass, without stopping, through three toll-barriers located 30 km apart on three sections of the road (named A, B, C).

The transportation cost of each link i of the network has been considered as a vector having two components: the monetary cost $c_i^m = 200$ lire per km and travel time t_i. We have considered t_i to be a function of flow travelling the link, assuming different functions for the various links of the network: $t_i = t_i(f_i)$.

The cost vector has been transformed into a scalar expressed in monetary units by multiplying the journey time by a coefficient v_t, which measure s the monetary value per unit time. Thus, the cost function of a link i is:

$$c_i(f_i) = c_i^m + v_t t_i(f_i) . \tag{29}$$

The monetary value of time is distributed randomly over the motorist population, so we have considered v_t to be a random, normally-distributed variable with mean VT and variance $\sigma^2 = (0.3 \cdot VT)^2$. The marginal maintenance cost of the road has been estimated at 2.8 lire per vehicle per km.

A year in the road's lifetime has been divided into a sequence of 8-hour time intervals τ, grouped into three sets j, characterised by different values of transportation demand per hour, namely, peak, off-peak and night periods. The toll vector π_j relative to interval $\tau \in j$ has three components, each of which defines the toll charged to drivers passing in either direction through one of the three barriers.

As link costs are random variables, we have computed the network equilibrium flows by means of a stochastic procedure, assigning transportation demand to the network through the MSA algorithm (Sheffi, 1985).

In order to assess the influence on the optimal toll values of the difficulty of financing road costs with public funds (which is measured by their marginal cost) and motorists' willingness to pay (as measured by the average monetary value of a unit time), the calculation has been repeated for three values of λ : $0.5, 1, 2$ and three values of VT : $100, 200, 300$ lire per minute (lpm).

Table I presents the optimum tolls on the three road sections for the three sets of time intervals and different combinations of parameters VT and λ. It can be noticed that, as would be expected, the optimal tolls increase in direct proportion to both the marginal cost of public funds (i.e., with the difficulty of raising public funds to finance road costs), as well as the unit-time monetary value (i.e., with drivers' willingness to pay). The optimal tolls are very different for the three sections, despite their being about the same length, and during the three sets of time intervals. These differences underscore that the optimal toll on a stretch of road is independent of its length, and therefore, its fixed costs. Such a finding stems from the absence of fixed costs in the expression of the objective function, as has already been mentioned in Section 2. On the contrary, the optimal toll does depend, given equal λ and VT, on the level of transportation demand and the costs of alternative routes, and thus on the network's characteristics.

Tab. I – Tolls (lire) on various road sections over various time period

Section	Period	$\lambda = 0,5$			$\lambda = 0,5$			$\lambda = 0,5$		
		VT = 100	VT = 200	VT = 300	VT = 100	VT = 200	VT = 300	VT = 100	VT = 200	VT = 300
A	P	0	0	0	0	511	977	1215	1864	2856
	M	0	254	333	0	491	769	0	671	1053
	N	0	153	198	100	362	566	199	483	815
B	P	320	576	961	412	684	1041	470	719	1043
	M	217	0	0	255	0	598	268	485	627
	N	0	190	0	232	0	0	255	0	587
C	P	1071	1227	1540	1370	2231	2820	1785	2451	3368
	M	1237	1795	2271	1491	2238	2698	1635	2628	3296
	N	1291	1826	2082	1474	2284	2818	1637	2563	3457

P=peak period

M=off–peak period

N=night period

Tab. II – Total travel time on the network and toll revenues (lire)

λ	VT=100		VT=200		VT=300	
	Time	Revenue	Time	Revenue	Time	Revenue
	Hours/day	Bil./year	Hours/day	Bil./year	Hours/day	Bil./year
0.5	646648	13.942	638501	20.283	636580	26.364
1.0	647312	16.371	640125	29.619	638408	44.365
2.0	651877	23.686	643016	39.849	641539	58.521
Without Toll	645731	—	637735	—	636051	—

Table II presents the total daily travel time throughout the network and the annual toll revenues collected when optimal tolls are imposed in the cases examined. Comparison of the calculated travel times when optimal tolls are imposed with those without tolls provides a measure of the distortion in the flow distribution caused by imposition of the toll. The table shows the influence of drivers' willingness to pay on travel times, and on the optimal partitioning of road costs between motorists and public financing. When drivers are rather reluctant to pay, toll revenues are lower and the increase in travel time higher than when their willingness to pay is high. For instance with $\lambda = 0.5$, when $VT=300$ lpm, we have optimal toll revenues of over 26 billion lire per year and an increase in travel time over the network equal to 529 hours per day. On the other hand, when $VT = 100$lpm, the optimal toll revenue is less than 14 billion lire per year and the increase in travel time is over 900 hours per day.

We have estimated that the total road costs of the FI-PI-LI road, i.e., maintenance costs and the amortisation costs of road work for the needed improvements, amount to 35 billion lire per year for a period of 20 years. The results in Table II show that, given scarce willingness on the part of drivers to pay ($VT = 100$), even if the difficulty of subsidising road costs with public funds is very high ($\lambda = 2$), the optimal toll revenue would be equal to 23.686 billion lire per year, so that 11.314 billion per year would be incumbent on public financing, involving damage to the economy of 22.628 billion per year. On the contrary, if drivers' willingness to pay is high ($VT = 300$), it is advantageous to charge all road costs to motorists, even if public funding is relatively available ($\lambda < 1$).

Table II shows that if both λ and VT are high ($VT = 200$ and $\lambda = 2$, or $VT = 300$ and $\lambda \geq 1$), the optimal toll revenues would be greater than the actual road costs. This means that the surplus toll revenues could be used to cover costs outside the road on which the tolls are collected, e.g., the costs of other roads in the network.

6. Conclusions.

This paper has shown the traditional theory of system optimization to be unrealistic, because it does not consider the real-life facts that, firstly, tolls can be charged on only some links of a road network, and secondly, the real reason for imposing road tolls is to be able to use the revenues as a means to cover at least part of road costs. Levying tolls on only some roads of a network will cause some of traffic to shift to alternative routes, thus modifying the total transportation cost throughout the network.

This modification may be a reduction or an increase depending on the price of the toll as well as other factors. In any event, there is a toll upper threshold beyond which the overall transportation cost of the network will inevitably increase. On the other hand, subsidising road costs through public financing requires imposing taxes on society as a whole and thus involves a loss of economy efficiency, as measured by the marginal cost of public funds. Therefore, raising money to cover road costs engenders a loss of social welfare, regardless of the source - be it motorists travelling the road or society as a whole. The partitioning of the costs burden between these two sources, and thereafter, the determination of the optimal tolls to impose must be performed in such a way as to minimise the social costs consequent to such fund-raising.

In the foregoing, we have presented a definition of optimal toll pricing for road networks and a method for its calculation. Both the theory and the results obtained through application of the method to a real case have revealed that optimal tolls are independent of the fixed road costs. On the other hand, they are markedly influenced by the difficulty in raising funds through public financing, a factor measured by the marginal cost of public funds, and by motorist willingness to pay, which varies according to the monetary value each driver attributes to travel time. In the case that both motorist willingness to pay and the marginal cost of public funds are high, the imposition of the optimal tolls would yield revenues greater than the road costs, so that the surplus toll revenues could be destined for use outside the road on which the tolls were imposed.

It seems worthwhile to stress that this result is a consequence of the principle on which the toll-pricing policy presented in the paper is founded: the aim of toll pricing is to achieve optimal social welfare, so that when public financing is in difficulty, while motorists are willing to pay, toll revenues could be destined for uses above and beyond the recovery of road costs.

References

[1] M.J. Beckmann, C.B. McGuire, C.B. Winsten, *Studies in the Economics of Transportation*, Yale University Press, New Haven, 1956.

[2] P. Bergendorff, D.W. Hearn, M.V. Ramana, *Congestion toll pricing of traffic networks*, In *Network Optimization*, P.M. Pardalos ed., Springer-Verlag series Lecture Notes in Economics and Mathematics (1957), 51–71.

[3] P. Ferrari, *Capacity constraints in urban transport networks*, Transpn. Res. 31B (1997), 291–301.

[4] D.W. Hearn, M.V. Ramana, *Solving congestion toll-pricing models*, In *Equilibrium and Advanced Transportation Modelling*, P. Marcotte and S. Nguyen (Eds), Kluwer Academic Publisher (1998), 109–114.

[5] J.J. Laffont, *The New Economics of Regulation: Ten Years After*, Econometrica 62 (1994), 507–537.

[6] J.A. Nelder, R. Mead, *A Simplex Method for Function Minimization*, Computer Journal 4 (1965), 308–313.

[7] Y. Sheffi, *Urban Transportation Networks*, Prentice-Hall, Englewood Cliffs, N.J., 1985.

[8] W. Spendley, G.R. Hext, F.R. Himswortf, *Sequential Application of Simplex Designs in Optimization and Evolutionary Operation*, Technometrics 4 (1962), 441–462.

ON THE STRONG SOLVABILITY OF A UNILATERAL BOUNDARY VALUE PROBLEM FOR NONLINEAR DISCONTINUOUS OPERATORS IN THE PLANE

Sofia Giuffrè

Department of Mathematics,
University of Catania
Viale A. Doria, 6 - 95125 Catania, Italy
e-mail: giuffre@dipmat.unict.it

Abstract   A uniqueness and existence theorem in the Sobolev space is proved for a unilateral boundary value problem for a class of nonlinear discontinuous operators in the plane. The operator is assumed to satisfy a suitable ellipticity condition, which allows us to apply nearness theory of mappings. Estimate

$$\int_\Omega \sum_{i,j=1}^2 \left(\frac{\partial^2 u}{\partial x_i \partial x_j} \right)^2 dx \leq \int_\Omega |\Delta u|^2 dx$$

$$\forall u \in W^{2,2}(\Omega) \ : \ u \geq 0, \ \frac{\partial u}{\partial n} \geq 0, \ u \cdot \frac{\partial u}{\partial n} = 0 \text{ on } \partial\Omega$$

having interest in itself, plays a fundamental role.

Keywords: Signorini problem, unilateral problems, theory of nearness.

F. Giannessi et al (eds.),
Equilibrium Problems: Nonsmooth Optimization and Variational Inequality Models, 119–127.
© 2001 *Kluwer Academic Publishers.*

1. Introduction.

In this paper we are concerned with the strong solvability of the following boundary value problem with unilateral boundary conditions

$$
\begin{cases}
\mathcal{A}\left(x, D^2 u\right) - \lambda u = f(x) & \text{a.e. in } \Omega \\[2mm]
u \geq 0, \ \dfrac{\partial u}{\partial n} \geq 0, \ u \cdot \dfrac{\partial u}{\partial n} = 0 & \text{on } \partial\Omega
\end{cases}
\tag{1.1}
$$

where the number λ is greater than zero and n is the unit outward normal to the boundary $\partial\Omega$.

Let us observe that when $\mathcal{A}\left(x, D^2 u\right) = \Delta u$ the problem

$$
\begin{cases}
\Delta u - \lambda u = f(x) & \text{a.e. in } \Omega \\[2mm]
u \geq 0, \ \dfrac{\partial u}{\partial n} \geq 0, \ u \cdot \dfrac{\partial u}{\partial n} = 0 & \text{on } \partial\Omega
\end{cases}
\tag{1.2}
$$

is the celebrated problem of Signorini, and, for each $f \in L^2(\Omega)$, it admits a unique solution $u \in W^{2,2}(\Omega)$ (see [1], [6]).

This problem represents the conceptual model of an elastic body Ω with boundary $\partial\Omega$ which is in contact with a rigid support body and is subject to volume force f. These forces produce a deformation of Ω and a displacement on $\partial\Omega$ with the normal component non negative. In a simplified model the displacement field u satisfies (2).

In the present paper we will prove that there exists a unique solution of problem (1) in the Sobolev space $W^{2,2}(\Omega)$.

We assume $\mathcal{A}(x, \xi)$, $(x, \xi) \in \Omega \times R^4$, to be a Carathéodory function, i.e. it is measurable in $x \in \Omega$ for all $\xi = \{\xi_{ij}\}_{i,j=1,2} \in R^4$ and continuous in $\xi \in R^4$ for a.a. $x \in \Omega$, and to satisfy the following condition introduced by S.Campanato (see [3]):

there exist three constants α, γ, $\delta > 0$, *with* $\gamma + \delta < 1$, *such that, for almost all* $x \in \Omega$, *for all* ξ, $\tau \in R^4$, *one has*

$$
\left| \sum_{i=1}^{2} \xi_{ii} - \alpha\left[\mathcal{A}\left(x, \xi + \tau\right) - \mathcal{A}\left(x, \tau\right)\right] \right| \leq \gamma \|\xi\| + \delta \left| \sum_{i=1}^{2} \xi_{ii} \right|
\tag{A}
$$

where $\|\xi\| = \left(\displaystyle\sum_{i,j=1}^{2} \xi_{ij}^2 \right)^{\frac{1}{2}}$.

Let us note that condition (A) does not imply the continuity of the mapping $\mathcal{A}(x, \xi)$ with respect to x, and, in view of Carathéodory's condition, we need only measurability of $\mathcal{A}(x,\xi)$ with respect to x.

Moreover condition (A) and the estimate (see Lemma 4.1)

$$\int_\Omega \sum_{i,j=1}^2 \left(\frac{\partial^2 u}{\partial x_i \partial x_j} \right)^2 dx \leq \int_\Omega |\Delta u|^2 \, dx,$$

$$\forall u \in W^{2,2}(\Omega) \, : \, u \geq 0, \; \frac{\partial u}{\partial n} \geq 0, \; u \cdot \frac{\partial u}{\partial n} = 0 \text{ on } \partial\Omega$$

ensure the possibility to apply the theory of nearness of mappings (see [3]).

The estimate above has interest in itself and probably can be extended to the case of R^n, $n > 2$.

Finally it is well known that Signorini problem (2), in a weak formulation, can be expressed by a variational inequality (see [1], [6]):

find $u \in K$ such that

$$a(u, v - u) \geq (f, v - u) \quad for \; v \in K$$

where $K = \{u \in H^1(\Omega) \, : \, u \geq 0 \text{ on } \partial\Omega\}$, $a(u,v) = \int_\Omega \sum_{i=1}^2 \frac{\partial u}{\partial x_i} \frac{\partial v}{\partial x_i} dx - \lambda \int_\Omega uv dx$ and $(f,v) = \int_\Omega f \, v \, dx$.

2. Basic assumptions and main results.

Let $\Omega \subset R^2$ be a bounded, convex and open set with C^2-smooth boundary $\partial\Omega$. We assume that $\partial\Omega$ is a closed curve and let

$$\begin{cases} x_1 = x_1(\varphi) \\ x_2 = x_2(\varphi) \end{cases}$$

be the normal parametrization of $\partial\Omega$, with φ being a curvilinear parameter, $\varphi \in [0, L]$.

Let us set $n = (x_2', -x_1')$ for the unit outward normal to $\partial\Omega$ and denote by $\chi(\varphi)$ the mean curvature of $\partial\Omega$, $\chi(\varphi) = x_2' x_1'' - x_1' x_2''$.

Consider a real valued function $\mathcal{A}(x, \xi) : \Omega \times R^4 \to R$, which satisfy Carathéodory's condition.

We are aimed at the investigation of the strong solvability of the following problem

$$\begin{cases} \mathcal{A}(x, D^2 u) - \lambda u = f(x) & \text{a.e. in } \Omega \\[2mm] u \geq 0, \; \frac{\partial u}{\partial n} \geq 0, \; u \cdot \frac{\partial u}{\partial n} = 0 & \text{on } \partial\Omega \end{cases} \tag{1.3}$$

where $D^2u = \{D_{ij}u\}_{i,j=1,2}$ denotes the Hessian matrix of u and λ is a positive constant.

As usual by strong solution $u \in W^{2,2}(\Omega)$, we mean a twice weakly differentiable function with L^2 summable derivatives.

Regarding the function $\mathcal{A}(x,\xi)$ we assume the ellipticity condition (A), and

$$\mathcal{A}(x,0) = 0.$$

The main result of the paper is the following

Theorem 1 *Let condition (A) be fulfilled. Then for each $f \in L^2(\Omega)$ and for each constant $\lambda > 0$, problem (3) is uniquely solvable in the Sobolev space $W^{2,2}(\Omega)$.*

3. Preliminary results.

In the proof of Theorem 1.1 we will use the nearness theory. Then we start recalling the definition of near operators.

Definition 1.1 *Let $\mathcal{B}$ be a set and $\mathcal{B}_1$ a real Banach space. Consider two mappings A and B defined on $\mathcal{B}$ with values in $\mathcal{B}_1$. The mapping A is said to be near to B if there exist two positive constants α and k, with $k \in (0,1)$, such that $\forall u, v \in \mathcal{B}$ it implies*

$$\|B(u) - B(v) - \alpha[A(u) - A(v)]\|_{\mathcal{B}_1} \le \|B(u) - B(v)\|_{\mathcal{B}_1}.$$

The following theorem holds.

Theorem 2 *The mapping $A : \mathcal{B} \to \mathcal{B}_1$ is injective, or surjective, or bijective if and only if it is near to a mapping $B : \mathcal{B} \to \mathcal{B}_1$, which is injective, or surjective, or bijective.*

We also recall the definition of monotone operators.

Definition 1.2 *Let $\mathcal{B}$ be a set, $\mathcal{B}_1$ a real Hilbert space, A, B two operators from $\mathcal{B}$ into $\mathcal{B}_1$. A is said to be monotone with respect to B if, $\forall u, v \in \mathcal{B}$, we have*

$$(A(u) - A(v), B(u) - B(v))_{\mathcal{B}_1} \ge 0.$$

We also need the following result.

Theorem 3 *Let A, B, C be operators from $\mathcal{B}$ into $\mathcal{B}_1$, with $\mathcal{B}_1$ being a real Hilbert space. If:*
A is near to B with costants α e k,
C is monotone with respect to B,

then the mapping $A + C$ is near to $B + \alpha C$ with the same constants α e k.

In order to obtain Theorem 1.1 we will use the following existence and uniqueness theorem.

Theorem 4 *For all $f \in L^2(\Omega)$, for all $\lambda > 0$, the Signorini problem*

$$
\begin{cases}
\Delta u - \lambda u = f(x) & \text{a.e. in } \Omega \\[2mm]
u \geq 0, \ \dfrac{\partial u}{\partial n} \geq 0, \ u \cdot \dfrac{\partial u}{\partial n} = 0 & \text{on } \partial\Omega
\end{cases}
$$

admits a unique solution $u \in W^{2,2}(\Omega)$.

4. Proof of the theorems.

We define $\mathcal{B}$ to be the closure in $W^{2,2}(\Omega)$ of the class $W = \{u \in C^2(\overline{\Omega}) \cap$

$C^3(\Omega) : u \geq 0, \ \dfrac{\partial u}{\partial n} \geq 0, \ u \cdot \dfrac{\partial u}{\partial n} = 0 \text{ on } \partial\Omega\}$

The main step in the proof is the following estimate.

Lemma 1.1 *Let $\Omega \subset R^2$ be an open, bounded and convex set with C^2-smooth boundary $\partial\Omega$. Then for each $u \in W$, it results*

$$
\int_\Omega \sum_{i,j=1}^2 \left(\frac{\partial^2 u}{\partial x_i \partial x_j} \right)^2 dx \leq \int_\Omega |\Delta u|^2 \, dx. \qquad (1.4)
$$

Proof. Let us set $p_i = \dfrac{\partial u}{\partial x_i}$, $i = 1, 2$, $p_{ij} = \dfrac{\partial^2 u}{\partial x_i \partial x_j}$, $i, j = 1, 2$. By the identity

$$
\sum_{i,k=1}^2 p_{ik}^2 + 2(p_{11}p_{22} - p_{12}^2) = (\Delta u)^2,
$$

in order to obtain estimate (4), we have to prove that

$$
\int_\Omega (p_{11}p_{22} - p_{12}^2) dx \geq 0.
$$

Taking into account the identity

$$
p_{11}p_{22} - p_{12}^2 = \frac{1}{2} \frac{\partial}{\partial x_1}(p_1 p_{22} - p_2 p_{12}) - \frac{1}{2} \frac{\partial}{\partial x_2}(p_1 p_{21} - p_2 p_{11})
$$

by means of the Gauss formula, we get

$$\int_\Omega (p_{11}p_{22} - p_{12}^2)dx = \frac{1}{2}\int_0^L \left[(p_1 p_{22} - p_2 p_{12})x_2' + (p_1 p_{21} - p_2 p_{11})x_1'\right]d\varphi.$$

Setting $u_0 = \dfrac{\partial u}{\partial n}$, let us consider the system $\forall \varphi \in [0, L]$

$$\begin{cases} p_1 x_1' + p_2 x_2' = \dfrac{du}{d\varphi}, \\ p_1 x_2' - p_2 x_1' = u_0. \end{cases} \tag{1.5}$$

It results

$$\begin{cases} p_1 = \dfrac{du}{d\varphi}x_1' + u_0 x_2', \\ p_2 = \dfrac{du}{d\varphi}x_2' - u_0 x_1', \end{cases}$$

and

$$p_1^2 + p_2^2 = \left(\frac{du}{d\varphi}\right)^2 + u_0^2.$$

If we substitute in the integral, we get

$$\int_\Omega (p_{11}p_{22} - p_{12}^2)dx = \frac{1}{2}\int_0^L \frac{du}{d\varphi}\left[p_{22}x_1'x_2' - p_{12}x_2'^2 - p_{11}x_1'x_2' + p_{12}x_1'^2\right]d\varphi$$

$$+ \frac{1}{2}\int_0^L u_0\left[p_{22}x_2'^2 + p_{12}x_1'x_2' + p_{11}x_1'^2 + p_{12}x_1'x_2'\right]d\varphi.$$

Differentiating the equations of system (5), it results

$$p_{11}x_1'^2 + p_{12}x_1'x_2' + p_{21}x_1'x_2' + p_{22}x_2'^2 = \frac{d^2u}{d\varphi^2} - [p_1 x_1'' + p_2 x_2'']$$

and

$$p_{11}x_1'x_2' + p_{12}x_2'^2 - p_{21}x_1'^2 - p_{22}x_1'x_2' = \frac{du_0}{d\varphi} - [p_1 x_2'' - p_2 x_1''].$$

Substituting in the integral, we have

$$\int_\Omega (p_{11}p_{22} - p_{12}^2)dx = \frac{1}{2}\int_0^L u_0\frac{d^2u}{d\varphi^2}d\varphi - \frac{1}{2}\int_0^L \frac{du_0}{d\varphi}\frac{du}{d\varphi}d\varphi$$

$$+ \frac{1}{2}\int_0^L \left\{\left(\frac{du}{d\varphi}\right)^2[x_1'x_2'' - x_2'x_1''] + u_0\frac{du}{d\varphi}[x_1'x_1'' + x_2'x_2'']\right\}d\varphi$$

$$+ \frac{1}{2}\int_0^L \left\{u_0^2[x_1'x_2'' - x_2'x_1''] - u_0\frac{du}{d\varphi}[x_1'x_1'' + x_2'x_2'']\right\}d\varphi.$$

Bearing in mind that $x_1' x_1'' + x_2' x_2'' = 0$ and $x_1' x_2'' - x_2' x_1'' = -\chi(\varphi)$, the integral becomes

$$\int_\Omega (p_{11} p_{22} - p_{12}^2) dx$$

$$= \frac{1}{2} \int_0^L u_0 \frac{d^2 u}{d\varphi^2} d\varphi - \frac{1}{2} \int_0^L \frac{du_0}{d\varphi} \frac{du}{d\varphi} d\varphi - \frac{1}{2} \int_0^L \left[\left(\frac{du}{d\varphi} \right)^2 + u_0^2 \right] \chi(\varphi) \, d\varphi$$

$$= \int_0^L u_0 \frac{d^2 u}{d\varphi^2} d\varphi - \frac{1}{2} \left[u_0 \frac{du}{d\varphi} \right]_0^L - \frac{1}{2} \int_0^L (p_1^2 + p_2^2) \chi(\varphi) \, d\varphi$$

$$= \int_0^L u_0 \frac{d^2 u}{d\varphi^2} d\varphi - \frac{1}{2} \int_0^L (p_1^2 + p_2^2) \chi(\varphi) \, d\varphi.$$

If we consider the first term at the right-hand side, since

$$u(x_1(\varphi), x_2(\varphi)) \geq 0 \quad \forall \varphi \in [0, L],$$

we have

$$\int_0^L u_0 \frac{d^2 u}{d\varphi^2} d\varphi = \int_{\{\varphi \in [0,L] : u(x_1(\varphi), x_2(\varphi)) > 0\}} u_0 \frac{d^2 u}{d\varphi^2} d\varphi +$$

$$+ \int_{\{\varphi \in]0,L[: u(x_1(\varphi), x_2(\varphi)) = 0\}} u_0 \frac{d^2 u}{d\varphi^2} d\varphi.$$

Since $u \in W$, if $u(x_1(\varphi), x_2(\varphi)) \neq 0$, with $\varphi \in [0, L]$, then

$$u_0(x_1(\varphi), x_2(\varphi)) = 0$$

and the first integral is equal to zero.

For what concerns the second integral, since $u(x_1(\varphi), x_2(\varphi)) \geq 0 \ \forall \varphi \in [0, L]$, each point $(x_1(\varphi^*), x_2(\varphi^*))$, with $\varphi^* \in]0, L[$, such that

$$u(x_1(\varphi^*), x_2(\varphi^*)) = 0$$

is a local minimum point.

Then $\dfrac{du}{d\varphi}(x_1(\varphi^*), x_2(\varphi^*)) = 0$, and $\dfrac{d^2 u}{d\varphi^2}(x_1(\varphi^*), x_2(\varphi^*)) \geq 0$.

Hence, taking into account that $u_0(x_1(\varphi), x_2(\varphi)) \geq 0 \ \forall \varphi \in [0, L]$, it follows

$$\int_0^L u_0 \frac{d^2 u}{d\varphi^2} d\varphi = \int_{\{\varphi \in]0,L[: u(x_1(\varphi), x_2(\varphi)) = 0\}} u_0 \frac{d^2 u}{d\varphi^2} d\varphi \geq 0.$$

Finally, bearing in mind that Ω is a convex set and hence $\chi(\varphi) \leq 0$ $\forall \varphi \in [0, L]$, we get

$$\int_\Omega (p_{11} p_{22} - p_{12}^2) dx$$

$$= \int_{\{\varphi \in]0,L[:u(x_1(\varphi),x_2(\varphi))=0\}} u_0 \frac{d^2 u}{d\varphi^2} d\varphi - \frac{1}{2} \int_0^L (p_1^2 + p_2^2) \, \chi(\varphi) \, d\varphi \geq 0.$$

Let us observe that estimate (4) holds true also for $u \in \mathcal{B}$.
In fact for $u_k \in W$, such that $u_k \to u$ in $W^{2,2}(\Omega)$, we have

$$\left\| D^2 u_k \right\|_{L^2(\Omega)} \leq \| \Delta u_k \|_{L^2(\Omega)} \, ,$$

and hence, when $k \to \infty$, we get

$$\left\| D^2 u \right\|_{L^2(\Omega)} \leq \| \Delta u \|_{L^2(\Omega)} \, .$$

Using estimate (4) and condition (A), it is possible to prove that the mapping $\mathcal{A}(x, D^2 u)$ is near to the Laplacian, both considered as operators from $\mathcal{B}$ in $L^2(\Omega)$. In fact:

$$\left(\int_\Omega \left| \Delta(u - v) - \alpha \left[\mathcal{A}\left(x, D^2 u \right) - \mathcal{A}\left(x, D^2 v \right) \right] \right|^2 dx \right)^{\frac{1}{2}}$$

$$\leq \left(\int_\Omega \left| \gamma \left\| D^2(u - v) \right\| + \delta \left| \Delta(u - v) \right| \right|^2 dx \right)^{\frac{1}{2}}$$

$$\leq \gamma \left(\int_\Omega \left\| D^2(u - v) \right\|^2 dx \right)^{\frac{1}{2}} + \delta \left(\int_\Omega |\Delta(u - v)|^2 dx \right)^{\frac{1}{2}}$$

$$\leq (\gamma + \delta) \left(\int_\Omega |\Delta(u - v)|^2 dx \right)^{\frac{1}{2}}$$

with $\gamma + \delta < 1$.
Moreover, by means of Theorem 3.2, to achieve that $\mathcal{A}(x, D^2 u) - \lambda u$ is near to $\Delta u - \alpha \lambda u$, it is enough to prove that $-\lambda u$ is a monotone operator with respect to Δu.
Thus, since from Theorem 3.3 it follows that the operator $\Delta u - \alpha \lambda u$ is a bijective one from $\mathcal{B}$ into $L^2(\Omega)$, we obtain that $\mathcal{A}(x, D^2 u) - \lambda u$ is bijective, that is our thesis.
Then we have just to prove that $-\lambda u$ is monotone with respect to Δu, that is

$$-\lambda \int_\Omega u \, \Delta u \, dx \geq 0.$$

By the identity

$$u\Delta u = [\frac{\partial}{\partial x_1}(u p_1) + \frac{\partial}{\partial x_2}(u p_2)] - (p_1^2 + p_2^2)$$

and the Green-Gauss formula, we get

$$-\lambda \int_\Omega u \, \Delta u \, dx = \lambda \int_\Omega (p_1^2 + p_2^2) dx - \lambda \int_0^L (up_1 x_2' - up_2 x_1') d\varphi.$$

Since $up_1 x_2' - up_2 x_1' = u \cdot \dfrac{\partial u}{\partial n} = 0$, the integral becomes

$$-\lambda \int_\Omega u \, \Delta u \, dx = \lambda \int_\Omega (p_1^2 + p_2^2) dx \geq 0.$$

References

[1] C. Baiocchi - A. Capelo, *Variational and Quasivariational Inequalities: applications to free boundary problems*, J. Wiley and Sons, Chichester (1984).

[2] H. Brezis, *Problèmes Unilatéraux*, J. Math. Pures et Appl. 51 (1972), pp. 1-168.

[3] S. Campanato, *On the condition of nearness between operators*, Ann. Mat. Pura Appl. 167 (1994), pp. 243-256.

[4] G. Fichera, *Boundary value problems in elasticity with unilateral constraints*, Handbuch der Physik VI a/2, Springer-Verlag, Berlin, Heidelberg New York (1972), pp. 347-389.

[5] J. L. Lions - G. Stampacchia, *Variational Inequalities*, Comm. Pure Appl. Math. 20 (1967), pp. 493-519.

[6] A. Signorini, *Questioni di elasticità non lineare e semilinearizzata*, Rend. Mat. 18 (1959), pp. 95-139.

[7] G. Talenti, *Problemi di derivata obliqua per equazioni ellittiche in due variabili*, Boll. Un. Mat. Ital. 22 (1967), pp. 505-526.

MOST LIKELY TRAFFIC EQUILIBRIUM ROUTE FLOWS ANALYSIS AND COMPUTATION

T. Larsson – J.T. Lundgren – C. Rydergren
Department of Mathematics,
Linköping University,
SE-581 83 Linköping, Sweden

M. Patriksson
Department of Mathematics,
Chalmers University of Technology,
SE-412 96 Göteborg, Sweden

Abstract   When analyzing equilibrium traffic flows it is usually the link flows and link travel demands that are of interest, but in some certain cases analyses require the knowledge of route flows. It is well known that the route flows are non–unique in the static and deterministic cases of traffic equilibrium. Furthermore, different assignment methods can generate different route flow output. We show how this non-uniqueness can affect the results in applications such as in the O–D estimation/adjustment problem, in the construction of induced O–D matrices, exhaust fume emission analyses and in link toll usage analyses. We state a model for finding, uniquely, the most likely route flows given the equilibrium link flows, and propose a solution algorithm for the problem based on partial dualization. We present computational results for the proposed algorithm and results from an application to exhaust fume emissions.

Keywords: Traffic equilibrium, route flows, entropy maximization.

F. Giannessi et al (eds.),
Equilibrium Problems: Nonsmooth Optimization and Variational Inequality Models, 129–159.
© 2001 *Kluwer Academic Publishers.*

1. Introduction.

Equilibrium traffic assignment based on the Wardrop conditions is frequently used in urban traffic planning to predict and analyze traffic flows. The equilibrium conditions are based on the assumption of rational route choice behaviour by the individual user and define a situation where the routes used by the travellers between each origin and destination are the shortest.

There are many assignment methods available for finding the unique link flow pattern that satisfies the user equilibrium condition. Standard solution methods for the traffic assignment problem often operate in the link flow space. One class of such methods is based on the method of Frank and Wolfe (see, e.g., Sheffi, 1985, Chapter 5, for an overview). Other methods, more recently applied to the assignment model, operate in the route flow space and therefore provide a route flow solution consistent with the equilibrium link flow solution (e.g., Larsson and Patriksson, 1992).

As is well known, the route flows are not uniquely determined in the deterministic assignment model; trips assigned to routes between origins and destinations can be swapped to other routes such that the link flows are unchanged. The route flow solution obtained depend therefore on the solution method applied, and for a given solution method also on the initiation of the solution procedure. Nevertheless, a route flow solution is needed as input to methods used in several applications.

2. Illustrative examples and applications.

In this section we present a comparison of the route flow output from two assignment algorithms and the result of the usage of these route flows in four potential applications. The assignment procedures used to compute equilibrium route flows are: (1) the Frank–Wolfe (FW) method supplied with a procedure to generate an equilibrium route flow and (2) the route flow based disaggregate simplicial decomposition (DSD) method of Larsson and Patriksson (1992). In all the examples the equilibrium flows have been determined to a very high accuracy with both methods.

2.1. Illustrative examples.

We here illustrate the differences in the route flow output between the FW and the DSD algorithms. We first study in particular the topology of the two sets of route flow solutions in O–D pair $(14, 10)$ in the Sioux Falls network (LeBlanc et al., 1975). Both the FW method and the DSD method generate two routes in this O–D pair. The travel demand from 14 to 10 is 2100 vehicles. The route flows in the FW solution are 837 for the route 14–11–10 and 1263 for the route 14–15–10 (see Figure 1). The route flows in the DSD solution are 368 and 1732, respectively.

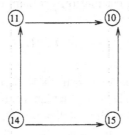

Figure 1: Sioux Falls O–D pair (14, 10).

We next give aggregated measures of the difference between the route flow solutions for applications of the FW and DSD methods to the network of Winnipeg (Florian and Nguyen, 1976). We first restrict the analysis to the 65 largest origin–destination (O–D) pairs (which provide 5% of the total travel demand) where the route flow solution from the DSD method includes five or more routes in each pair. Comparing this route flow solution to the one from the FW method for these O–D pairs we have that, on average, 53% of the O–D demand use identical routes. If we calculate the same measure for the 270 largest O–D pairs (which provide 21% of the total travel demand) which have two or more equilibrium routes in each pair, then the corresponding number is 74%. If we instead analyze the O–D disaggregated link flow solution for the seven links with the largest flow and with 258 O–D pairs or more present on each of the links we have that 92% of the flow originate from identical O–D pairs. This high percentage can be explained by the fact that most of the O–D pairs present on these links only have one route between the origin and destination in both of the route flow solutions.

2.2. Applications.

We give four examples of potential applications where a route flow solution is used as input and present results based on the route flow output from the DSD and the FW method. In the examples the test network of Sioux Falls is used.

Origin–destination matrix estimation.   The estimation or adjustment of O–D trip matrices is a well known problem in traffic planning. The estimation is often based on traffic count observations (e.g., Cascetta and Nguyen, 1988, and Fisk, 1988). Assume that we wish to estimate an O–D matrix consisting of N elements (O–D pairs) where we have observed traffic counts $\bar{v}_i$ for M links. The fundamental equation in a standard O–D matrix estimation problem can be expressed as

$$\sum_{j=1}^{N} p_{ij}t_j = \bar{v}_i, \quad i = 1, ..., M,$$

where p_{ij} is the proportion of trips in O–D pair j using link i and t_j is the number of trips in O–D pair j which is to be determined. In the case of a congested network the coefficients p_{ij} are functions of t_j and can only be determined in an iterative procedure. Usually, an equilibrium assignment for a given t_j is made alternatively with an adjustment of the O–D matrix based on the coefficients p_{ij}. In the procedure the coefficients p_{ij} are extracted from a route flow solution (or an O–D disaggregated link flow solution) and the result is therefore dependent of the route flow solution obtained.

Construction of an induced origin–destination matrix.   Most urban areas can be divided into a number of subareas, each with different characteristics, for example, a heavily congested core area and its surrounding arterials. To analyze each of the subareas separately we need consistent induced O–D matrices for each of the subareas. A transfer decomposition approach for calculating an induced O–D matrix is presented in Hearn (1984) and Barton et al. (1989). The induced O–D matrix obtained by this approach is non-unique. Another approach (described in Drive project v1054), which uses route flow information, to obtain O–D matrices for each of the subnetworks is to use a two step procedure: (1) perform an equilibrium assignment in the complete network using an algorithm that provides route flow information, and (2) use the route flow information to calculate the induced O–D matrices. Also by this approach the induced O–D matrices are non-unique and are dependent of the route flow solution used.

We give an example using the Sioux Falls network where we are interested in a consistent O–D matrix for the sub-network of nodes 14, 15, 23 and 22. We introduce the new nodes A, B, C, D, E and F to represent the area outside the defined sub-area (see Figures 2 and 3). Two route flow solutions are studied, the route flow output from the FW method and the DSD method. The number of trips from the node A to the

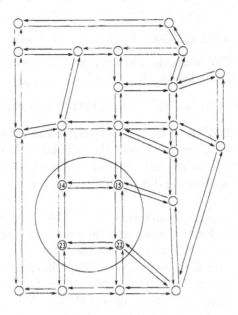

Figure 2: The Sioux Falls network.

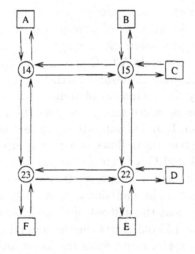

Figure 3: Sub-area in the Sioux Falls network.

nodes

$$\{14, 15, 23, 22, B, C, D, E, F\}$$

are

$$\{4112, 740, 527, 2914, 0, 459, 0, 0, 1024\}$$

and
$$\{4592, 767, 586, 2353, 0, 420, 0, 0, 1058\}$$
for the DSD and FW methods, respectively. The number of O–D trips from node 14 to the nodes {A, B, C} are

$$\{3974, 2226, 2900\} \quad \text{and} \quad \{4582, 1618, 2900\}$$

for the DSD and FW methods, respectively. (The number of trips to nodes 15, 23, 22 are unchanged from the original O–D matrix and the number of trips to the other nodes are zero.) Clearly, the two route flow solutions induce significantly different O–D matrices.

Exhaust fume emission analysis.   The application that gave the first inspiration for this work was the application of exhaust fume emission analysis and especially the analysis of exhaust fume emission from vehicles starting with cold engines. Experiments have shown that cold engines in combination with a cold catalytic converter emit large amounts of, for example, nitrogen oxides (NO_x). When analyzing the spatial distribution of emissions from cold starts the emission distribution is given as a function of the travelled distance, for which an equilibrium route flow information is needed (e.g., Série and Joumard, 1997 and Edwards, 1998). An example of a simplified exhaust fume emission analysis is given using the network in Figure 4. Assume that a group of 10 vehicles travel from node A to C and another group of 10 vehicles travels from node B to node C. Assume further that the equilibrium link flows are 10 on every link. One set of route flows consistent with the equilibrium link flows is where group one uses the route that includes the upper link between B and C and group two uses only the lower link. Another set of consistent route flows is where group one uses only the lower link between B and C and group two use the upper link. Let us assume that each link has the length of 200 meters and that the effects of the cold starts result in an emission only during the first 200 meters travelled. We observe that the first set of route flows results in an emission corresponding to 10 cold starts on the lower link between B and C and for the second set of route flows the lower link between node B and C is free from emission originating from cold starts. The conclusion is that the analysis made in this network is very sensitive to the set of route flows used.

Link toll usage analysis.   The old remedy for traffic congestion was to expand the traffic network. Experiments have shown that this approach is undermined by the level of suppressed travel demand as new roads may result in even more traffic. Congestion pricing and especially

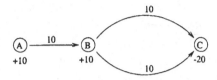

Figure 4: Example network for emission analysis.

link tolls is a viable alternative for improving the efficiency of the road system. Other motives for congestion pricing are environmental protection and/or infrastructure financing (see The Economist, December 6, 1997 for a discussion on these subjects). When implementing a congestion pricing scheme where link tolls are introduced, there may be several reasons for analyzing which traveller groups are affected by the link tolls. One example is in the case of reimbursement of link toll charges to the travellers (see e.g., Small, 1992). To target these reimbursements an analysis of the origin of the travellers using the tolled links is made (Rossi et al., 1989). In such analyses O–D disaggregated link flows (or route flows) are needed as input.

We give an illustrative example where we analyze the origin of the travellers on a specific link in the network of Sioux Falls. Let us assume that we would like to determine the origin of the travellers using the link $(14, 15)$, marked in Figure 5. The equilibrium link flow on $(14, 15)$ is 9035 vehicles. The FW and the DSD methods are used to obtain two route flow solutions. According to both route flow solutions the vehicles that use the link are O–D pairs with origin nodes in the set $\{11, 12, 13, 14, 23, 24\}$. In the FW route flow solution the flows from these origins on link $(14, 15)$ are $\{1043, 356, 95, 6266, 1156, 119\}$, respectively and in the DSD route flow solution the corresponding flows are $\{1356, 344, 59, 6643, 536, 96\}$.

In this section we have given four examples of applications where a more detailed description of the equilibrium flow than the link flow solution is needed as input. Two of the applications, the O–D matrix estimation and the link toll usage analysis, need to have the link flow disaggregated by commodity, that is, O–D link flows. The other two applications, the construction of an induced O–D matrix and the exhaust fume emission analyses need an even more disaggregated flow description, that is, a route flow solution. For the construction of an induced O–D matrix an O–D link flow solution gives us unique row and column sums of the O–D matrix, but a route flow solution is necessary to determine (uniquely) the actual O–D matrix. In the exhaust fume emission

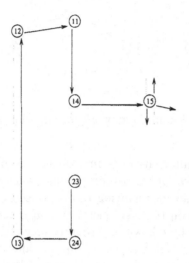

Figure 5: Link example network.

analysis we need the route flow information to calculate the travelled distance. However, if the exhaust fume emissions are given as a function of the travelled time it is sufficient to have an O–D link flow solution.

These applications, together with the numerical examples, illustrate how the non-uniqueness of the route flows can affect the results in each of the applications and illustrate the need for a procedure that provides a specified route flow solution (or in some cases a specified O–D link flow solution) that is reasonable according to some specified criterion.

In this paper we state an entropy problem (Rossi et al., 1989), which as its optimal solution gives a unique set of reasonable route flows (see Section 3). For this problem we propose an exact solution method based on a partial dualization approach.

In Section 3 necessary notation for the analysis of equilibrium flows is introduced, and previous work on finding a reasonable and uniquely defined route flow solution (the most likely route flow solution) is reviewed. We establish that the solution to the model stated for finding the most likely route flows is a limiting case of the stochastic user equilibrium route flows when the travellers' cost perception dispersion tends to zero. The relations between the solution to the problem of finding the most likely O–D link flow according to Akamatsu, (1997) and the solution to the most likely route flow problem are discussed. We also discuss a procedure for finding the most likely route flow solution from the most likely O–D link flow solution. In Section 4, we present a solution procedure for the problem of finding the most likely route flows. In Section 5 promising experimental results for medium scale networks are reported and, in Section 6, results from an application to exhaust fume emissions is presented.

3. Most likely equilibrium flows.

3.1. Preliminaries.

Consider a network describing an urban region with sets of nodes, $\mathcal{N}$, and directed links, $\mathcal{A}$. Between every O–D pair of nodes, $(p, q) \in \mathcal{C}$, there is a fixed trip demand, d_{pq}. Each link, $a \in \mathcal{A}$, is assumed to be associated with a travel cost (or time) which depend on the total number of travellers, f_a, on link a, defined by a positive, continuous, and strictly increasing function, c_a. The link travel costs are assumed to be additive, that is, each route cost is assumed to be the sum of the travel cost for the links defining the route. The solution to the assignment model defines a situation where each traveller minimizes his/her own travel cost, and results in the Wardrop equilibrium where the travel cost on the routes used are equal and equal to or less than those which would be experienced by a single vehicle on any unused route.

The equilibrium conditions are formulated using route flows, h_{pqr}, that is, the flow between origin p and destination q on route $r \in \mathcal{R}_{pq}$, where $\mathcal{R}_{pq}$ denotes the set of routes in this specific O–D pair. A set of non-negative route flows satisfying the O–D demand can be described by the following constraints which define the route flow polytope, $\mathcal{H}^r$:

$$\sum_{r \in \mathcal{R}_{pq}} h_{pqr} 1 = d_{pq}, \qquad \forall (p, q) \in \mathcal{C}, \tag{1a}$$

$$h_{pqr} \geq 0, \qquad \forall r \in \mathcal{R}_{pq}, \quad \forall (p, q) \in \mathcal{C}. \tag{1b}$$

To relate routes and links we define $\delta_{pqra} = 1$ if route r between p and q includes link a, and 0 otherwise. We can now express the relation between link and route flows as

$$f_a(h) = \sum_{(p,q) \in \mathcal{C}} \sum_{r \in \mathcal{R}_{pq}} \delta_{pqra} h_{pqr}, \qquad \forall a \in \mathcal{A}. \tag{2}$$

The cost of using route r between p and q can, by the additivity assumption and (2), be expressed as

$$c_{pqr}(h) = \sum_{a \in \mathcal{A}} \delta_{pqra} c_a(f_a).$$

The Wardrop equilibrium condition can now be expressed as

$$h_{pqr} > 0 \implies c_{pqr} = \pi_{pq}, \qquad \forall r \in \mathcal{R}_{pq},$$

$$h_{pqr} = 0 \implies c_{pqr} \geq \pi_{pq}, \quad \forall r \in \mathcal{R}_{pq},$$

where π_{pq} denotes the minimum route cost in O–D pair (p,q). Denote the equilibrium routes in O–D pair (p,q) by

$$\mathcal{R}^*_{pq} = \{r \in \mathcal{R}_{pq} | c_{pqr}(h) = \pi_{pq}\}. \tag{3}$$

The equilibrium link flow solution, f^*, can be shown to be found as the optimal solution to the problem (e.g., Patriksson, 1994, Theorem 2.1)

$$\min_{h \in \mathcal{H}^r} \sum_{a \in \mathcal{A}} \int_0^{f_a(h)} c_a(s)\, ds. \tag{4}$$

The relation between equilibrium link and route flows is

$$\sum_{(p,q) \in \mathcal{C}} \sum_{r \in \mathcal{R}^*_{pq}} \delta_{pqra} h_{pqr} f^*_a, \quad \forall a \in \mathcal{A},$$

where the set of equilibrium route flow solutions is defined by the polytope, $(\mathcal{H}^r)^*$,

$$\sum_{r \in \mathcal{R}^*_{pq}} h_{pqr} = d_{pq}, \qquad \forall (p,q) \in \mathcal{C}, \tag{5a}$$

$$\sum_{(p,q) \in \mathcal{C}} \sum_{r \in \mathcal{R}^*_{pq}} \delta_{pqra} h_{pqr} = f^*_a, \qquad \forall a \in \mathcal{A}, \tag{5b}$$

$$h_{pqr} \geq 0, \qquad \forall r \in \mathcal{R}_{pq}. \quad \forall (p,q) \in \mathcal{C}. \tag{5c}$$

This set is usually not a singleton, as was demonstrated in Section 2. We consider the following model for finding a unique set of "reasonable" equilibrium route flows. The problem is defined using route flow variables and the optimal solution defines the set of *most likely route flows*, denoted h^*.

$$\max - \sum_{(p,q) \in \mathcal{C}} \sum_{r \in \mathcal{R}^*_{pq}} h_{pqr} \ln h_{pqr} \tag{6a}$$

subject to

$$\sum_{r \in \mathcal{R}^*_{pq}} h_{pqr} = d_{pq}, \qquad \forall (p,q) \in \mathcal{C}, \tag{6b}$$

$$\sum_{(p,q) \in \mathcal{C}} \sum_{r \in \mathcal{R}^*_{pq}} \delta_{pqra} h_{pqr} = f^*_a, \qquad \forall a \in \mathcal{A}, \tag{6c}$$

$$h_{pqr} \geq 0, \qquad \forall r \in \mathcal{R}^*_{pq}, \quad \forall (p,q) \in \mathcal{C}, \tag{6d}$$

In problem (6) the route flow entropy is maximized under the restriction that the route flows are equilibrium flows.

A derivation of problem (6) was presented in Rossi et al. (1989).

The derivation of the model is based on the minimum information principle in Snickars and Weibull (1977). The equilibrium link flow solution is defined as a macro state which arises as a result of the route choices of individual travellers. The route choices between minimum and equal cost routes of a set for individual travellers define a set of micro states. All micro states consistent with the macro state are argued to be equally probable, since the travellers are indifferent to which route they use. Using well known principles from information theory they argue that the most likely route flow solution is the one which corresponds to the largest number of micro states among those which are consistent with the observed macro state. After an approximation using the Stirling formula the problem (6) is obtained for determining this (unique) micro state.

The indifference of the travellers and the assumption that all micro states are equally probable can be shown to also follow from the *efficiency principle* on which an alternative derivation based on behavioural assumptions can be made (Smith, 1978 and Smith, 1983).

In Rossi et al. (1989) a procedure to obtain the most likely route flows described as a two-step procedure is presented. First, the construction of the entropy program requires that the link flow solution is calculated and all equilibrium routes are enumerated. Second, the entropy program is solved. Rossi et al. present a numerical result for a small-scale problem where the general-purpose nonlinear programming code MINOS was used. The solution to problem (6) is analyzed for a small scale network in Bell and Iida (1997, Section 5.5).

3.2. An alternative derivation.

The definition of a stochastic user equilibrium extends the Wardrop conditions (the deterministic user equilibrium). To reflect a variation in the user perception of cost a random component, here assumed to be Gumbel distributed, is added to the route cost. The stochastic user equilibrium can be defined as the optimal solution to the following optimization model (Fisk, 1980),

$$\min \sum_{a \in \mathcal{A}} \int_0^{f_a(h)} t_a(s)\, ds + \frac{1}{\theta} \sum_{(p,q) \in \mathcal{C}} \sum_{r \in \mathcal{R}_{pq}} h_{pqr} \ln h_{pqr}, \qquad (7a)$$

subject to

$$(1a), (1b) \text{ and } (2), \qquad\qquad (7b)$$

where $\theta > 0$ is the dispersion parameter. Denote the optimal route flow solution to problem (7) for a fixed θ by $h^*(\theta)$. The following theorem tells us that the optimal solution to the stochastic user equilibrium problem when $\theta \to \infty$ coincides with the most likely route flow solution, h^*.

Theorem 1 $\lim_{\theta \to \infty} h^*(\theta) = h^*$.

Proof. See Appendix A.

Akamatzu et al. have in a set of papers developed a *stochastic user equilibrium* model stated in O–D disaggregated link flows using an "entropy decomposition" technique (Akamatsu et al. 1990, Akamatsu 1996, Akamatsu 1997). The "entropy decomposition" technique is applicable also to the problem of finding the most likely O–D link flows; such a model is given in Akamatsu (1997). The model is stated in O–D link flow variables, f_{pqa}, that is, the O–D flow in pair $(p,q) \in C$ on link $a = (i,j) \in \mathcal{A}$:

$$\min \sum_{(p,q)\in C} \left(\sum_{a \in \mathcal{A}} f_{pqa} \ln f_{pqa} - \right.$$

$$\left. - \sum_{j \in \mathcal{N}} \left(\sum_{i \in T(j)} f_{pq(i,j)} \right) \ln \left(\sum_{i \in T(j)} f_{pq(i,j)} \right) \right) \tag{8a}$$

subject to

$$\sum_{(p,q)\in C} f_{pqa} = f_a^*, \qquad \forall a \in \mathcal{A}, \tag{8b}$$

$$\sum_{i \in T(j)} f_{pq(i,j)} - \sum_{i \in S(j)} f_{pq(j,i)} =$$

$$= \begin{cases} d_{pq}, & \text{if } j = q, \\ -d_{pq}, & \text{if } j = p \qquad j \in \mathcal{N}, \quad (p,q) \in C, \\ 0, & \text{otherwise}, \end{cases} \tag{8c}$$

$$f_{pqa} \geq 0, \qquad \forall a \in \mathcal{A}, \quad \forall (p,q) \in C. \tag{8d}$$

where $T(j)$ denote the set of nodes i such that $(i,j) \in \mathcal{A}$ and $S(j)$ denote the set of nodes i such that $(j,i) \in \mathcal{A}$. Problem (8) is convex and has a strictly convex objective function (Akamatsu, 1997), and the optimal solution to the problem defines uniquely the most likely O–D link flow solution. The relation between the most likely O–D link flow solution and the most likely route flow solution is made clear by the following theorem.

Theorem 2 *The most likely O–D link flows defined by the optimal solution to problem (8), are equal to the O–D link flows defined by*

$$\sum_{r \in \mathcal{R}_{pq}^*} \delta_{pqra} h_{pqr}^*$$

where h^*_{pqr} is the optimal solution to problem (6).

Proof. See Appendix B.

The model (8) can in certain situations be more favourable than (6). Such situations may arise when an equilibrium solution disaggregated by origin and destination (or just by origin or by destination) is sufficient. This is the case for the O–D matrix estimation and link toll usage analysis applications (cf. Section **??**). The efficiency of the solution method for (8) sketched in Akamatsu (1997) is an open question.

Given the most likely O–D link flow solution from (8) it is possible, with a simple procedure, to obtain the most likely route flow solution. Such a procedure include an enumeration of a restricted set of routes in the network and is discussed at the end of Appendix B.

Another approach is taken in Janson (1993) where a procedure for finding an approximate most likely O–D link flow solution is presented. The approach is based on the fact that different O–D link flow solutions are obtained when varying the initialization of the Frank–Wolfe method. A fixed number of user equilibria are obtained from different initiations to obtain a set of O–D link flow solutions. Each of the O–D link flow solutions is incorporated as a column in a restricted entropy problem. The solution to the restricted entropy problem is a set of O–D link flows which are more likely than any solution found by solving only one equilibrium problem. Computational results for a small scale network are presented. Note, however, that the most likely O–D link flows defined by the optimal solution to the model stated in Janson (1993) are *not* consistent with the most likely route flows obtained as the optimal solution to problem (6). This is pointed out in Akamatsu (1997).

4. Solution procedure for the entropy program.

In this section we present a solution procedure for the problem of finding the most likely route flows, problem (6). We assume that the assignment problem (4) is solved and that the equilibrium link flow f^* is known. The procedure for finding the most likely route flows is a two phase procedure: (1) generate the equilibrium routes, $\mathcal{R}^*_{pq}$, from which an instance of the entropy model is constructed; and (2) solve the entropy program instance.

To generate the equilibrium routes, that is, to determine the sets $\mathcal{R}^*_{pq}$, we use the link flow solution obtained from the assignment problem and the equilibrium link costs $c_a(f^*_a)$. Construct sub-networks $\mathcal{G}^*_p = (\mathcal{N}, \mathcal{A}^*_p)$ of equilibrium links, one for each origin p, in the following way. Identify the set of equilibrium node potentials y^*_i, $i \in \mathcal{N}$ by solving a shortest route problem from source node p where the link costs are defined as

$c_a(f_a^*)$. The set of equilibrium links $\mathcal{A}_p^*$ then consists of links a such that $y_{t(a)}^* - y_{h(a)}^* = c_a(f_a^*)$, where $t(a)$ and $h(a)$ denote the tail and head of arc a, respectively. The set of equilibrium routes $\mathcal{R}_{pq}^*$, for all O–D pairs $(p,q) \in \mathcal{C}$ is obtained by an enumeration of the routes in the sub-network $\mathcal{G}_p^*$. The route enumeration is made using a depth-first search procedure with root node p. From the sets of equilibrium routes and the set of equilibrium link flows we construct an instance of the entropy program.

In phase 2, we solve the entropy program. An overview of solution algorithms for entropy programs can be found in, for example, Lamond and Stewart (1981) and a method that work on the dual problem directly is found in, for example, Fang and Tsao (1994).

We present a solution approach for the entropy program (6) that works on the dual side of the problem. The approach is based on the dualization of the constraints (6c) in problem (6). Let $\beta \in \mathbb{R}^{|\mathcal{A}|}$ be the vector of dual variables for the link flow equality constraints (6c). After transforming problem (6) into a minimization problem we construct the dual objective function

$$\varphi(\beta) = \min L(h, \beta) \tag{9a}$$

subject to

$$(6b) \quad \text{and} \quad (6d), \tag{9b}$$

with

$$L(h, \beta) = \sum_{(p,q) \in \mathcal{C}} \sum_{r \in \mathcal{R}_{pq}^*} h_{pqr} \ln h_{pqr} + \sum_{a \in \mathcal{A}} \beta_a \left(\sum_{(p,q) \in \mathcal{C}} \sum_{r \in \mathcal{R}_{pq}^*} h_{pqr} \delta_{pqra} - f_a^* \right).$$

Problem (9) separates in one problem for each pair $(p,q) \in \mathcal{C}$. We define

$$\varphi_{pq}(\beta) = \min \sum_{r \in \mathcal{R}_{pq}^*} \left[h_{pqr} \ln h_{pqr} + h_{pqr} \sum_{a \in \mathcal{A}} \delta_{pqra} \beta_a \right] \tag{10a}$$

subject to

$$\sum_{r \in \mathcal{R}_{pq}^*} h_{pqr} = d_{pq}, \tag{10b}$$

$$h_{pqr} \geq 0, \qquad \forall r \in \mathcal{R}_{pq}^*. \tag{10c}$$

This problem is explicitly solved by applying its Karush–Kuhn–Tucker conditions, which gives the solution

$$h_{pqr}(\beta) = d_{pq} \frac{e^{-\sum_{a \in \mathcal{A}} \beta_a \delta_{pqra}}}{\sum_{r \in \mathcal{R}_{pq}^*} e^{-\sum_{a \in \mathcal{A}} \beta_a \delta_{pqra}}}.$$

This expression is recognized as a logit model for route choice (restricted to equilibrium links) where the link costs are β_a. The dual problem then is

$$\sup \varphi(\beta) = -\sum_{a \in \mathcal{A}} \beta_a f_a^* + \sum_{(p,q) \in \mathcal{C}} \varphi_{pq}(\beta). \tag{11}$$

The dual objective function φ is continuously differentiable and concave and the dual problem, thus, is an unconstrained differentiable, concave maximization problem. Under the assumption that problem (6) has a positive feasible solution, the supremum in problem (11) will be attained for some vector $\beta = \beta^*$ of Lagrange multipliers (see, e.g., Bazaraa et al., 1993, Theorem 5.3.1). Further, strong duality holds, and the optimal solution to problem (6) is

$$h_{pqr}^* = d_{pq} \frac{e^{-\sum_{a \in \mathcal{A}} \beta_a^* \delta_{pqra}}}{\sum_{r \in \mathcal{R}_{pq}^*} e^{-\sum_{a \in \mathcal{A}} \beta_a^* \delta_{pqra}}}. \tag{12}$$

If a route flow is enforced to be zero in an optimal solution, this route can (in principle) be excluded from the set $\mathcal{R}_{pq}^*$ (see, e.g., Erlander and Stewart, 1990). Otherwise, the supremum in problem (11) will not be reached and some β_a^* will attain 'infinite values'.

The dual problem is solved using a conjugate gradient method. The gradient of φ at β is

$$\nabla \varphi(\beta) = \left[\sum_{(p,q) \in \mathcal{C}} \sum_{r \in \mathcal{R}_{pq}^*} \delta_{pqra} h_{pqr}(\beta) - f_a^* \right]_{a \in \mathcal{A}}$$

that is, the i:th component of $\nabla \varphi(\beta)$ is the difference between the link flow solving (10), calculated as a sum of the route flows, and the equilibrium link flow for link a. The norm of the gradient is used in the termination criterion for the algorithm. The route flows are, for numerical reasons, calculated as

$$h_{pqr}(\beta) = d_{pq} \frac{e^{-\sum_{a \in \mathcal{A}} \beta_a \delta_{pqra} - \gamma_{pq}}}{\sum_{r \in \mathcal{R}_{pq}^*} e^{-\sum_{a \in \mathcal{A}} \beta_a \delta_{pqra} - \gamma_{pq}}},$$

where

$$\gamma_{pq} = \min_{r \in \mathcal{R}_{pq}^*} \sum_{a \in \mathcal{A}} \beta_a \delta_{pqra}.$$

Now, we introduce an iteration counter k. The search direction is computed in iteration $k = 0$ as $d^k = \nabla\varphi(\beta^k)$ and for $k > 0$ as

$$d^k = \nabla\varphi(\beta^k) + \frac{\nabla\varphi(\beta^k)^T\nabla\varphi(\beta^k)}{\nabla\varphi(\beta^{k-1})^T\nabla\varphi(\beta^{k-1})}d^{k-1}.$$

An approximate step length is computed based on a one dimensional Newton-step as

$$t^k = -\frac{\nabla\varphi(\beta^k)^T d}{d^{k^T}\nabla^2\varphi(\beta^k)d^k},$$

where the Hessian is (see, e.g., Bertsekas, 1995, Chapter 6, Equation 1.11)

$$\nabla^2\varphi(\beta) = -\nabla g(h(\beta))^T\nabla^2_{hh}L(h,\beta)^{-1}\nabla g(h(\beta)) =$$

$$= -\nabla g(h(\beta))^T\text{diag}\,[h_{pqr}(\beta^k)]\nabla g(h(\beta^k)) = -\Delta\,\text{diag}\,[h_{pqr}(\beta^k)]\Delta^T,$$

where g is the vector function of the left-and-sides in (6c) and $\Delta^T = (\delta_{pqra})$. A sufficient ascent is ensured by the use of an Armijo backtracking line search (with an acceptance parameter 0.2). The dual feasible solution $\beta = 0$ is used as a starting solution. The solution method is terminated when $\|\nabla\varphi(\beta)\| \le \varepsilon$ for some value of $\varepsilon > 0$.

5. Experimental results.

Three test networks have been used for the evaluation of the solution method described in Section 4 and for the analysis of the properties of the most likely route flow solution. The networks studied are the networks of Sioux Falls (SD, USA), Winnipeg (Canada) and Linköping (Sweden), see Table 1 for their sizes. In the computational experiments, the route flow solutions obtained from the FW and the DSD methods are compared to the most likely route flow solution, the number of routes in each of the route flow solutions are analysed, and the topological differences between the route flow solutions are studied. The test results are presented separately for each of the three test networks.

Network	# Nodes	# Links	# O–D pairs
Sioux Falls	24	76	528
Winnipeg	1052	2836	4344
Linköping	335	882	12372

Table 1: Traffic equilibrium networks.

The network equilibrium problems are solved to a relative error between lower and upper bounds of the optimal objective value of 10^{-6}

for the DSD algorithm and to $5 \cdot 10^{-5}$, $3 \cdot 10^{-5}$ and $7 \cdot 10^{-6}$ for the Sioux Falls, Winnipeg and Linköping networks, respectively for the FW method. The lower required accuracy for the FW algorithm is chosen since the method has a slower convergence rate. The number of routes generated by the DSD method is highly dependent of how accurate each of the master problems (see, Larsson and Patriksson, 1992) are solved. In our experiments, we have solved each of the master problems to a relative error between lower and upper bounds of the optimal objective value of 10^{-6}.

In order to implement the two-phase procedure, we introduce two tolerances: (1) a link is included in the subnetworks $\mathcal{A}_p^*$ of equilibrium links if the reduced cost for a link divided by the link cost is smaller than 0.01. This tolerance is crucial for the equilibrium route enumeration, if the tolerance is zero, only the shortest route generated in the last iteration of the equilibrium assignment method will be an equilibrium route. If however the tolerance is large, routes considered as equilibrium route may include non-equilibrium links, and this may lead to cycles in the sub-networks, thus making the enumeration impossible. (2) In the enumeration procedure, an enumerated route is considered to be an equilibrium route if the sum of reduced costs divided by the sum of link travel costs is smaller than 0.1.

Precautions have been made to ensure that problem (6) have a feasible solution; with the tolerances used, a small number of routes supplied by the DSD method are excluded from the resulting equilibrium sub-network, $\mathcal{A}_p^*$. These routes are however added to the set of equilibrium routes during the route enumeration procedure. The number of added routes are 0, 80, 93 for the evaluation networks of Sioux Falls, Winnipeg and Linköping, respectively. (This number is highly dependent of the solution accuracy required in the assignment solution.)

In the tables below a route is considered significant in the equilibrium solution if

$$h_{pqr} > \frac{d_{pq}}{100|\mathcal{R}_{pq}^*|},$$

that is, if the route flow is larger than one percent of the flow obtained if the demand were uniformly spread over all equilibrium routes in the O–D pair (p, q).

All problems are solved on a SUN Ultrasparc 170E with 192 MB memory.

5.1. The Sioux Falls network.

For the Sioux Falls network the termination criterion for the entropy problem is set to $\varepsilon = 10^{-6}$. The computational time for the route enumeration and the solution of the entropy problem is 0.3 and 20 CPU seconds, respectively. In Table 2 the number of equilibrium routes, the number of equilibrium routes per O–D pair, and the entropy value for the route flow solution are presented for the DSD, FW, and the most likely route flow solution (the number in the parenthesis is the number of enumerated equilibrium routes).

	Entropy value	# Routes		# Routes/O–D pair
DSD	42.17	718		1.37
FW	54.75	785		1.49
EP	54.80	762	(770)	1.45

Table 2: Sioux Falls.

The entropy objective function in problem (6) gives a measure of probability for a given route flow solution. The entropy values for the three route flow solutions indicate that the route flow solution generated by the FW method is close to the most likely route flow relative to the solution to the DSD method. Note however, that in the route flow solution from the FW method a number of non-equilibrium routes are included, and the entropy value may therefore be deceptive.

The entropy value obtained as the objective function value in problem (6) is an aggregated measure. In an individual O–D pair the most likely route flow need not have a higher entropy value than a route flow solution from another method. This is the case in O–D pair (16, 8) were the route flow solution from the FW method have two routes with significant flow and the entropy value for this pair is -1.19 while the most likely route flow solution also has two routes with significant flow and the entropy value is -1.24. In the example in Section ?? two route flows for O–D pair (14, 10) for the DSD and the FW method are presented. The entropy in this O–D pair is -0.146 (-0.583) for the route flow solution from the FW (DSD) method. The most likely route flow solution has 838 and 1262 vehicles for the routes 14–11–10 and 14–15–10, respectively. This route flow set has an entropy value of -0.145 (calculated in thousands of vehicles).

The maximal number of significant routes in any O–D pair is 2 and 8 in the DSD and most likely route flow solution, respectively. The largest difference between the number of significant O–D link flows on one link in the DSD and the most likely route flow solution is 7 and the number of

O–D pairs with a significant O–D link flow on a link is on average 28.72 and 29.53 in DSD and the most likely route flow solutions, respectively. The small difference in the averaged numbers can be explained by the fact that a large number of the O–D pairs only use one route between the origin and the destination.

We also compute an aggregated measure of the difference between the route flow solutions. The analysis is restricted to the 72 O–D pairs where the DSD method has generated two or more routes in each pair (these pairs provide 21% of the total travel demand). When comparing the most likely route flow solution with the route flow solution from the DSD method for these O–D pairs, on average, 82% of the O–D demand use identical routes. The corresponding number for the most likely route flow and the solution to the FW method is 97%.

5.2. The Winnipeg network.

For the Winnipeg network the termination criterion for the entropy program is set to $\varepsilon 3D1.0$. This termination criterion implies that the left-hand-side in (6c) have a relative difference less than 0.2% from the equilibrium link flows when considering equilibrium link flows with a flow larger then 10 units (less than 10% of the links have a flow smaller than 10 units). The computational time for the route enumeration and the solution of the entropy problem is 125 and 638 CPU seconds, respectively.

	Entropy value	# Routes		# Routes/O–D pair
DSD	−178355	8957		2.06
FW	−168212	16494		3.80
EP	−166648	18111	(22338)	4.17

Table 3: Winnipeg.

The entropy values and route data are given in Table 3 (the number in parenthesis is the number of enumerated equilibrium routes).

Next, we compare the topological differences between the most likely route flow and the route flow solution from the DSD method for one O–D pair with a relatively large travel demand. The set of equilibrium links in O–D pair number (1, 15) in the Winnipeg network is given in Figure 6 (the links in the network are aggregates of sequences of links in the original network). The route enumeration procedure gives that 110 routes in this network are equilibrium routes. The DSD method gives a route flow solution in which 6 equilibrium routes have a significant flow. These equilibrium routes can be found in the network in Figure 7 and in Table 4 we list the routes and the route flows from the DSD method and the corresponding route flows in the most likely route flow solution. Comparing the two route flow solutions we observe that there is a large difference in the route flows for the two solutions in this O–D pair. We note that in the most likely route flow solution, only about 60% of the O–D demand use the routes that carry all the flow in the DSD route flow solution.

Equilibrium DSD routes	Flow, DSD	Flow, EP
1 – 2 – 3 – 6 – 12 – 13 – 14 – 15	33.52	5.52
1 – 2 – 4 – 5 – 7 – 9 – 11 – 10 – 14 – 15	0.98	0.03
1 – 3 – 6 – 9 – 12 – 13 – 14 – 15	0.67	1.95
1 – 3 – 6 – 12 – 13 – 14 – 15	1.68	13.77
1 – 2 – 3 – 4 – 5 – 8 – 10 – 14 – 15	2.86	0.19
1 – 3 – 4 – 5 – 7 – 9 – 11 – 13 – 14 – 15	0.29	1.93
Sum	40.00	23.40

Table 4: Two set of equilibrium routes in the order they are generated in the DSD scheme.

The routes in the most likely route flow solution are depicted in Figure 8. The most likely route flow solution uses 51 significant routes.

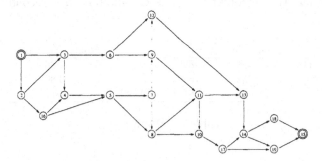

Figure 6: Network in which equilibrium routes are enumerated

An aggregated measure of the difference between the route flow solutions is given where we restrict the analysis to the 270 largest O–D pairs (these pairs provide 21% of the total travel demand) where the DSD method have generated two or more routes in each pair. Comparing the most likely route flow solution to the route flow solution from the DSD method, we have that, on average, 75% of the O–D demand in these O–D pairs use identical routes. The corresponding number for the most likely route flow compared with the solution to the FW method is 94%.

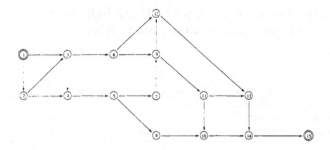

Figure 7: Network of generated routes in DSD solution.

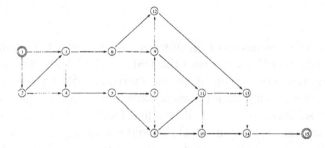

Figure 8: Network of significant equilibrium routes in max entropy solution.

The largest number of routes used in one O–D pair is 9 and 396 in the DSD and most likely route flow solutions, respectively. The largest difference between the number of significant O–D link flows on one link in the DSD and the most likely route flow solution is 113. The number of O–D pairs with a significant O–D link flow on a link is on average 47.61 and 47.78 in the DSD and the most likely route flow solutions, respectively. Also for the Winnipeg network we conclude that the small difference in the averaged numbers can be explained by the fact that a large number of the O–D pairs have only one route between the origin and the destination.

5.3. The Linköping network.

For the Linköping network the termination criterion for the entropy program is set to $\varepsilon = 2.0$. The computational time for the route enumeration and the solution of the entropy problem is 64 and 970 CPU seconds, respectively. The number of equilibrium routes with significant route flow, the number of routes per O–D pair and the entropy value are given in Table 5. Although the network is, in some areas, highly congested the number of routes per O–D pair indicates that the number of possible route alternatives in the network is low.

	Entropy value	# Routes		# Routes/O–D pair
DSD	-757620	18672		1.51
FW	-743075	21193		1.71
EP	-743069	21447	(21894)	1.73

Table 5: Linköping.

We restrict the analysis of the difference of the route flow solutions to the 382 largest O-D pairs where the DSD method have generated two or more routes in each pair (these pairs provide 13% of the total travel demand). When comparing the most likely route flow solution with the route flow solution from the DSD method for these O–D pairs we have that, on average, 81% of the O–D demand use identical routes. For the same O–D pairs and the route flow solution from the FW method and the most likely route flow solution we found that 97% of the demand travel on identical routes.

The maximal number of used routes in any O–D pair is 5 and 19, respectively, in the DSD and the most likely route flow solutions. The largest difference in the number of significant O–D link flows on one link is 165 between the DSD and the most likely route flow solutions. The number of O–D pairs with a significant O–D link flow on a link is 238.6 and 241.2 in the DSD and the max entropy solutions, respectively, on average.

From the numerical experiments we conclude that the solution procedure can be used to obtain the most likely route flow solution in reasonable computing time for medium sized networks. The difference between the route flow solution from the DSD method and the most likely route flow solution is significant in certain O–D pairs of our test networks. The FW method produces, for our test networks, a route flow solution very similar to the most likely route flows in most O–D pairs.

For large networks, especially those with a grid structure, there is a possibility that the number of routes found in the path enumeration procedure becomes very large. To avoid this and still use the presented approach the networks need to be aggregated. The establishment of a proper principle for aggregation the network is a subject for future studies.

6. An application: Exhaust fume emission analysis.

In this section an exhaust fume emission analysis is made for the Linköping network. In the experiments made we analyze the spatial distribution on the links of cold start dependent excess from vehicles starting with cold engines and catalytic converters. The data for vehicle emissions, the travel demand and the link delay functions used in the analysis was provided by the Swedish National Road and Transport Research Institute (VTI). Figure 9 shows the central part of Linköping, which is the set of links studied in detail. The average exhaust fume

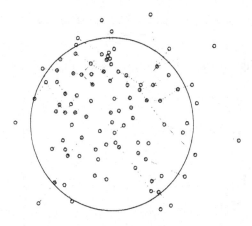

Figure 9: The central part of Linköping.

emission of nitrogen oxides (NO_x) arising from a cold engine (in contrast to that from a warm engine) is assumed to be described by an exponential function of the travelled distance only, see Figure 10.

In Figure 11 the emission distribution of the links in the central part of Linköping is shown. The emission distribution is computed based on the most likely route flow solution. From this figure we may observe that some links have a very high level of NO_x compared to other links.

Experiments where the NO_x emission calculation is based on the route flow output from the DSD method have shown that the difference to the emission obtained by the most likely route flow is very small. (Significant differences are found only on links where the emission are relatively small compared to other links.) Due to the structure of the network the results do not differ very much. However, in the analysis made, there are differences in the resulting link emission and these differences are likely to be larger in a network where the traveller has a larger number of alternative routes to choose from, although lack of sufficient data prevent us from making analyses of other city centres.

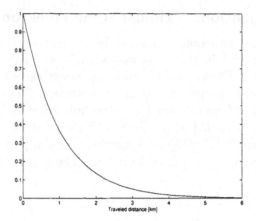

Figure 10: The NO_x emission distribution.

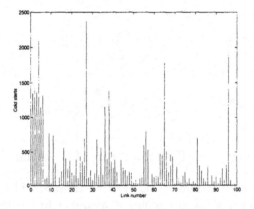

Figure 11: NO_x emission on the links in the central part of Linköping according to the most likely route flow solution.

Appendix A: Relation between the stochastic user equilibrium and the most likely route flows.

Proof. (**Theorem 1**). The proof is stated for two arbitrary functions, f and g, and is based on the proof in Polyak (1987, Chapter 6.1, Theorem 4), modified to the case where g is strictly convex.

Assume that $f : \mathbb{R}^n \to \mathbb{R}$ is convex and $g : \mathbb{R}^n \to \mathbb{R}$ is strictly convex, and let $X_f^* = \operatorname{argmin}_{x \in X} f(x)$, where $X \subset \mathbb{R}^n$ is a nonempty, closed and bounded set. Define the trajectory $\{x_\varepsilon | \varepsilon > 0\}$ by $x_\varepsilon = \operatorname{argmin}_{x \in X} \{f(x) + \varepsilon g(x)\}$ and let $\{x^*\} = \operatorname{argmin}_{x \in X_f^*} g(x)$. To prove the theorem we show that $x_\varepsilon \to x^*$ when $\varepsilon \to 0^+$.

For any fixed $\varepsilon > 0$, $f + \varepsilon g$ is strictly convex. Since X is compact and convex, x_ε exists and is unique, and the trajectory $\{x_\varepsilon | \varepsilon > 0\}$ is well-defined and bounded.

From the definition of x_ε, and since $x^* \in X$ we have that

$$f(x_\varepsilon) + \varepsilon g(x_\varepsilon) \leq f(x^*) + \varepsilon g(x^*). \tag{13}$$

Further, since $x^* \in X_f^*$ and $x_\varepsilon \in X$ we also have the relation $f(x^*) \leq f(x_\varepsilon)$. From these two inequalities, $g(x_\varepsilon) \leq g(x^*)$, follows.

Extract a subsequence $\{x_{\varepsilon_i}\}$, $\varepsilon_i \downarrow 0$, converging to a point $\tilde{x} \in X$. Since g is continuous, $\lim_{i \to \infty} g(x_{\varepsilon_i}) = g(\tilde{x})$ holds, which gives that,

$$g(\tilde{x}) \leq g(x^*). \tag{14}$$

Taking the limit corresponding to the subsequence $\{x_{\varepsilon_i}\}$ in (13) gives that $f(\tilde{x}) \leq f(x^*)$, and therefore $\tilde{x} \in X_f^*$. By relation (14), the definition of x^*, and the strict convexity of g, we then obtain that $\tilde{x} = x^*$. It follows that $x_{\varepsilon_i} \to x^*$.

Since the subsequence was arbitrary, $\{x_\varepsilon\} \to x^*$ follows.

Now, identify x with the link flow variables $f_a(h)$, f with

$$\sum_{a \in A} \int_0^{f_a(h)} t_a(s) \, ds$$

and g with

$$\sum_{(p,q) \in \mathcal{C}} \sum_{r \in \mathcal{R}_{pq}} h_{pqr} \ln h_{pqr}$$

to complete the proof.

Appendix B: Relation between the models for finding the most likely O–D link flows and the most likely route flows.

Proof. (**Theorem 2**). The proof is made in two steps. In the first we show that problems (6) and (8) have the same feasible set with respect to O–D link flows, that is there can not exist a set of O–D link flows that is feasible in problem (8) and that has a positive O–D link flow on a non-equilibrium route. In the second step we show that the optimal solutions to the two problems (6) and (8) coincide in terms of the O–D link flows.

We show that the link flow in O–D pair (p,q) on link $a \in \mathcal{A}$ is zero whenever link a is not included in an equilibrium route from p to q. The Wardrop conditions can be stated as (e.g., Patriksson, 1992, Theorem 3.14a)

$$c(h^*)^T(h - h^*) \geq 0, \quad \forall h \in \mathcal{H}^r.$$

Let $\bar{h} \in \{h \in \mathcal{H}^r \mid \sum_{(p,q)\in\mathcal{C}} \sum_{r\in\mathcal{R}_{pq}} \delta_{pqra} h_{pqr} = f_a^*\}$. From the additivity assumption we have that

$$c(h^*)^T(\bar{h} - h^*) = \sum_{a\in\mathcal{A}} c_a(f_a^*) \left(\sum_{(p,q)\in\mathcal{C}} \sum_{r\in\mathcal{R}_{pq}} \delta_{pqra} \bar{h}_{pqr} - f_a^* \right) =$$

$$= c(f^*)^T(f^* - f^*) = 0.$$

Hence,

$$c(h^*)^T(h - \bar{h}) = c(h^*)^T(h - h^*) - c(h^*)^T(\bar{h} - h^*) =$$

$$= c(h^*)^T(h - h^*) \geq 0, \quad \forall h \in \mathcal{H}^r.$$

This implies that $\bar{h} \in (\mathcal{H}^r)^*$, that is, if $c_{pqr} > \pi_{pq}$ then $\bar{h} = 0$. We conclude that a link not included in any equilibrium route from p to q has a zero O–D link flow in pair (p,q), and that the feasible set of problems (6) and (8) therefore coincide with respect to O–D link flows.

Problem (8) can thus be equivalently stated as

$$\min \sum_{(p,q)\in\mathcal{C}} \left(\sum_{a\in\mathcal{A}} f_{pqa} \ln f_{pqa} - \sum_{j\in\mathcal{N}} \left(\sum_{i\in T'(j)} f_{pq(i,j)} \right) \right.$$

$$\left. \cdot \ln \left(\sum_{i\in T'(j)} f_{pq(i,j)} \right) \right) \qquad (15a)$$

subject to

$$\sum_{(p,q)\in\mathcal{C}} f_{pqa} = f_a^*, \qquad \forall a \in \mathcal{A}, \qquad (15b)$$

$$\sum_{i\in T'(j)} f_{pq(i,j)} - \sum_{i\in\mathcal{S}'(j)} f_{pq(j,i)} =$$

$$= \begin{cases} d_{pq}, & \text{if } j = q, \\ -d_{pq}, & \text{if } j = p, \qquad j \in \mathcal{N}, \quad (p,q) \in \mathcal{C}, \qquad (15c) \\ 0, & \text{otherwise,} \end{cases}$$

$$f_{pqa} \geq 0, \qquad \forall a \in \mathcal{A}, \quad \forall(p,q) \in \mathcal{C}. \qquad (15d)$$

where $T'(j)$ denote nodes i such that $(i,j) \in \mathcal{A}$ and is included in an equilibrium route from (p,q) and $\mathcal{S}'(j)$ denote nodes i such that $(j,i) \in \mathcal{A}$ and is included in an equilibrium route from (p,q).

In step two we show that the optimal solutions to problems (6) and (15) coincide in terms of the O–D link flows. One O–D link flow solution, feasible in problem (15), can be calculated from the most likely route flows, h^*, as

$$f_{pqa}(h^*_{pqr}) = \sum_{r \in \mathcal{R}^*_{pq}} \delta_{pqra} h^*_{pqr}, \qquad \forall (p,q) \in \mathcal{C}, \quad \forall a \in \mathcal{A},$$

where

$$h^*_{pqr} = d_{pq} \frac{e^{-\sum_{a \in \mathcal{A}} \beta^*_a \delta_{pqra}}}{\sum_{r \in \mathcal{R}^*_{pq}} e^{-\sum_{a \in \mathcal{A}} \beta^*_a \delta_{pqra}}}, \qquad \forall r \in \mathcal{R}^*_{pq}, \quad \forall (p,q) \in \mathcal{C} \qquad (16)$$

with β^* being the vector of Lagrange multipliers for constraints (6c). Consider an arbitrary node j, and define $\mathcal{R}^j_{pq}$ as the set of routes in $\mathcal{R}^*_{pq}$ that pass node j. Any route in $\mathcal{R}^j_{pq}$ passing node $i \in T'(j)$ can be partitioned into three parts, from p to i, from i to j, and from j to q. Hence,

$$\mathcal{R}^j_{pq} = \mathcal{R}^j_{pi} \bigcup \left(\bigcup_{i \in T'(j)} (i,j) \right) \bigcup \mathcal{R}^j_{jq}.$$

Define k as the denominator in expression (16) and define

$$k_j = \sum_{r \in \mathcal{R}^j_{pq}} e^{-\beta_r} = \sum_{i \in T'(j)} \sum_{r_1 \in \mathcal{R}^j_{pi}} \sum_{r_2 \in \mathcal{R}^j_{jq}} e^{-\beta_{r_1}} e^{-\beta_{ij}} e^{-\beta_{r_2}} =$$

$$= \left(\sum_{r \in \mathcal{R}^j_{jq}} e^{-\beta_r} \right) \sum_{i \in T'(j)} \left(e^{-\beta_{ij}} \sum_{r \in \mathcal{R}^j_{pi}} e^{-\beta_r} \right), \qquad (17)$$

where $\beta_r = \sum_{a \in r} \beta_a$. The last equality in (17) is adopted from Van Vliet (1981). Define $\mathcal{A}(j) = \{(i,j) | i \in T(j)\}$ and $\mathcal{A}'(j) = \{(i,j) | i \in T'(j)\}$. The total O–D flow in pair (p,q) into node j then is

$$\sum_{a \in \mathcal{A}'(j)} f_{pqa}(h^*) = \sum_{a \in \mathcal{A}'(j)} \sum_{r \in \mathcal{R}^j_{pq}} \delta_{pqra} h^*_{pqr} 3D \frac{k_j}{k} d_{pq}.$$

Choose an arbitrary node $\bar{i} \in T'(j)$. The flow in O–D pair (p,q) on link $\bar{a} = (\bar{i}, j)$ is,

$$f_{pq\bar{a}}(h^*) = \sum_{r \in \mathcal{R}^j_{p\bar{i}} \times \{(\bar{i},j)\} \times \mathcal{R}^j_{jq}} \delta_{pqr\bar{a}} h^*_{pqr} =$$

$$= \frac{d_{pq}}{k} \left[\left(\sum_{r \in \mathcal{R}^j_{p\bar{i}}} e^{-\beta_r} \right) e^{-\beta_{\bar{i}j}} \left(\sum_{r \in \mathcal{R}^j_{jq}} e^{-\beta_r} \right) \right].$$

Introducing the notation $\mu_i = -\ln\left(\sum_{r\in\mathcal{R}^j_{pi}} e^{-\beta_r}\right)$, we now have that

$$\frac{f_{pq\bar{a}}(h^*)}{\sum_{a\in\mathcal{A}'(j)} f_{pqa}(h^*)} = \frac{\left(\sum_{r\in\mathcal{R}^j_{p\bar{i}}} e^{-\beta_r}\right) e^{-\beta_{\bar{i}j}}}{\sum_{i\in\mathcal{T}'(j)}\left(e^{-\beta_{ij}}\sum_{r\in\mathcal{R}^j_{pi}} e^{-\beta_r}\right)} =$$

$$= \frac{e^{-\mu_{\bar{i}}} e^{-\beta_{\bar{i}j}}}{\sum_{i\in\mathcal{T}'(j)} e^{-\beta_{ij}} e^{-\mu_i}}. \qquad (18)$$

By observing that

$$1 = \sum_{l\in\mathcal{A}'(j)} \frac{f_{pql}}{\sum_{a\in\mathcal{A}'(j)} f_{pqa}} = \frac{1}{e^{-\mu_j}} \sum_{i\in\mathcal{T}'(j)} \frac{e^{-\beta_{ij}}}{e^{-\mu_i}},$$

and by taking the logarithm of expression (18) we have that,

$$\ln f_{pq\bar{a}} - \ln\left(\sum_{a\in\mathcal{A}'(j)} f_{pqa}\right) + \beta_{\bar{i}j} + \mu_{\bar{i}} - \mu_j = 0$$

which holds for all $j \in \mathcal{N}$ and all $\bar{i} \in \mathcal{T}'(j)$.

This expression is however exactly the Karush–Kuhn–Tucker conditions for the problem (15) with $\beta \in \mathbb{R}^{|\mathcal{A}|}$ and $\mu \in \mathbb{R}^{|\mathcal{N}|}$ interpreted as the Lagrange multipliers for the constraint sets (15b) and (15c).

Since the O–D link flows $f_{pqa}(h^*_{pqr}) = \sum_{r\in\mathcal{R}^*_{pq}} \delta_{pqra} h^*_{pqr}$, for all $(p,q) \in \mathcal{C}$, and $a \in \mathcal{A}$ is feasible in problem (15), and fulfill the Karush–Kuhn–Tucker conditions for this problem, we conclude that they are optimal.

We conclude this appendix by describing a procedure for finding the most likely route flows given the most likely O–D link flows. The procedure is similar to a backward pass in the Dial assignment method. Given the subnetwork of equilibrium links for origin p, denoted $\mathcal{G}^*_p(\mathcal{N}, \mathcal{A}^*_p)$ (see Section 4), perform an enumeration of the routes between p and q, starting with node q. In each intermediate node calculate the route choice portions as

$$\frac{f^*_{pqa}}{\sum_{a\in\mathcal{A}^*_p} f^*_{pqa}},$$

where f^*_{pqa} is the optimal solution to problem (8). The most likely route flow on a specific route is then obtained as the product of these quantities along the route.

References

[1] T. Akamatsu, Y. Tsuchiya, T. Shimazaki, *Parallel distributed processing on neural network for some transportation equilibrium assignment problems*, In *Proceedings of the 11th International Symposium on the Theory of Traffic Flow and Transportation*, Ed. M. Koshi, Yokohama, Elsevier, NY, (1990), 369–386.

[2] T. Akamatsu, *Cyclic flows, Markov process and stochastic traffic assignment*, Transportation Research 30 (1996), 369–386.

[3] T. Akamatsu, *Decomposition of path choice entropy in general transport networks*, Transportation Science 31 (1997), 349–362

[4] R.R. Barton, D.W. Hearn, S. Lawphongpanich, *The equivalence of transfer and generalized Benders decomposition methods for traffic assignment*, Transportation Research 23B (1989), 61–73.

[5] M.S. Bazaraa, H.D. Sherali, C.M. Shetty, *Nonlinear Programming: Theory and Algorithms*, John Wiley & Sons, New York, NY, 1993.

[6] M.G.H. Bell, Y. Iida, *Transportation Network Analysis*, John Wiley & Sons, Chichester, England, 1997.

[7] D.P. Bertsekas, *Nonlinear Programming.* Athena Scientific, Belmont, MA., 1995.

[8] E. Cascetta, S. Nguyen, *A unified framework for estimating or updating origin/destination matrices from traffic counts*, Transportation Research 22B (1988), 437–455

[9] *Drive project v1054*, 1st. Deliverable, Report on model requirements, 1989.

[10] S. Erlander, N.F. Stewart, *The Gravity Model in Transportation Analysis*, VSP BV, Utrecht, The Netherlands, 1990.

[11] H. Edwards, *Estimation of excess cold start emissions on links in traffic networks*, (In Swedish) Working paper, VTI, S-581 95 Linköping, Sweden, 1998.

[12] S.C. Fang, H.S.J. Tsao, *A quadratic convergent global algorithm for the linearly-constrained minimum cross-entropy problem*, European Journal of Operations Research 79 (1994), 369–378.

[13] C.S. Fisk, *Some developments in equilibrium traffic assignment*, Transportation Research 14B (1980), 243–255.

[14] C.S. Fisk, *On combing maximum entropy trip matrix estimation with user optimal assignment*, Transportation Research 22B (1988), 69–79.

[15] M. Florian, S. Nguyen, *An application and validation of equilibrium trip assignment methods*, Transportation Science 10 (1976), 374–390.

[16] D.W. Hearn, *Practical and theoretical aspects of aggregation problems in transportation models*, In *Transportation Planning Models*, Proceedings of the course given at the International Center for Transportation Studies, Amalfi, Italy, October 11–16, 1982, Ed. M. Florian, North-Holland, Amsterdam (1984), 257–287.

[17] B.N. Janson, *Most likely origin–destination link uses from equilibrium assignment*, Transportation Research 27B (1993), 333–350.

[18] B. Lamond, N.F. Stewart, *Bregman's balancing method*, Transportation Research 15B (1981), 239–248.

[19] T. Larsson, M. Patriksson, *Simplicial decomposition with disaggregate representation for the traffic assignment problem*, Transportation Science 26 (1992), 4–17.

[20] L.J. LeBlanc, E.K. Morlok, W.P. Pierskalla, *An efficient approach to solving the road network equilibrium traffic assignment problem*, Transportation Research 9 (1975), 309–318.

[21] M. Patriksson, *The Traffic Assignment Problem—Models and Methods*, VSP BV, Utrecht, The Netherlands, 1994.

[22] B.T. Polyak, *Introduction to Optimization*, Optimization Software, Inc., Publications Division, N.Y., 1987.

[23] T.F. Rossi, S. McNeil, C. Hendrickson, *Entropy model for consistent impact fee assessment*, Journal of Urban Planning and Development/ASCE 115 (1989), 51–63.

[24] E. Sérié, R. Joumard, *Modelling of cold start emissions for road vehicles*, INRETS report LEN 9731, 1997.

[25] K.A. Small, *Using revenues from congestion pricing*, Transportation 19 (1992), 359–381.

[26] T.E. Smith, *A cost-efficiency of spatial interaction behaviour*, Regional Science and Urban Economics 8 (1978), 137–168.

[27] T.E. Smith, *A cost-efficiency approach to the analysis of congested spatial-interaction behavior*, Environment and Planning 15A (1983), 435–464.

[28] F. Snickars, J.W. Weibull, *A minimum information principle: Theory and practice*, Regional Science and Urban Economics 7 (1977), 137–168.

[29] Y. Sheffi, *Urban Transportation Networks: Equilibrium Analysis with Mathematical Programming Methods*, Prentice–Hall, Englewood Cliffs, NJ, 1985.

[30] *The Economist*, December 6 1997, 21–24.

[31] D. Van Vliet, *Selected node-pair analysis in Dial's assignment algorithm*, Transportation Research 15B (1981), 65–68.

EXISTENCE OF SOLUTIONS TO BILEVEL VARIATIONAL PROBLEMS IN BANACH SPACES

Maria Beatrice Lignola
Dipartimento di Matematica e Applicazioni "R. Caccioppoli"
Università degli Studi di Napoli "Federico II"
Via Claudio, 21 - 80125 Napoli

Jacqueline Morgan
Dipartimento di Matematica e Applicazioni "R. Caccioppoli"
Università degli Studi di Napoli "Federico II"
Compl. universitario Monte S. Angelo
Via Cintia - 80126 Napoli

1. Introduction.

Multilevel Optimization, which became in last years a very extended research field, studies the optimization problems with equilibrium conditions as constraints. A large number of problems, arising when modelling socio-economic and engineering systems ([2], [19], ...) are included in this field:

- The basic *leader-follower* strategy, considered by von Stackelberg [27], in which one player (called the leader) selects his optimal strategy anticipating the reaction of the other player (called the follower).

In particular, if X is the action set of the leader L, $Y(x) \subseteq Y$ is the action set of the follower F for any $x \in X$, l and f are the cost functions of L and F respectively, many situations may be modelized in one of the following way:

$$\text{Inf}_{x \in X} \, \text{Inf}_{y \in \text{Argmin } f_{(x, \cdot)}} \, l(x, y)$$

or

$$\text{Inf}_{x \in X} \, \text{Sup}_{y \in \text{Argmin } f_{(x, \cdot)}} \, l(x, y).$$

F. Giannessi et al (eds.),
Equilibrium Problems: Nonsmooth Optimization and Variational Inequality Models, 161–174.
© 2001 *Kluwer Academic Publishers.*

The first of these formulations corresponds to a pessimistic model, the second one to an optimistic one (see, for example, [10] and [20]).
Here the lower level is described by the parametric optimization problem:

$$P(x) \qquad \inf_{y \in Y(x)} f(x, y).$$

- The Bilevel Programming Problem (BPP) is a problem of type (2) in which the set-valued function Y is described by a finite number of inequalities:

$$Y(x) = \{y \in Y : g_i(x, y) \leq 0 \quad i = 1, ..., n\}.$$

- The Mathematical Program with Equilibrium Constraints (MPEC) is an optimization problem in which the constraints are defined by a parametric Variational Inequality or Complementarity System. This subject had been extensively studied in [19], in which also a complete list of references can be found.
We point out that (MPEC) generalizes a large number of interesting problems, since, as it is well known, variational inequalities and minimum problems are equivalent under some convexity assumptions.

In [12] we introduced a new multilevel problem considering at the first level a very general parametric problem and we gave existence results for solutions in the setting of finite dimensional spaces. In order to extend these results to infinite dimensional spaces, throughout the paper, we consider a real reflexive Banach space E, a topological space X, a function h from $X \times E \times E$ to $\mathbf{R}$, a function ϕ from $X \times E \times E$ to $\mathbf{R} \cup \{+\infty\}$ and a function f from $X \times E$ to $\mathbf{R} \cup \{+\infty\}$.
The Bilevel Variational Problem (BVP for short) consists in finding a pair $(\hat{x}, \hat{y}) \in X \times E$ such that:

$$\text{(BVP)} \qquad \hat{y} \in T(\hat{x}) \quad \text{and} \quad f(\hat{x}, \hat{y}) = \inf_{x \in X} \inf_{y \in T(x)} f(x, y),$$

where, for all $x \in X$, $T(x)$ is the solution set to the following parametric Variational Problem:

$$\text{(VP)}(x) \quad \text{find } y_0 \in E \text{ such that:}$$

$$h(x, y_0, z) + \phi(x, y_0, y_0) \leq \phi(x, y_0, z) \quad \text{for every } z \in E .$$

We recall that, in the case in which the problem (VP)(x) does not depend from the parameter x, one obtains the problem introduced by U. Mosco in [24]:

$$\text{(VP)} \quad \text{find } y_0 \in E \text{ such that } h(y_0, z) + \phi(y_0, y_0) \leq \phi(y_0, z) \text{ for every}$$

$z \in E$.

Many problems can be described by a (VP), as it had been illustrated in [1]:

- Variational Inequalities - (VI) (see [24], [9], [7])
Consider an operator A from E to E^*, a closed convex subset K of E and take the functions $h(y, z) = \langle Ay, y - z \rangle$ and $\phi(y, z) = \psi_K(y)$, where ψ_K is the indicator function of the set K. Then (VP) becomes a Variational Inequality:

$$\text{(VI)} \quad \text{find } y_0 \in K \text{ such that } \langle Ay_0, y_0 - z \rangle \leq 0 \text{ for all } z \in K.$$

- Quasi Variational Inequalities - (QVI) (see [1] and [24])
Consider an operator A from E to E^* and a set-valued function S from E to E which has closed and convex values. Taking h as in (VI) and $\phi(y, z) = \psi_{S(y)}(z)$, where $\psi_{S(y)}$ is the indicator function of the set $S(y)$, (VP) becomes a Quasi Variational Inequality :

$$\text{(QVI)} \quad \text{find } y_0 \in E \text{ such that } y_0 \in S(y_0) \text{ and } \langle Ay_0, y_0 - z \rangle \leq 0,$$
$$\text{for every } z \in S(y_0).$$

Moreover, as shown in [14], also the following problems can be interpretated as a (VP):

- Generalized Variational Inequalities - (GVI) (see [28] [26])
Let G be a set-valued operator from E to E^* and take

$$h(y, z) = \sup_{y^* \in G(y)} \langle y^*, y - z \rangle$$

and ϕ as in (VI). Then (VP) becomes:

$$\text{(GVI)} \quad \text{find } y_0 \in K \text{ such that } \sup_{y^* \in G(y_0)} \langle y^*, y_0 - z \rangle \leq 0 \text{ for all } z \in K.$$

- Generalized Quasi-Variational Inequalities - (GQVI) (see [4] [14])
Let G be a set-valued operator from E to E^* and S be a set-valued function from E to E. Taking h as in (GVI) and ϕ as in (QVI) the problem (VP) is the following:

$$\text{(GQVI)} \quad \text{find } y_0 \in E \text{ such that}$$
$$y_0 \in S(y_0) \quad \text{and} \quad \sup_{y^* \in G(y_0)} \langle y^*, y_0 - z \rangle \leq 0 \text{ for all } z \in S(y_0) .$$

As already mentioned , aim of this paper is to give existence results of solutions to problems (BVP) in infinite dimensional spaces. To this end, differently from the finite dimensional case, some monotonicity assumptions on the function h are required in order to obtain the closedness of the set-valued function T, a property which is fundamental for our purpose.

We recall that a real-valued function g on $E \times E$ is called *monotone* if

$$g(y, z) + g(z, y) \geq 0 \quad \text{for all } y \in E, \ z \in E$$

and it is called *pseudomonotone* if

$$g(y, z) \leq 0 \quad \text{implies} \quad g(z, y) \geq 0.$$

A single-valued operator A is monotone if

$$\langle Ay - Az, y - z \rangle \geq 0 \quad \text{for all } y \in E, \ z \in E,$$

and it is pseudomonotone if

$$\langle Ay, y - z \rangle \leq 0 \quad \text{implies}$$

$$\langle Az, y - z \rangle \leq 0 \quad \text{for all } y \in E, \ z \in E.$$

A set-valued function M from X to E is *sequentially closed* on X if the graph of M is a sequentially closed subset of $X \times E$.

In the following, the set-valued function T will be supposed non-empty and weakly compact-valued (for results concerning the existence of solutions to (VP) see [1] for monotone functions and [6] for pseudomonotone functions).

2. A general existence result.

In this section we give a sufficient condition for the existence of solutions to problems (BVP), which will be further applied to Optimization Problems with constraints defined by one of the equilibrium problems mentioned in Section 1.

Theorem A. Let (X, τ) be sequentially compact and assume that:

A_0) f is sequentially lower semicontinuous on $(X \times E, \ \tau \times w)$ (where w stands for the weak topology) and it is equicoercive on E, that is: for every convergent sequence $(x_n)_n$, $x_n \in X$, and every sequence

$(y_n)_n$, $y_n \in E$, such that $f(x_n, y_n) < c$, $\forall n \in I\!N$, there exists a weakly convergent subsequence of $(y_n)_n$;

$A_1)$ ϕ is sequentially lower semicontinuous on $(X \times E \times E, \tau \times w \times w)$;

$A_2)$ for every (x, y, z) and every sequence $(x_n, y_n)_n$ converging in $(\tau \times w)$ to (x, y), there exists a sequence $(z_n)_n$ strongly converging to z such that
$$\phi(x, y, z) \geq \limsup_n \phi(x_n, y_n, z_n);$$

$A_3)$ for every x and y, $h(x, y, \cdot)$ is concave and $\phi(x, y, \cdot)$ is convex;

$A_4)$ for every x and z, $h(x, \cdot, z)$ is sequentially lower semicontinuous on the segments;

$A_5)$ for every x and y, $h(x, y, y) = 0$;

$A_6)$ for every $x \in X$ and every sequence $(x_n)_n$ converging to x, if $(y_n)_n$ is a sequence of solutions to $(VP)(x_n)$ weakly converging to a point y, then y is a solution to the following problem:

find $y_0 \in E$ such that:
$$-h(x, z, y_0) + \phi(x, y_0, y_0) \leq \phi(x, y_0, z) \ \forall z \in E. \tag{1}$$

Then the set-valued map T is weakly sequentially closed on $(X \times E, \tau \times w)$ and the problem (BVP) has at least a solution.

Proof. In order to achieve the closedness of T, it is sufficient to prove that every solution to the problem (1) is also a solution to $(VP)(x)$ and to apply $A_6)$. Assume that it fails to be true and let $\hat{x}$ be a point of X and $\hat{y}$ be a point of E such that

$$-h(\hat{x}, z, \hat{y}) + \phi(\hat{x}, \hat{y}, \hat{y}) \leq \phi(\hat{x}, \hat{y}, z) \quad \forall z \in E \tag{2.1}$$

which does not solve $(VI)(\hat{x})$. So there exists $\hat{z} \in E$ for which

$$h(\hat{x}, \hat{y}, \hat{z}) + \phi(\hat{x}, \hat{y}, \hat{y}) > \phi(\hat{x}, \hat{y}, \hat{z}). \tag{2.2}$$

Take $z_n = t_n \hat{y} + (1 - t_n)\hat{z}$ where $(t_n)_n$ is a sequence of $[0, 1]$ converging to 1.
¿From concavity assumption on h:

$$h(\hat{x}, z_n, z_n) \geq t_n h(\hat{x}, z_n, \hat{y}) + (1 - t_n)h(\hat{x}, z_n, \hat{z}) \tag{2.3}$$

while convexity of $\phi(\hat{x}, \hat{y}, \cdot)$ implies:

$$\phi(\hat{x}, \hat{y}, z_n) \leq t_n \phi(\hat{x}, \hat{y}, \hat{y}) + (1 - t_n)\phi(\hat{x}, \hat{y}, \hat{z}). \tag{2.4}$$

Since (2.1) holds,

$$-t_n\phi(\hat{x},\hat{y},z_n) \le t_n h(\hat{x},z_n,\hat{y}) - t_n\phi(\hat{x},\hat{y},\hat{y}),$$

which together with (2.4) gives:

$$(1-t_n)\phi(\hat{x},\hat{y},z_n) \le t_n\phi(\hat{x},\hat{y},\hat{y}) + (1-t_n)\phi(\hat{x},\hat{y},\hat{z}) + $$
$$+ t_n h(\hat{x},z_n,\hat{y}) - t_n\phi(\hat{x},\hat{y},\hat{y})$$

so, from (2.3),

$$(1-t_n)\phi(\hat{x},\hat{y},z_n) + (1-t_n)h(\hat{x},z_n,\hat{z}) - (1-t_n)\phi(\hat{x},\hat{y},\hat{z}) \le$$

$$\le h(\hat{x},z_n,z_n) = 0 . \tag{2.5}$$

In light of A_1), A_4) and (2.2) we get existence of $m \in I\!N$ such that

$$h(\hat{x},z_n,\hat{z}) + \phi(\hat{x},\hat{y},z_n) > \phi(\hat{x},\hat{y},\hat{z}) \quad \forall n \ge m$$

and this contradicts (2.5). It follows that T is sequentially closed.
By classical arguments, lower semicontinuity of the marginal function $\underset{y\in T(x)}{\text{Inf}}\ f(x,y)$ can be obtained. Indeed, if $x_0 \in X$, $(x_n)_n$ converges to x_0 in X and c is a real number such that $c > \underset{n}{\lim\inf}\ \underset{y\in T(x_n)}{\text{Inf}}\ f(x_n.y)$, there exists a subsequence $(n_k)_k$ and $(y_{n_k})_k$ such that $y_{n_k} \in T(x_{n_k})$ and $f(x_{n_k},y_{n_k}) < c$.
¿From A_0), $(y_{n_k})_k$ has a subsequence, still denoted by (y_{n_k}), which converges to $y_0 \in E$. Since T is closed, $y_0 \in T(x_0)$ and to conclude it is enough to appeal to sequential lower semicontinuity of f.   ∎

3. Monotone case.

We start by giving a result concerning OPVIC (Optimization Problems with Variational Inequality Constraints).

Let A be an operator from $X \times E$ to E^* and K be a set-valued function from X to E. The problem (OPVIC) is the following:

(OPVIC)   find $(\hat{x},\hat{y}) \in X \times E$ such that

$$\hat{y} \in T(\hat{x}) \quad \text{and} \quad f(\hat{x},\hat{y}) = \underset{x\in X}{\text{Inf}}\ \underset{y\in T(x)}{\text{Inf}}\ f(x,y),$$

where $T(x)$ is the solution set to the parametric variational inequality (VI)(x):

(VI)(x)   find $y_0 \in K(x)$ such that

$$\langle A(x, y_0), y_0 - z \rangle \leq 0 \quad \text{for all} \quad z \in K(x).$$

Proposition B. Let (X, τ) be sequentially compact and assume that A_0) and the following assumptions hold:

B_1) the set-valued function K is convex-valued and sequentially closed from (X, τ) to (E, w) and sequentially lower semicontinuous from (X, τ) to (E, s);

B_2) for every $x \in X$, the operator $A(x, \cdot)$ is monotone and uniformly bounded;

B_3) for every $x \in X$, $A(x, \cdot)$ is hemicontinuous on E;

B_4) for every (x, y) and every $(x_n)_n$ converging to x, there exists a sequence $(y_n')_n$ strongly converging to y such that $(A(x_n, y_n'))_n$ strongly converges to $A(x, y)$.

Then there exists a solution to the problem (OPVIC).

Proof. Let us define $\phi(x, y, z) = \phi(x, y) = \psi_{K(x)}(y)$ (the indicator function of the set $K(x)$), $h(x, y, z) = \langle A(x, y), y - z \rangle$, for all $(x, y, z) \in X \times E \times E$ and observe that assumptions A_1) to A_5) are easily satisfied.

In order to prove A_6), let $(x_n)_n$ be a sequence in X converging to x_0 and $(y_n)_n$ be a sequence of solutions to (VI)(x_n), that is:

$$y_n \in K(x_n) \quad \text{and} \quad \langle A(x_n, y_n), y_n - z \rangle \leq 0 \quad \forall z \in K(x_n).$$

If $(y_n)_n$ weakly converges to y_0, from B_1) we have that $y_0 \in K(x_0)$. Let $z \in K(x_0)$ and let $(z_n)_n$ be a sequence strongly converging to z, such that $z_n \in K(x_n)$, for n sufficiently large. Let $(z_n')_n$ be the sequence defined in B_4) strongly converging to z and such that the sequence $(A(x_n, z_n'))_n$ strongly converges to $A(x_0, z)$. We have:

$$-h(x_0, z, y_0) = \langle A(x_0, z), y_0 - z \rangle = \lim_n \langle A(x_n, z_n'), y_n - z_n' \rangle .$$

Since $A(x_n, \cdot)$ is monotone and $z_n \in K(x_n)$ for every $n \in \mathbb{N}$:

$$\begin{aligned}
0 &\geq \langle A(x_n, y_n), y_n - z_n \rangle = \langle A(x_n, y_n), y_n - z_n' \rangle + \langle A(x_n, y_n), z_n' - z_n \rangle \\
&\geq \langle A(x_n, z_n'), y_n - z_n' \rangle + \langle A(x_n, y_n), z_n' - z_n \rangle .
\end{aligned}$$

Being $A(x, \cdot)$ uniformly bounded,

$$\lim_n \langle A(x_n, y_n), z_n' - z_n \rangle = 0 ,$$

so that

$$-h(x_0, z, y_0) = \lim_n \langle A(x_n, z_n'), y_n - z_n' \rangle \leq 0 \quad \forall z \in K(x_0),$$

and the proof is complete.   ∎

For (OPQVIC), that is for bilevel problems with quasi-variational inequality constraints, we have:

Proposition C: Let (X, τ) be sequentially compact and assume that $A_0)$, $B_2)$, $B_3)$, $B_4)$ and the following assumption are satisfied:

$C_1)$ the set-valued function S from $(X \times E, \tau \times w)$ to (E, w) is sequentially closed and sequentially lower semicontinuous from $(X \times E, \tau \times w)$ to (E, s).

Then there exists a solution to (OPQVIC).

Proof. Let us define $\phi(x, y, z) = \psi_{S(x,y)}(z)$, $h(x, y, z) = \langle A(x, y), y - z \rangle$, for all $(x, y, z) \in X \times E \times E$ and observe that as in Proposition B the crucial part consists in proving that assumption $A_6)$ is satisfied. To this end, let $(x_n)_n$ be a sequence in X converging to x_0 and $(y_n)_n$ be a sequence of solutions to (QVI)(x_n), that is:

$$y_n \in S(x_n, y_n) \quad \text{and} \quad \langle A(x_n, y_n), y_n - z \rangle \leq 0 \quad \forall z \in S(x_n, y_n).$$

If $(y_n)_n$ weakly converges to y_0, from $C_1)$ $y_0 \in S(x_0, y_0)$. Let $z \in S(x_0, y_0)$ and let $(z_n)_n$ be a sequence strongly converging to z, such that $z_n \in S(x_n, y_n)$, for n sufficiently large, which exists since S is lower semicontinuous. Arguing as in Proposition B it is easy to prove that $-h(x_0, y_0, z_0) \leq 0$ and condition $A_6)$ is satisfied.   ∎

Remark 1.   Results concerning bilevel problems with constraints defined by (GQVI) cannot be obtained in an analogous way. Indeed, in this case, assumptions $B_2)$ and $B_3)$ would amount respectively to monotonicity and lower semicontinuity on the segments of the set-valued operator $G(x, \cdot)$. But a multifunction which satisfies both the two properties is necessarily single-valued. Nevertheless, in finite dimensional spaces, the existence of solutions to (OPGVIC) and (OPGQVIC) may easily be obtained: see Corollary 4.3 in [12].

4. Pseudomonotone case.

In this section we give some results involving pseudomonotone operators pointing out that the price that we have to pay in order to weaken the monotonicity assumption is to strenghten the "continuity" condition B_4) and, moreover, that the case in which E is infinite dimensional has to be distinguished from the finite dimensional one.

We have:

Proposition D: Let (X, τ) be sequentially compact, E be finite dimensional and assume that A_0), B_1), B_3) and the following conditions are satisfied:

D_1) for every $x \in X$, $\text{int} K(x) \neq \emptyset$;

D_2) for every $x \in X$, the operator $A(x, \cdot)$ is pseudomonotone;

D_4) for every $x \in X$ and every sequence (x_n) converging to x, the sequence $(A(x_n, y))_n$ strongly converges to $A(x, y)$ for every $y \in E$.

Then there exists a solution to the problem (OPVIC).

Proof. In order to prove that condition A_6) is satisfied, let $(x_n)_n$ be a sequence converging to $x_0 \in X$, $(y_n)_n$ be a sequence of solutions to (VI)(x_n), for every $n \in \mathbb{N}$, and assume that $(y_n)_n$ weakly converges to $y_0 \in E$. From B_1), $y_0 \in K(x_0)$ and our aim is to prove that $\langle A(x_0, y_0), y_0 - z \rangle \leq 0$ for every $z \in K(x_0)$. First, we consider points $z \in \text{int} K(x_0)$. Each of these points has to belong to $K(x_n)$ for n sufficiently large: indeed, being $\text{int} K(x_0) \neq \emptyset$ and the set-valued function K being sequentially lower semicontinuous, we have

$$\text{int } K(x_0) \subseteq \bigcup_{m \in \mathbb{N}} \bigcap_{n \geq m} \text{int } K(x_n)$$

(see [16]).
So, there exists $m \in N$ such that $\langle A(x_n, y_n), y_n - z \rangle \leq 0$ for every $n \geq m$ and from D_2)

$$\langle A(x_n, z), y_n - z \rangle \leq 0 \quad \text{for every } n \geq m ,$$

which in turn implies, because of D_4), $\langle A(x_0, z), y_0 - z \rangle \leq 0$. Now, it is sufficient to observe that, in our assumptions, Minty's lemma holds, so that $\langle A(x_0, y_0), y_0 - z \rangle \leq 0$ for every $z \in \text{int} K(x_0)$.
When we consider a point z belonging to the boundary of $K(x_0)$, we can approximate it by a sequence of points $(z_n)_n$ lying on a segment contained in the interior of $K(x_0)$. Therefore we have from B_3):

$$\langle A(x_0, z), y_0 - z \rangle = \lim_n \langle A(x_0, z_n), y_0 - z_n \rangle \leq 0$$

and this concludes the proof.   ∎

When E is not a finite dimensional space, an additional assumption is needed: in fact in [16] it has been shown that D_1) is not sufficient to conclude that for every sequence $(x_n)_n$ converging in X to x_0,

$$\operatorname{int} K(x_0) \subseteq \bigcup_{m \in \mathbb{N}} \bigcap_{n \geq m} \operatorname{int} K(x_n).$$

So, we have:

Proposition E: Let (X, τ) be sequentially compact and assume that A_0), B_1), B_3), D_1), D_2), D_4) and the following assumption are satisfied:

D_5) for every convergent sequence $(x_n)_n$ there exists $m \in \mathbb{N}$ such that
$$\operatorname{int} \bigcap_{n \geq m} K(x_n) \neq \emptyset.$$

Then there exists a solution to (OPVIC).

Proof. As in Proposition D, taking into account that, from D_4) and D_5), we can obtain that

$$\operatorname{int} K(x_0) \subseteq \bigcup_{m \in \mathbb{N}} \bigcap_{n \geq m} \operatorname{int} K(x_n) .$$

                                                                                 ∎

We observe that, when the constraints are described by one inequality, assumption D_5) amounts to a type of "uniform Slater condition". Indeed, if $K(x) = \{y \in E : g(x, y) \leq 0\}$, it is easy to see that condition D_5) is satisfied when g is upper semicontinuous and there exists a point y_0 such that $g(x, y_0) < 0$ for every $x \in X$.

When K is described by n inequalities, for $n > 1$, some additional assumptions are needed in order to obtain sequential lower semicontinuity of K, see [13].

We point out that, in line with Proposition C a result for (OPQVIC) can be easily obtained, while for (OPGVIC) and (OPGQVIC) the conclusion of Remark 1 still holds, since pseudomonotonicity jointly with lower semicontinuity also force a multi-valued operator to be single-valued.

We conclude this Section by showing that in general assumption D_4) cannot be weakened to B_4).

Example 2. Let $X = \mathbb{N} \cup \{+\infty\}$, $E = \mathbb{R}$, $K_n = K_\infty = [0, +\infty[$ and consider the following operators:

$$A_n(y) = e^{-(ny)^2} - 1 \text{ for every } n \in I\!N \text{ and } y \in I\!R.$$

It is obvious that $A_\infty(y) = -1$ for every $y \in I\!R$ satisfies $B_4)$ (for $y = 0$ $y_n = \dfrac{1}{n^3}$ is such that $(A_n(y_n))_n$ converges to $A_\infty(0)$) and it does not satisfy condition $D_4)$. It is also easy to see that the operators A_n are pseudomonotone (but not monotone) and $y_n = 0$ for every $n \in I\!N$ is a solution to $\langle A_n y, y - z \rangle \leq 0 \quad \forall z \in K_n$ while $y_0 = 0$ is not a solution to the variational inequality

$$\langle A_\infty y_0, y_0 - z \rangle \leq 0 \quad \forall z \in K.$$

5. Open problems.

After introducing (BVP) and proving that these problems have solutions under suitable assumptions, some questions arise naturally.
- Under which conditions the solutions to (BVP) are stable, that is, if one perturbes a problem, the solutions to the perturbed problems converge to a solution of the original one?
- How constructing converging algorithms for such problems?
For what stability is concerned, we recall that it has been already investigated for Stackelberg problems in [18], [15], for (BPP) in [13] and for (OPVIC) in [17]. In all cases only results for approximate solutions had been obtained. In fact, in general, stability of exact solutions cannot be expected. Thus, the first step will consist in defining an appropriate concept of approximate solution to (BVP).
In order to construct algorithms for these problems, a guideline could be the algorithms proposed in [21],[25] and [8] for optimization problems with constraints defined by means of a variational inequality.
We wish to recall that in order to verify the convergence of algorithms, one can prove, as in optimization([5]), in non zero sum-games ([3]), in zero-sum games ([23]), in (BPP) ([22]) and in (OPVIC) ([11]), that the problem is *well − posed* is some suitable sense. So, another question consists in defining an appropriate concept of well-posedness and finding classes of functions in which (BVP) are well-posed.

References

[1] C. Baiocchi, A. Capelo, *Variational and quasivariational inequalities, applications to free boundary problems*, John Wiley and Sons, New-York, 1984.

[2] T. Basar, G.J. Olsder, *Dynamic Noncooperative Games*, Academic Press, New York, Second Edition, 1995.

[3] E. Cavazzuti, J. Morgan, *Well-posed saddle point problems, Optimization, Theory and Algorithms*, J.B.Hiriart-Urruty, W.Oettli, J.Stoer Eds., Marcel Dekker, New-York (1978), 61–76.

[4] X.P. Ding, K.K. Tan, *Generalized variational inequalities and generalized quasi-variational inequalities*, Journal of Mathematical Analysis and Applications 148 (1990), 497–508.

[5] A.L. Dontchev, T. Zolezzi, *Well-Posed Optimization Problems*, Lecture Notes in Mathematics 1543 (1993), Springer-Verlag, Berlin.

[6] N. Hadjisavvas, S. Schaible, *Quasimonotonicity and pseudo monotonicity in variational inequalities and equilibrium problems, Generalized convexity, generalized monotonicity:recent results (Luminy, 1996)*, Nonconvex Optim.Appl. 27 (1998), Kluwer Acad. Publ., Dordrecht, 257-275.

[7] P.T. Harker, J.S. Pang, *Finite-dimensional variational inequality and nonlinear complementarity problems: a survey of theory, algorithms and application*, Mathematical Programming 48 (1990), 161–220.

[8] P.T. Harker, S.C. Choi, *Penalty function approach for mathematical programs with variational inequality constraints*, Information and Decision Technologies 17 (1991), 41–50.

[9] D. Kinderlehrer, G. Stampacchia, *An introduction to variational inequality and their applications*, Academic press, New-York, 1980.

[10] G. Leitmann, *On Generalized Stackelberg Strategies*, Journal of Optimization, Theory and Applications 26 (1978), 637–643.

[11] M.B. Lignola, J. Morgan, *Well-Posedness for Optimization Problems with constraints defined by Variational Inequality having a unique solution*, to appear on Journal of Global Optimization.

[12] M.B. Lignola, J. Morgan, *Existence of solutions to generalized bilevel programming problem, Multilevel Programming-Algorithms and Applications*, Eds A. Migdalas, P.M. Pardalos and P. Varbrand, Kluwer Academic Publishers (1998), 315-332.

[13] M.B. Lignola, J. Morgan, *Stability for regularized bilevel programming problem*, Journal of Optimization, Theory and Applications 93 (1997), n.3, 575–596.

[14] M.B. Lignola, J. Morgan, *Convergence of Solutions of Quasivariational Inequalities and Applications*, Topological Methods in Nonlinear Analysis 10 (1997), 375–385.

[15] M.B. Lignola, J. Morgan, *Topological existence and stability for Stackelberg problems*, Journal of Optimization Theory and Applications 84 (1995), n.1, 145–169.

[16] M.B. Lignola, J. Morgan, *Semicontinuity and Episemicontinuity: Equivalence and Applications*, Bollettino dell'Unione Matematica Italiana 8B (1994), 1–16.

[17] M.B. Lignola, J. Morgan, *Approximate solutions to variational inequalities and Applications*, Equilibrium Problems with Side Constraints Langrangean Theory and Duality, Eds F. Giannessi and A. Maugeri Le Matematiche 49 (1994), 281–293.

[18] P. Loridan, J. Morgan, *New results on approximate solutions in two level optimization*, Optimization 20 (1989), 819–836.

[19] Z.Q. Luo, J.S. Pang, D. Ralph, *Mathematical Programs with Equilibrium Constraints*, Cambridge University Press, 1996.

[20] L. Mallozzi, J. Morgan, , *Hierarchical Systems with Weighted Reaction Set, Nonlinear Optimization and Applications*, Ed. Di Pillo and F. Giannessi, Plenum Press New York and London (1996), 271–283.

[21] P. Marcotte, D.L. Zhu, *Exact and Inexact Penalty Methods for the Generalized Bilevel Programming Problem*, Mathematical Programming 74 (1996), 141–157.

[22] J. Morgan, *Constrained well-posed two-level optimization problem, Non-smooth optimization and related topics*, F.H.Clarke, V.F. Demyanov, F.Giannessi, Eds., Plenum press, New-York (1989), 307–325.

[23] M. Margiocco,F. Patrone, L. Pusillo, *A new approach to Tikhonov well-posedness for Nash Equilibria*, Optimization 40 (1997), 385–400.

[24] U. Mosco, *Implicit variational problems and quasi variational inequalities*, Proc. Summer School (Bruxelles, 1975), *Nonlinear Operators and the Calculus of Variations*, Lecture Notes in Math. 543 (1976), Springer, Berlin, 83–156.

[25] J.V. Outrata, *On optimization problems with variational inequality constraint*, Siam J. on Optimization 4 (1994), 334–357.

[26] M.H. Shih, K.K. Tan, *The Ky Fan Minimax Principle with Convex Sections and Variational Inequalities, Differential Geometry, Calculus of Variations, and Their Applications*, edited by G.M. Rassias and T.M. Rassias, Marcel Dekker, New York (1985), 471–481.

[27] H. von Stackelberg, *Marktform und Gleichgewicht*, Julius Springer, Vienna, 1934.

[28] C.L. Yen, *A minimax inequality and its applications to variational inequalities*, Pacific J. Math. 97 (1981), 477–481.

[29] M.B. Lignola, J. Morgan, *Convergences for Variational Inequalities and Generalized Variational Inequalities*, Atti Seminario Matematico Fisico Universitá di Modena 45 (1997), 377–388.

ON THE EXISTENCE OF SOLUTIONS TO VECTOR OPTIMIZATION PROBLEMS

Giandomenico Mastroeni
Department of Mathematics
University of Pisa, Pisa, Italy

Massimo Pappalardo
Department of Applied Mathematics
University of Pisa, Pisa, Italy

Abstract   The existence of solutions to a Vector Optimization Problem is carried out by means of the image space analysis. Classic existence results are revisited and presented under suitable compactness assumptions on the image of the Vector Optimization Problem.

Keywords: Vector optimization, optimality conditions, image space.

AMS Classification: 90C, 49J, 65K.

1. Introduction.

In recent years the study of vector optimization problems has widely grown, both from a theoretical point of view and as it concerns applications to real problems. It has been shown that the theory of Vector Optimization can be based on the image space analysis and theorems of the alternative or separation theorems [1,4,5,6,7,8].

The present paper contributes to develop the study of the existence of solutions to a Vector Optimization Problem (for short, VOP) by performing the analysis in the image space associated to VOP, defined by the product space where the image of the objective and constraint functions run. Indeed, a first proposal in this direction was done in [11]. We will see (Sect. 2) that the VOP can be equivalently formulated in the

F. Giannessi et al (eds.),
Equilibrium Problems: Nonsmooth Optimization and Variational Inequality Models, 175–185.
© 2001 *Kluwer Academic Publishers.*

image space so that the existence of solutions can be obtained by means of cone-compactness assumptions on the image of VOP (Sect. 3). The notion of cone-compactness will be analysed more in details (Sect. 4) pointing out the main topological properties and the most important results, existing in in the literature, that guarantee the cone-compactness of the image of the VOP.

We recall the main notations and definitions that will be used in the sequel. Let $M \subseteq \mathbb{R}^\ell$. $intM$ and clM will denote the interior and the closure of M, respectively. $\mathbb{R}^\ell_+ := \{x \in \mathbb{R}^\ell : x \geq 0\}$.

Let $D \subseteq \mathbb{R}^m$ be a convex cone; the *positive polar* of D is the set $D^* := \{x^* \in \mathbb{R}^m : \langle x^*, x \rangle \geq 0, \forall x \in D\}$. D is said *pointed* iff $D \cap (-D) = \{0\}$. $\pi(\mathbb{R}^\ell; K)$ will denote the projection of the set $K \subseteq \mathbb{R}^\ell \times \mathbb{R}^m$ on $\mathbb{R}^\ell$.

2. Image space and separation.

Let the cone $C \subset \mathbb{R}^\ell$ be given. In the following it will be assumed that C is convex, pointed, with apex at the origin, and with $int\, C \neq \emptyset$.

Assume we are given the vector–valued functions $f : \mathbb{R}^n \to \mathbb{R}^\ell$, $g : \mathbb{R}^n \to \mathbb{R}^m$, and the subset $X \subseteq \mathbb{R}^n$. We will consider the following VOP:

$$(2.1) \qquad \min_{C\setminus\{0\}} f(x) \quad , \quad \text{subject to } x \in K := \{x \in X : g(x) \geq 0\},$$

where $\min_{C\setminus\{0\}}$ marks vector minimum with respect to the cone $C\setminus\{0\}$; $y \in K$ is a (global) vector minimum point (for short, v.m.p.) of (2.1), iff

$$(2.2) \qquad\qquad f(y) \not\geq_{C\setminus\{0\}} f(x) , \quad \forall x \in K,$$

where the inequality means $f(y) - f(x) \notin C\setminus\{0\}$. At $C = \mathbb{R}^\ell_+$, (2.1) becomes the classic *Pareto vector problem*.

We observe that (2.2) is satisfied iff the system (in the unknown x):

$$(2.3) \qquad f(y) - f(x) \geq_{C\setminus\{0\}} 0 , \quad g(x) \geq 0 , \quad x \in X$$

is impossible.

Consider the sets:

$$\mathcal{H} := \{(u,v) \in \mathbb{R}^\ell \times \mathbb{R}^m : u \geq_{C\setminus\{0\}} 0 , v \geq 0\} = (C\setminus\{0\}) \times \mathbb{R}^m_+ ,$$

$$\mathcal{K}(y) := \{(u,v) \in \mathbb{R}^\ell \times \mathbb{R}^m : u = f(y) - f(x) , v = g(x), x \in X\}.$$

In what follows, when there is no fear of confusion, $\mathcal{K}(y)$ will be denoted merely by $\mathcal{K}$. $\mathcal{H}$ and $\mathcal{K}$ are subsets of $\mathbb{R}^{\ell+m}$, which is called *image space*; $\mathcal{K}$ is called the *image* of (2.1).

Definition 2.1. The VOP

$$(2.4) \qquad \max_{C\backslash\{0\}} u \ , \ \text{s.t.} \ (u,v) \in \mathcal{K} \cap (\mathbb{R}^\ell \times \mathbb{R}^m_+) \ ,$$

is called *image problem* associated to (2.1).

Now, observe that system (2.3) is impossible iff

$$(2.5) \qquad \mathcal{H} \cap \mathcal{K} = \emptyset \ .$$

Proposition 2.1. i) Suppose that $\mathcal{H} + cl\mathcal{H} = \mathcal{H}$. Then (2.5) holds iff

$$(2.6) \qquad \mathcal{H} \cap [\mathcal{K} - cl\mathcal{H}] = \emptyset \ ,$$

ii) (2.1) has a v.m.p. iff , for every fixed $y \in K$, (2.4) has a vector maximum point.

Proof. i) It is proved in [2], Theorem 1.1.

ii) Let $\bar{y}$ be a solution of (2.1). We have to prove that $\forall y \in K, \exists \bar{u} \in \mathcal{K}(y) \cap (\mathbb{R}^\ell \times \mathbb{R}^m_+)$ such that $\bar{u}$ is a solution of (2.4).

Since $\bar{y}$ is a v.m.p. of (2.1) we have

$$f(\bar{y}) - f(x) \not\geq_{C\backslash\{0\}} 0 \ , \ \forall x \in K.$$

Let $u' := f(\bar{y})$. The previous condition is equivalent to the following one:

$$u' - u \not\geq_{C\backslash\{0\}} 0 \ , \forall u \in f(K).$$

Since

$$(2.7) \qquad f(K) = f(y) - \pi(\mathbb{R}^\ell; \mathcal{K}(y) \cap (\mathbb{R}^\ell \times \mathbb{R}^m_+))$$

we obtain

$$u' - f(y) + u \not\geq_{C\backslash\{0\}} 0, \ \forall u \in \pi(\mathbb{R}^\ell; \mathcal{K}(y) \cap (\mathbb{R}^\ell \times \mathbb{R}^m_+)).$$

Putting $\bar{u} := f(y) - f(\bar{y}) \in \pi(\mathbb{R}^\ell; \mathcal{K}(y) \cap (\mathbb{R}^\ell \times \mathbb{R}^m_+))$, we have that $\bar{u}$ is a vector maximum point of (2.4).

Viceversa let $\bar{u} = f(y) - f(y^*)$ be an optimal solution of (2.4) for any fixed $y \in K$. Therefore

$$(2.8) \qquad u - \bar{u} \not\geq_{C\backslash\{0\}} 0, \ \forall u \in \pi(\mathbb{R}^\ell; \mathcal{K}(y) \cap (\mathbb{R}^\ell \times \mathbb{R}^m_+)).$$

Recalling (2.7) we have that (2.8) is equivalent to

$$f(y) - f(x) - \bar{u} \not\geq_{C\backslash\{0\}} 0, \ \forall x \in K,$$

that is

$$f(y^*) - f(x) \not\geq_{C\backslash\{0\}} 0, \ \forall x \in K,$$

so that y^* is a v.m.p. of (2.1). $\qquad\qquad\qquad\qquad\qquad\qquad \square$

Remark 2.1. In Proposition 2.1 i) we have assumed that $\mathcal{H} = \mathcal{H} + cl\mathcal{H}$. In [2] a wide discussion about this hypothesis has been developed and many examples have shown that it is not too restrictive.

$\mathcal{E}(y) := \mathcal{K} - cl\mathcal{H}$ is called *conic extension* of the image and often enjoys more properties than $\mathcal{K}$ itself.

Remark 2.2. We observe that the sets of the optimal solutions of the problems (2.1) and (2.4) are related in the following way: $\bar{y}$ is a v.m.p. of (2.1) iff $\bar{u} := f(y) - f(\bar{y}), \bar{v} := g(\bar{y})$ is a vector maximum point of (2.4).

Following the approach developed in [4,5,6,10] it is of interest to deepen the analysis of the image problems associated to (2.1).
Besides of (2.4), the introduction of the conic extension of $\mathcal{K}(y)$ allows to define the extended image problem associated to (2.1) defined by

$$(2.9) \qquad \max_{C\backslash\{0\}} u \ , \quad \text{s.t.} \ (u,v) \in \mathcal{E}(y) \cap (\mathbb{R}^\ell \times \mathbb{R}^m_+) \ ,$$

Proposition 2.2 The following statements are equivalent:
i) (2.1) has a v.m.p. ;
ii) for every fixed $y \in K$, (2.4) has a vector maximum point;
3i) for every fixed $y \in K$, (2.9) has a vector maximum point.
Proof. i) $\Longleftrightarrow$ ii). It is proved in ii) of Proposition 2.1.
3i) $\Longrightarrow$ ii). Let $\bar{u}$ be a solution of (2.9), then there exists $(\tilde{u}, \tilde{v}) \in \mathcal{K}(y)$ such that $\bar{u} = \tilde{u} - h$ with $h \in C\backslash\{0\}$ and

$$u - (\tilde{u} - h) \not\geq_{C\backslash\{0\}} 0, \quad \forall (u,v) \in \mathcal{E}(y) \cap (\mathbb{R}^\ell \times \mathbb{R}^m_+)$$

In particular

$$u - (\tilde{u} - h) \not\geq_{C\backslash\{0\}} 0, \quad \forall (u,v) \in [\mathcal{K}(y) - (h,0)] \cap (\mathbb{R}^\ell \times \mathbb{R}^m_+)$$

that is

$$u - \tilde{u} \not\geq_{C\backslash\{0\}} 0, \quad \forall (u,v) \in \mathcal{K}(y) \cap (\mathbb{R}^\ell \times \mathbb{R}^m_+)$$

so that $\tilde{u}$ is a vector maximum point of (2.4).
ii) $\Longrightarrow$ 3i). Let $\bar{u}$ be a solution of (2.4).
Then

$$(2.10) \qquad u - \bar{u} \not\geq_{C\backslash\{0\}} 0 \ , \forall (u,v) \in \mathcal{K}(y) \cap (\mathbb{R}^\ell \times \mathbb{R}^m_+)$$

We will prove that

$$(2.11) \qquad u - \bar{u} \not\geq_{C\backslash\{0\}} 0 \ , \forall (u,v) \in \mathcal{E}(y) \cap (\mathbb{R}^\ell \times \mathbb{R}^m_+)$$

Ab absurdo suppose that $\exists (u^*, v^*) \in \mathcal{E}(y) \cap (\mathbb{R}^\ell \times \mathbb{R}^m_+)$ such that (2.11) does not hold, that is

$$u^* = \tilde{u} - h_1, \quad v^* = \tilde{v} - h_2, \ (\tilde{u}, \tilde{v}) \in \mathcal{K}(y), \ (h_1, h_2) \in cl\mathcal{H},$$

and $u^* - \bar{u} = c \in C\backslash\{0\}$. Therefore $\tilde{u} - \bar{u} = c + h_1 \in C\backslash\{0\}$ and $0 \leq v^* = \tilde{v} - h_2$, so that $\tilde{v} \geq h_2 \geq 0$, which contradicts (2.10). $\quad\square$

Remark 2.3. The problems (2.4) and (2.9) have the same set of optimal solutions.

3. Existence of a vector minimum point.

The existence of a v.m.p. of the problem (2.1) is closely related to the properties of the image $f(K)$ of the feasible set. The most important ones are those of cone compactness [9,10] that we will briefly recall.

Definition 3.1. Let D be a cone in $\mathbb{R}^\ell$. A set $Y \subset \mathbb{R}^\ell$ is said to be D–compact, iff, for any $y \in Y$, the set $(y - clD) \cap Y$ is compact. As

shown in the Remark 3.2.2 of [10] a D–compact set is not necessarily compact. Consider, for example, the set $Y = \{y \in \mathbb{R}^2 : y_1 + y_2 \geq 0\}$ and $D = \mathbb{R}^2_+$.
The following theorem is a first important existence result for (2.1).

Theorem 3.1. Let C be a convex cone in $\mathbb{R}^\ell$ such that clC is pointed. If $f(K)$ is nonempty and C–compact then (2.1) admits a v.m.p..
Proof. We recall that, if C is a convex cone,then C is acute iff clC is pointed (see Proposition 2.1.4 of [10]). The proof follows from Theorem 3.2.3 of [10], putting $Y = f(K)$, $D = C$ and taking into account that clC is pointed. $\quad\square$

In the image space analysis a key role is played by the conic extension $\mathcal{E}(y)$ of the image $\mathcal{K}(y)$. Since the conic extension $\mathcal{E}(y)$ involves the image of both feasible and unfeasible points, in order to investigate the existence of the v.m.p. of (2.1) we consider the following subset of $\mathcal{E}(y)$:

$$\mathcal{E}_u(y) := \mathcal{E}(y) \cap \{(u, v) \in clC \times \{0\}\}.$$

Proposition 3.1

$$\mathcal{H} \cap \mathcal{E}(y) = \emptyset \iff \mathcal{H} \cap \mathcal{E}_u(y) = \emptyset .$$

Proof. Since $\mathcal{E}_u(y) \subseteq \mathcal{E}(y)$ we only have to prove that

$$\mathcal{H} \cap \mathcal{E}_u(y) = \emptyset \implies \mathcal{H} \cap \mathcal{E}(y) = \emptyset .$$

Ab absurdo, suppose that $\mathcal{H} \cap \mathcal{E}(y) = \emptyset$. Then $\exists (u^*, v^*) \in \mathcal{K}(y)$ such that $u^* - h_1 \geq_{C \setminus \{0\}} 0$, $v^* - h_2 \geq 0, h_1 \in clC, h_2 \geq 0$. Therefore $v^* \geq h_2 \geq 0$ and the point $(u^*, v^*) - (h_1, v^*) = (u^* - h_1, 0) \in \mathcal{E}_u \cap \mathcal{H}$, which contradicts the hypothesis.   □

Proposition 3.2. Let $f(K)$ be closed. If, $\forall y \in K$, $\mathcal{E}_u(y)$ is a bounded set, then (2.1) admits a v.m.p..

Proof. We observe that

$$\mathcal{E}_u(y) := \{(u, v) \in \mathbb{R}^\ell \times \mathbb{R}^m : u = f(y) - f(x) - h_1, v = g(x) - h_2, x \in X,$$

$$h_1 \in clC, h_2 \geq 0\} \cap (clC \times \{0\}),$$

so that $\pi(\mathbb{R}^\ell; \mathcal{E}_u(y)) = \{u \in \mathbb{R}^\ell : u = f(y) - f(x) - h_1, x \in K, h_1 \in clC\} \cap (clC) = [f(y) - f(K) - clC] \cap clC \supseteq [f(y) - f(K)] \cap clC$. Since $\mathcal{E}_u(y)$ is a bounded set then, $\forall y \in K$, $\pi(\mathbb{R}^\ell; \mathcal{E}_u(y))$ is a bounded set; therefore $[f(y) - f(K)] \cap clC$ is compact, since it is a closed subset of $\pi(\mathbb{R}^\ell; \mathcal{E}_u(y))$.

It is easy to show that the set $[f(y) - clC] \cap f(K)$ is compact; indeed it is obtained by a linear transformation of the set $[f(y) - f(K)] \cap clC$. Therefore the set $f(K)$ is C–compact. Applying Theorem 3.1 we complete the proof.   □

In Proposition 3.2 $f(K)$ is assumed to be closed. This condition, together with the boundedness of $\mathcal{E}_u(y)$, ensures the C–compactness of $f(K)$. In fact, the conic extension $\mathcal{E}(y)$ provides a regularization of $f(K)$ that no longer preserves the closedness of $f(K)$ even though $\mathcal{E}(y)$ turns out to be closed. To overcome this difficulty we can define a further extension of the set $\mathcal{K}(y)$ that does not involve the objective functions, namely,

$$\mathcal{E}'(y) := [\mathcal{K}(y) - (0 \times \mathbb{R}^m_+)],$$

and the corresponding

$$\mathcal{E}_v(y) := \mathcal{E}'(y) \cap \{(u, v) \in clC \times \{0\}\}.$$

Another possibility consists in considering the image $\mathcal{K}(y)$ and the subset of the feasible images

$$\mathcal{K}_v(y) := \mathcal{K}(y) \cap \{(u, v) \in clC \times \mathbb{R}^m_+\}.$$

We observe that in both sets $\mathcal{E}_v(y)$ and $\mathcal{K}_v(y)$ the objective functions have not been altered, so that it is possible to consider the closedness assumptions directly on the above sets. Adapting again the proof of the Proposition 3.2 we obtain the following result.

Proposition 3.3. If one of the following conditions hold:
(i) $\forall y \in K$, $\mathcal{E}_v(y)$ is a compact set;
(ii) $\forall y \in K$, $\mathcal{K}_v(y)$ is a compact set;
then (2.1) admits a v.m.p..

Proof. (i) We have that

$$\mathcal{E}_v(y) := \{(u,v) \in \mathbb{R}^\ell \times \mathbb{R}^m : u = f(y) - f(x),$$

$$v = g(x) - h, x \in X, \ h \geq 0\} \cap (clC \times \{0\}).$$

Therefore $\pi(\mathbb{R}^\ell; \mathcal{E}_v(y)) = \{u \in \mathbb{R}^\ell : u = f(y) - f(x), \ x \in K, \} \cap clC = [f(y) - f(K)] \cap clC$. Since $\mathcal{E}_v(y)$ is a compact set then, $\forall y \in K$, $\pi(\mathbb{R}^\ell; \mathcal{E}_v(y))$ is a compact set; with the same arguments used in the proof of the Proposition 3.2 we have that $[f(y) - clC] \cap f(K)$ is compact so that $f(K)$ is C–compact. By Theorem 3.1 we complete the proof of part (i).
(ii) $\mathcal{K}_v(y) :=$

$$\{(u,v) \in \mathbb{R}^\ell \times \mathbb{R}^m : u = f(y) - f(x), \ v = g(x), x \in X\} \cap (clC \times \mathbb{R}^m_+\}.$$

$\pi(\mathbb{R}^\ell; \mathcal{K}_v(y)) = \{u \in \mathbb{R}^\ell : u = f(y) - f(x), \ x \in K, \} \cap clC = [f(y) - f(K)] \cap clC$. Since $\mathcal{K}_v(y)$ is a compact set then, $\forall y \in K$, $\pi(\mathbb{R}^\ell; \mathcal{K}_v(y))$ is a compact set; with the same arguments used in the proof of part (i) we complete the proof of Proposition 3.3. □

4. About the cone–compactness.

The existence results shown in the previous section are essentially obtained by proving that the set $f(K)$ is C–compact. It is, therefore, of interest to consider conditions on the objective function f and on the constraint function g that guarantee the C–compactness of $f(K)$. A first obvious remark is that this condition holds when K is a compact set and f is a continuous function on K. In order to weaken the previous assumptions it is necessary to deepen the analysis of the notion of D–compactness of a set: this has led to consider a slightly weaker property, the D–semicompactness [3,12,13] and to express the D–compactness of a set Y in terms of the D–boundedness and D–closedness of Y, generalizing, in this way, the classic topological results.

Definition 4.1. Let Y be a nonempty set in $\mathbb{R}^\ell$, and let D be a cone in $\mathbb{R}^\ell$. Then Y is said to be
(i) D–closed iff $Y + \mathrm{cl}\,C$ is closed;
(ii) D–bounded iff $Y^+ \cap (-\mathrm{cl}\,D) = \{0\}$, where $Y^+ := \{y' \in \mathbb{R}^\ell : \text{there}$ exist sequences $\{\alpha_k\} \subset (\mathbb{R}_+ \setminus \{0\})$ and $\{y_k\} \subset Y$ s.t. $\alpha_k \to 0$ and $\alpha_k y_k \to y'\}$.

Definition 4.2. Let Y be a nonempty set in $\mathbb{R}^\ell$, and let D be a cone in $\mathbb{R}^\ell$. Then, Y is said to be D–semicompact iff every open cover of Y of the form

$$\{(y^\gamma - \mathrm{cl}\,D)^c : y^\gamma, \; \gamma \in \Gamma\}$$

has a finite subcover, where Γ is some index set and the superscript c denotes the complement of a set.

The following theorem states the equivalence between the concepts of D–compactness, D–semicompactness and the ones of D–closedness and boundedness under the hypotheses that cl D be a pointed and convex cone and Y be a closed and convex set. It is shown in Proposition 3.2.1 of [10] that if Y is D–compact then Y is D–semicompact. In the hypotheses of closedness and convexity of the set Y the previous concepts are equivalent.

Theorem 4.1. Let cl D be a pointed and convex cone in $\mathbb{R}^\ell$, Y be a nonempty, closed and convex set in $\mathbb{R}^\ell$. Then, the following statements are equivalent:
(i) Y is D–compact;
(ii) Y is D–closed and D–bounded;
(3i) Y is D–semicompact.
Proof. We observe that Y is D–closed, D–bounded, D–semicompact iff, respectively, Y is cl$D$–closed, cl$D$–bounded, cl$D$–semicompact; hence Theorem 4.1 follows from Proposition 3.2.3 of [10].   □

Remark 4.1. In the previous theorem we can obviously drop the hypotheses of closedness and convexity on Y, replacing Y with cl conv Y.

Theorem 4.2. [10] Let C be a convex cone in $\mathbb{R}^\ell$ such that cl$C$ is pointed. If $f(K)$ is nonempty and C–semicompact then (2.1) admits a v.m.p..

By means of the new concept of D-semicontinuity of a vector function it is possible to extend the classic Weierstrass theorem for the existence of an optimal solution of a constrained extremum problem.

Definition 4.3. [3] Let D be a cone in $\mathbb{R}^\ell$. A function $f : \mathbb{R}^n \longrightarrow \mathbb{R}^\ell$ is said to be D-semicontinuous if

$$f^{-1}(y - clD) = \{x \in \mathbb{R}^n : y - f(x) \in clD\}$$

is closed, $\forall y \in \mathbb{R}^\ell$.

Theorem 4.3. Let X be a compact set in $\mathbb{R}^\ell$, f be a C-semicontinuous function and $-g : X \longrightarrow \mathbb{R}^m$ be a l.s.c. function on X. Then the set $f(K)$ is C-semicompact.

Proof. It is simple to prove that, in the considered hypotheses, the set K is compact so that the statement follows from Lemma 3.2.6 of [10]. □

Remark 4.2. It is known that $f : \mathbb{R}^n \longrightarrow \mathbb{R}^\ell$ is $\mathbb{R}^\ell_+$-semicontinuous iff each f_i, i=1,...,ℓ is lower semicontinuous.

In the results of Section 3, we have only considered classic hypotheses of boundedness and compactness on the image $\mathcal{K}(y)$ and on tha extended image $\mathcal{E}(y)$. This is due to the dependance of these sets on the point $y \in K$. If we define the image $\mathcal{K}$ independently on the point y, then we can directly assume C-compactness or C-semicompactness hypotheses on this set, since these assumptions involve the point y themselves. In fact let

$$\tilde{\mathcal{K}} := \{(u, v) \in \mathbb{R}^\ell \times \mathbb{R}^m : u = f(x) , v = g(x), x \in X\}.$$

Proposition 4.1. Let C be a convex cone in $\mathbb{R}^\ell$ such that clC is pointed. If $\tilde{\mathcal{K}}$ is $(C \times \mathbb{R}^m_+)$-compact then (2.1) admits a v.m.p..

Proof. The hypotheses imply that the set $[y - (clC \times \mathbb{R}^m_+) \cap \tilde{\mathcal{K}}$ is compact, $\forall y \in \tilde{\mathcal{K}}$, so that $[f(y) - clC] \cap f(K) = \pi(\mathbb{R}^\ell; y - (clC \times \mathbb{R}^m_+) \cap \tilde{\mathcal{K}})$ is compact, that is $f(K)$ is C-compact. Theorem 3.1 allows to prove our statement. □

It would be interesting to estabilish, in the image space, existence results involving semicompactness assumptions. To this end we should deepen the analysis of the topological properties of the notion of D-semicompactness, proving, in particular, under which assumptions the projection on a subspace of a D-semicompact set is still semicompact.

Further developments of the analysis on the existence of the solutions to a VOP can be made considering scalarization methods. Solutions to a

VOP are obtained solving parametric scalar optimization problems $P(\lambda)$

$$\min \ h_\lambda(x) \quad s.t. \quad x \in K,$$

where $h_\lambda(x) := \sum_{i=1}^{\ell} \lambda_i f_i(x), \ \lambda \in C^*$.

References

[1] G. Bigi and M. Pappalardo, *Regularity conditions in Vector Optimization*, Jou. of Optimiz. Theory Appls. 102, No.1, (1999), 83–96.

[2] M. Castellani, G. Mastroeni and M. Pappalardo, *On regularity for generalized systems and applications*, In *Nonlinear Optimization and Applications*, G. Di Pillo et al. (Eds.), Plenum Press, New York (1996), 13–26.

[3] H.W. Corley, *An existence result for maximization with respect to cones*, Jou. of Optimiz. Theory Appls. 31 (1980), 277–281.

[4] F. Giannessi, *Theorems of the alternative, quadratic programs and complementarity problems*, In *Variational Inequalities and complementarity problems*, R.W. Cottle et al. Eds., J. Wiley (1980), 151–186.

[5] F. Giannessi, *Theorems of the alternative and optimality conditions*, Jou. Optimiz. Theory Appls., Plenum, New York, Vol.42, No.11, 1984, 331–365.

[6] F. Giannessi, *Vector Variational Inequalities and Vector Equilibria, Mathematical Theories*, F. Giannessi, Eds., Kluwer Academic Publishers,Dordrecht, Boston, London, 2000.

[7] F. Giannessi, G. Mastroeni and L. Pellegrini, *On the Theory of Vector Optimization and Variational Inequalities. Image Space Analysis and Separation*, In *Vector Variational Inequalities and Vector Equilibria, Mathematical Theories*, F. Giannessi, Eds., Kluwer Academic Publishers,Dordrecht, Boston, London, 2000.

[8] F. Giannessi and L. Pellegrini, *Image space Analysis for Vector Optimization and Variational Inequalities. Scalarization*, In *Advances in Combinatorial and Global Optimization*, A. Migdalas, P. Pardalos and R. Burkard, Eds., Worlds Science.

[9] Hartley, *On cone-efficiency, cone-convexity and cone compactness*, SIAM Jou. Appl. Math. 34, N.2, (1978), 211–222.

[10] Y. Sawaragi, H. Nakayama and T. Tanino, *Theory of multiobjective Optimization* Academic Press, New York, 1985.

[11] F. Tardella, *On the image of a constrained extremum problem and some applications to the existence of a minimum*, Jou. of Optimiz. Theory Appls. 60, No.1, (1989), 93–104.

[12] F. Tardella, *Some topological properties in optimization theory*, Jou. of Optimiz. Theory Appls. 60, No.1, (1989), 105–116.

[13] D.H. Wagner, *Semicompactness with respect to a Euclidean cone*, Canad. Jour. Math. 29 (1977), 29–36.

EQUILIBRIUM PROBLEMS AND VARIATIONAL INEQUALITIES

Antonino Maugeri

Dipartimento di Matematica

Università di Catania

v.le A. Doria, 6 95125 Catania – Italia

e-mail: maugeri@dmi.unict.it

Abstract   We show that many equilibrium problems fulfill the common laws expressed by a set of conditions and that the equilibrium solution is obtained as a solution to a Variational Inequality. In particular we study the traffic equilibrium problem in the continuum case and we solve the problem to express this problem by means of a Variational Inequality.

Keywords: Variational Inequality, Signorini Problem, Obstacle Problem, Traffic Equilibrium Problem, Continuum Traffic Equilibrium Problem, Duality, Lagrangean Theory.

AMS classification: 90C, 49J, 65K.

1. Introduction.

Many equilibrium problems arising from various fields of science may be expressed, under general conditions, in a unified way:

$$\begin{cases} \mathcal{B}u\,\mathcal{L}u = 0 \\ \mathcal{B}u \geq 0 \; \mathcal{L}u \geq 0 \\ u \in S \end{cases} \tag{1.1}$$

where $\mathcal{B}$ and $\mathcal{L}$ are suitable operators defined in a suitable functional class S.

The kind of equilibrium described by the structure (1.1), in general, is different from the one obtained by minimization of a cost functional

F. Giannessi et al (eds.),

Equilibrium Problems: Nonsmooth Optimization and Variational Inequality Models, 187–205.

© 2001 *Kluwer Academic Publishers.*

or of an energy integral. Moreover the structure (1.1) leads, in general, to a Variational Inequality on a convex, closed subset $\mathbb{K}$ of S.

This happens for Unilateral Problems in continuum mechanics (the celebrated Signorini Problem), for the Obstacle Problem, for the Discrete and Continuum Traffic Equilibrium Problem, for the Spatial Price Problem, for the Financial Problem, etc. (see for many of these problems [17]).

The equilibrium given by the structure (1.1) may be considered as an equilibrium from a "local point of view". It is different, in general, from the one which we call "global", obtained by minimizing the usual functionals, and represents a complementary aspect which makes clear unknown features of equilibrium problems.

The reason for which the past decades have witnessed an exceptional interest for the equilibrium problems of the type (1.1) rests on the fact that the Variational Inequality Theory, which in general expresses the equilibrium conditions (1.1), provides a powerful methodology, that in these last years has been improved by studying the connections with the Separation Theory and the Gap Function, the Lagrangean Theory and the Duality and many related Computational Procedures.

In the present paper some of the most recent results and the open problems are focused.

2. The Signorini problem.

The celebrated Signorini problem (see [19], [9]) has been the first problem whose formulation is included in the structure (1.1). It can be summarized, in a weak formulation, as follows

$$
\begin{cases}
\mathcal{L}u - f = -\sum_{i,j=1}^{n} \frac{\partial}{\partial x_i}\left(a_{ij}\frac{\partial u}{\partial x_j}\right) + \sum_{i=1}^{n} b_i\frac{\partial u}{\partial x_i} + cu - f = 0 \text{ in } \Omega \\
u \geq 0 \text{ on } \partial\Omega \\
\frac{\partial u}{\partial \nu} = \sum_{i,j=1}^{n} a_{ij}\frac{\partial u}{\partial x_j}n_i \geq 0 \text{ on } \partial\Omega \\
u\frac{\partial u}{\partial \nu} = 0 \text{ on } \partial\Omega \\
u \in H^1(\Omega) \text{ on } \partial\Omega
\end{cases}
\tag{2.1}
$$

where n_i are the components of the outer normal to $\partial\Omega$ and it represents the conceptual model of an elastic body Ω with boundary $\partial\Omega$ which is in contact with a rigid support body and is subject to volume forces f. These forces produce a deformation of Ω and a displacement u on

$\partial\Omega$ either with the conormal component non negative if the body has a contact with $\partial\Omega$ (when $u = 0$) or with this component zero if the body has not contact with $\partial\Omega$ (when $u > 0$). The usual variational technique allows one to prove that problem (2.1) can be expressed by the variational inequality (see [3])

$$a(u, v - u) \geq (f, v - u) \; \forall v \in \mathbb{K} \tag{2.2}$$

where

$$a(u, v) = \int_\Omega \left(\sum_{ij} a_{ij} \frac{\partial u}{\partial x_i} \frac{\partial v}{\partial x_j} + \sum_{i=1}^n b_i u \frac{\partial v}{\partial x_i} + cuv \right) dx$$

$$(f, v) = \int_\Omega fv \, dx$$

$$K = \{v \in H^1(\Omega) : v(x) \geq 0 \text{ on } \partial\Omega\}. \tag{2.3}$$

The assumptions that it is necessary to consider in order to give sense to (2.2) are the following:
Ω open, bounded connected subset of $\mathbb{R}^n$, with boundary $\partial\Omega$ of class $C^{0,1}$;
$a_{ij} \in L^\infty(\Omega)$; $b_i \in L^n(\Omega)$ if $n > 2$, $b_i \in L^{2+\epsilon}(\Omega)$ with $\epsilon > 0$ if $n = 2$, $b_i(x) \in L^2(\Omega)$ if $n = 1$; $c \in L^{\frac{n}{2}}(\Omega)$ if $n > 2$, $c \in L^{1+\epsilon}(\Omega)$ if $n = 2$ ($\epsilon > 0$), $c \in L^1(\Omega)$ if $n = 1$. Moreover, we assume that the operator L is elliptic:

$$\sum_{i,j=1}^n a_{ij}(x)\xi_i\xi_j \geq \alpha \sum_{i=1}^n \xi_i^2 \; \forall \xi \in \mathbb{R}^n, \quad \alpha > 0.$$

Then assuming that the bilinear form $a(u, v)$ is coercive on $H^1(\Omega)$

$$a(u, u) \geq \nu \|u\|_{H^1(\Omega)}^2 \; \forall u \in H^1(\Omega) \; \nu > 0 \tag{2.4}$$

the existence of a solution to the variational inequality (2.2) is ensured (see [9], [10], [13]). For example (2.4) is true if $b_i = 0$ and $c \in L^\infty(\Omega)$ and $c(x) \geq c_0 > 0$.

Problem (2.2) has been the prototype and the landmark of the theory of problems with unilateral conditions on the boundary. Starting from problem (2.1), several important and deep generalizations have been obtained.

It is worth remarking that problem (2.2) cannot be obtained from the minimization of the functional

$$\min_{u \in \mathbb{K}} \frac{1}{2} a(u, u) - (f, u) \tag{2.5}$$

if the form $a(u, v)$ is not symmetric.

In order to describe in an effective way important aspects and features of the linear and non linear elasticity problems, problem (2.2) has to be generalized in a suitable way for what concerns the form $a(u, v)$, the convex $\mathbb{K}$ and the coerciveness condition (2.4).

We refer to [10], [11], [2], [5] for some of the possible generalizations. We only mention how the coerciveness can be weakened when the bilinear form

$$a(u, v) = \int_{\Omega} \sum_{i,j=1}^{n} a_{ij}(x) \frac{\partial u}{\partial x_i} \frac{\partial v}{\partial x_j} \, dx$$

is symmetric, the ellipticity condition holds and $\mathbb{K}$ is a more general closed, convex subset of $H^1(\Omega)$.

We may use a result due to [2] (see also [9], [10]) that makes use of the concepts of recession function and recession cone.

Definition 1 *Let $F(u) : H^1(\Omega) \to] - \infty, +\infty]$ be a proper, convex and weakly lower semicontinuous function. The recession function F^{∞} of F is defined by*

$$F^{\infty}(u) = \lim_{\lambda \to \infty} \frac{1}{\lambda} F(u_0 + \lambda u) \tag{2.6}$$

where u_0 is any element of $\mathrm{dom}\, F = \{u \in H^1(\Omega) : F(u) < \infty\}$.

Definition 2 *Let $\mathbb{K}$ be a non–empty, convex, closed subset of $H^1(\Omega)$. We call recession cone of $\mathbb{K}$ the set (closed, convex cone)*

$$\mathbb{K}^{\infty} = \cap_{\lambda > 0} \frac{1}{\lambda} (\mathbb{K} - u_0) \tag{2.7}$$

where u_0 is an element of $\mathbb{K}$ and $\mathbb{K} - u_0$ denotes the set $\{u - u_0, u \in \mathbb{K}\}$.

Remark 1. If

$$F(u) = \frac{1}{2} a(u, u) = \frac{1}{2} \int_{\Omega} \sum_{i,j=1}^{n} a_{ij}(x) \frac{\partial u}{\partial x_i} \frac{\partial u}{\partial x_j} \, dx$$

and

$$\sum_{i,j=1}^{n} a_{ij}(x) \xi_i \xi_j \geq \alpha \sum_{i=1}^{n} \xi_i^2 \quad \alpha > 0, \quad \forall \xi \in \mathbb{R}^n,$$

it results

$$F^{\infty} = \begin{cases} 0 & \text{if } F(u) = 0 \\ +\infty & \text{if } F(u) \neq 0 \end{cases} \tag{2.8}$$

as it can be easily seen by observing that

$$F(u_0 + \lambda u) = \frac{1}{2}\lambda^2 a(u,u) + \lambda a(u,u_0) + \frac{1}{2}a(u_0,u_0).$$

Moreover $F(u) = 0$ means that

$$0 = \int_\Omega \sum_{i,j=1}^n a_{ij}(x)\frac{\partial u}{\partial x_i}\frac{\partial u}{\partial x_j}\,dx \geq \alpha \int_\Omega \sum_{i=1}^n \left(\frac{\partial u}{\partial x_i}\right)^2 dx$$

and hence we obtain

$$u = c \quad c \in \mathbb{R}.$$

Then

$$Ker F = \{u \in H^1(\Omega) : F(u) = 0\} = \{u : u = c,\ c \in \mathbb{R}\}. \qquad (2.9)$$

Furthermore, if we set $L(u) = (f,u) = \int_\Omega fu\,dx$ with $f \in L^2(\Omega)$, it results

$$Ker(F^\infty - L) = \{u \in H^1(\Omega) : F^\infty(u) = L(u)\} =$$
$$= \{u \in H^1(\Omega) : L(u) = 0 \text{ if } F(u) = 0\} =$$
$$= Ker F \cap Ker L = \{u : u = c,\ c\int_\Omega f\,dx = 0\} =$$

$$= \begin{cases} \{0\} & \text{if } \int_\Omega f\,dx \neq 0 \\ \\ \mathbb{R} & \text{if } \int_\Omega f\,dx = 0. \end{cases} \qquad (2.10)$$

Remark 2. If $\mathbb{K}$ is a cone with vertex at the origin, as in the case:

$$\mathbb{K} = \{u \in H^1(\Omega) : u(x) \geq 0 \text{ on } \partial\Omega\} \qquad (2.11)$$

we obtain $\mathbb{K} = \mathbb{K}^\infty$, because $\frac{1}{\lambda}\mathbb{K} = \mathbb{K}\ \forall \lambda > 0$.
Remark 3. Let us set

$$G(u) = F(u) - L(u) + \chi_{\mathbb{K}}(u) \qquad (2.12)$$

where $\chi_{\mathbb{K}}(u)$ is the indicator function of $\mathbb{K}$, that is,

$$\chi_{\mathbb{K}}(u) = \begin{cases} 0 & \text{if } u \in \mathbb{K} \\ \\ +\infty & \text{if } u \notin \mathbb{K}. \end{cases}$$

Let us prove that $G(u)$ fulfills the following compactness property (see [2] Lemma 4.11):

"If $\{\lambda_n\}$, $\{x_n\}$ and u are such that $\lambda_n \to +\infty$, x_n weakly converges to u and $G(\lambda_n u_n) \leq c$ for some constant c, then u_n strongly converges to u."

In fact, F being positively homogeneous of degree 2, sequentially l.s.c. and non negative, we get

$$0 \leq F(u) \leq \liminf_{n\to\infty} F(u_n) \leq \limsup_{n\to\infty} F(u_n) = \limsup_{n\to\infty} \frac{1}{\lambda_n^2} F(\lambda_n u_n) \leq$$

$$\leq \limsup_{n\to\infty} \frac{1}{\lambda_n^2} [G(\lambda_n u_n) + L(\lambda_n u_n)] \leq \limsup_{n\to\infty} \frac{1}{\lambda_n^2} [c + L(\lambda_n u_n)] = 0.$$

Hence $F(u) = 0$ and $F(u_n - u) = F(u_n) \to 0$. On the other hand, $\|u_n - u\|_{L^2(\Omega)} \to 0$ and, by virtue of the estimate

$$\alpha \|u_n - u\|_{H^1(\Omega)}^2 \leq \alpha \|u_n - u\|_{L^2(\Omega)}^2 + F(u_n - u)$$

the assertion follows.

We may recall the existence result due to G. Fichera and C. Baiocchi (see [2], [9], [11]):

Theorem 1 *Let $F : H^1(\Omega) \to [0, +\infty[$ be a proper, convex, seqentially l.s.c. functional, $L : H^1(\Omega) \to \mathbb{R}$ a linear, continuous functional, $\mathbb{K}$ a non-empty, convex closed subset of $H^1(\Omega)$. Let the compactness condition of Remark 3 hold for the functional G defined as in (2.12). Assume that*

$$F^\infty(u) \geq L(u) \quad \forall u \in \mathbb{K}^\infty \tag{2.13}$$

$$Ker(F^\infty - L) \cap \mathbb{K}^\infty \text{ is a subspace.} \tag{2.14}$$

Then there exists $u \in \mathbb{K}$ such that

$$F(u) - L(u) \leq F(v) - L(v) \quad v \in \mathbb{K}. \tag{2.15}$$

Taking into account Remarks 1 and 3, from Theorem 1, we easily get the following:

Corollary 1 *Let $F(u) = \frac{1}{2}a(u,u)$, $a(u,u)$ being a bilinear, continuous positive form defined on $H^1(\Omega) \times H^1(\Omega)$ and $L(u) = \int_\Omega fu\,dx$ with $f \in L^2(\Omega)$.*

Assume that

$$c \int_\Omega f(x)\,dx \leq 0 \quad \forall c \in \mathbb{R}, \ c \in \mathbb{K}^\infty \tag{2.16}$$

and that

$$\{c \in \mathbb{R} : c \int_\Omega f(x)\,dx = 0, \quad c \in \mathbb{K}^\infty\} \text{ is a subspace.} \tag{2.17}$$

Then there exists $u \in \mathbb{K}$ such that

$$F(u) - L(u) \leq F(v) - L(v) \quad \forall v \in \mathbb{K}.$$

Remark 4. $\mathbb{K}^\infty$ being a cone, (2.17) is equivalent to assume that

$$\text{"If } c \in \left\{ c \in \mathbb{R} : c \int_\Omega f(x)\, dx = 0, \quad c \in \mathbb{K}^\infty \right\}, \text{ then } -c \in \mathbb{K}^* \text{"}.$$

If coerciveness condition (2.4) is fulfilled, for example when

$$a(u, v) = \int_\Omega \left(\sum_{i,j=1}^n a_{ij}(x) \frac{\partial u}{\partial x_i} \frac{\partial v}{\partial x_j} + c(x) uv \right) dx$$

with $c(x) \geq c_0 > 0$, it turns out to be $Ker F = \{0\}$ and conditions (2.13), (2.14) are automatically verified (see also (2.16) and (2.17)).

Remark 5. The non symmetric, noncoercive case can be found in G. Fichera [9], [10]. Applications to noncoercive optimization problems have been considered by A. Auslender (see [3]).

Remark 6. Making use of the conjugate convex function, it is possible to consider the dual problem in the symmetric case (see, for example, [5] I.1.3). However, we can consider a dual formulation in the general case.

Setting

$$\psi(v) = a(u, v - u) - (f, v - u) \quad v \in \mathbb{K},$$

where $u \in \mathbb{K}$ is the solution to the Variational Inequality (2.2), it is easy to see that

1 $\psi(v) \geq 0 \quad \forall v \in \mathbb{K};$

2 $\min_{v \in \mathbb{K}} \psi(v) = \psi(u) = 0.$

$\psi(v)$ is an example of a so called Gap function.

Then the problem

$$\min_{v \in \mathbb{K}} \psi(v) \tag{2.18}$$

is equivalent to problem (2.2).

We can prove the following results.

Lemma 1 *Problem (2.18) is equivalent to the Problem*

$$\min_{v \in H^1(\Omega)} \sup_{\mu \in C} \{\psi(v) - \langle \mu, v \rangle\} \tag{2.19}$$

where

$$C = \{\mu \in H^{-\frac{1}{2}}(\partial\Omega) : \mu \geq 0 \text{ on } \partial\Omega\} \tag{2.20}$$

(see [7], Lemma 2.1).

Let us consider the dual problems

$$\max_{\mu \in C} \inf_{v \in H^1(\Omega)} [\psi(v) - \langle \mu, v \rangle] \tag{2.21}$$

and

$$\max_{\Lambda \in \Delta} \Lambda \tag{2.22}$$

where

$$\Delta = \{\Lambda \in \mathbb{R} : \psi(v) - \langle \mu, v \rangle \geq \Lambda \quad \forall v \in H^1(\Omega), \forall \mu \in H^{-\frac{1}{2}}(\partial\Omega)\}. \tag{2.23}$$

The following results hold.

Lemma 2 $\overline{\mu} \in C$ *is a maximal solution to the dual problem (2.21) if and only if*

$$\overline{\Lambda} = \max_{\mu \in C} \inf_{v \in H^1(\Omega)} [\psi(v) - \langle \mu, v \rangle]$$

is a solution to the problem (2.22) (see [7], Lemma 2.3).

Lemma 3 *If the primal problem (2.18) is solvable, then the dual problem (2.21) is also solvable and the extremal values of the problems are equal (see [7], Lemma 2.4).*

Let us consider the Lagrangean function

$$L : H^1(\Omega) \times C \Rightarrow \mathbb{R}$$

defined by setting

$$L(v, \mu) = \psi(v) - \langle \mu, v \rangle. \tag{2.24}$$

Then the following theorem holds.

Theorem 2 *A point $(u, \overline{\mu}) \in H^1(\Omega) \times C$ is a saddle point of the Lagrangean function L if and only if u is a solution to the primal problem (2.18), $\overline{\mu}$ is a solution to the dual problem (2.8) and the extremal values of the two problems are equal.*

We can deduce some consequences from Theorem 2. In fact, from Theorem 2 it follows that, if u is a solution to the Variational Inequality and $\overline{\mu}$ is a solution to the dual problem (2.8), it results

$$\psi(u) - \langle \overline{\mu}, u \rangle = 0 \tag{2.25}$$

and hence

$$\langle \overline{\mu}, u \rangle = 0. \tag{2.26}$$

Moreover, from

$$L(v, \bar{\mu}) \geq L(u, \bar{\mu}) = 0 \ \forall v \in H^1(\Omega),$$

we get

$$a(u, v - u) - (f, v - u) - \langle \bar{\mu}, v \rangle \geq 0 \ \forall v \in H^1(\Omega). \tag{2.27}$$

Choosing $v = u \pm \varphi \ \forall \varphi \in H^1(\Omega)$, from (2.27), and taking into account (2.26), we get

$$a(u, \varphi) - (f, \varphi) - \langle \mu, \varphi \rangle = 0. \tag{2.28}$$

Taking into account that (see (2.1))

$$a(u, \varphi) - (f, \varphi) = \langle \mathcal{L}u - f, \varphi \rangle + \int_{\partial \Omega} \frac{\partial u}{\partial \nu} \varphi \, ds = \int_{\partial \Omega} \frac{\partial u}{\partial \nu} \varphi \, ds$$

from (2.28) we obtain:

$$\bar{\mu} = \frac{\partial u}{\partial \nu} \tag{2.29}$$

and hence it is clear the meaning of the Lagrangean multiplier $\bar{\mu}$.

3. The obstacle problem.

Let ψ be a given function in $H^2(\Omega)$. The obstacle problem fulfills the general structure (1.1):

$$\begin{cases} (\mathcal{L}u - f)(u - \psi) = 0 \\ \mathcal{L}u - f \geq 0, \ u - \psi \geq 0 \\ u \in H^2(\Omega) \end{cases} \tag{3.1}$$

and it can be considered as an unilateral problem with unilateral constraint on Ω.

Assuming $\psi(x) \leq 0$ on $\partial \Omega$ and $f(x) \in L^2(\Omega)$, setting

$$\mathbb{K} = \{u \in H_0^1(\Omega) : u \geq \psi \text{ a.e. } x \in \Omega\},$$

the weak solution to the problem (3.1) is given by the solution to the Variational Inequality (see [3], [13], [5], [4]):

$$\text{Find } u \in \mathbb{K} : a(u, v - u) \geq (f, v - u) \quad \forall u \in \mathbb{K} \tag{3.2}$$

where

$$a(u, v) = (\mathcal{L}u, v) = \int_\Omega \sum_{i,j=1}^n a_{ij}(x) \frac{\partial u}{\partial x_i} \frac{\partial v}{\partial x_j} \, dx +$$

$$+ \int_\Omega a_0(x)u(x)v(x)\, dx, \tag{3.3}$$

and the following assumption holds:

$$\begin{cases} a_{ij}, a_0(x) \in L^\infty(\Omega), a_0(x) \geq 0 \text{ and} \\[2mm] \sum_{i,j=1}^n a_{ij}(x)\xi_i\xi_j \geq \alpha \sum_{i=1}^n \xi_i^2 \ \ \alpha > 0, \ \forall \xi \in \mathbb{R}^n. \end{cases} \tag{3.4}$$

Existence and regularity results can be found in the papers [13], [5], [4] (see also the references there quoted). We may follow the duality theory presented in Section n. 2 in order to give a dual formulation of (3.2) in the asymmetric case.

Setting

$$\psi(v) = a(u, v - u) - (f, v - u) \ v \in \mathbb{K} \tag{3.5}$$

with $u \in \mathbb{K}$ the unique solution to (3.2), we can prove also in this case the following equivalence results.

Lemma 4 *The problems*

$$\min_{v \in \mathbb{K}} \psi(v) \tag{3.6}$$

and

$$\min_{v \in H_0^1(\Omega)} \sup_{\mu \in C} \{\psi(v) - \langle \mu, v - \psi \rangle\} \tag{3.7}$$

with

$$C = \{\mu \in H^{-1}(\Omega), \ \mu \geq 0\} \tag{3.8}$$

are equivalent.

Lemma 5 *Let us consider the dual problems*

$$\max_{\mu \in C} \inf_{v \in H_0^1(\Omega)} \{\psi(v) - \langle \mu, v - \psi \rangle\} \tag{3.9}$$

and

$$\max_{\Lambda \in \Delta} \Lambda \tag{3.10}$$

where

$$\Delta = \{\Lambda \in \mathbb{R} : \psi(v) + \langle \mu, v - \psi \rangle \geq \Lambda \ \forall v \in H_0^1(\Omega), \ \forall \mu \in H^{-1}(\Omega)\}. \tag{3.11}$$

Then $\overline{\mu}$ is a maximal solution to the dual problem (3.9) if and only if

$$\overline{\Lambda} = \max_{\mu \in C} \inf_{v \in H_0^1(\Omega)} \{\psi(v) - \langle \mu, v - \psi \rangle\}$$

is a solution to the problem (3.10).

Lemma 6 *If the primal problem (3.6) is solvable, then the dual problem (3.9) is also solvable and the extremal values of the problems are equal.*

We can consider the Lagrangean function

$$L : H_0^1(\Omega) \times C \Rightarrow \mathbb{R}$$

$$L(v, \mu) = \psi(v) - \langle \mu, v - \psi \rangle \tag{3.12}$$

and a similar theorem to Theorem 2 holds true.

Theorem 3 *A point $(u, \overline{\mu}) \in H_0^1(\Omega) \times C$ is a saddle point of the Lagrangean function L if and only if u is a solution to the primal problem (3.6), $\overline{\mu}$ is a solution to the dual problem (3.9) and the extremal values of the two problems are equal.*

Also in this case we can deduce some interesting consequences from Theorem 3.

We have

$$\psi(v) - \langle \overline{\mu}, v - \psi \rangle \geq \psi(u) - \langle \overline{\mu}, u - \psi \rangle = 0 \ \forall v \in H_0^1(\Omega). \tag{3.13}$$

Hence

$$\langle \overline{\mu}, u - \psi \rangle = 0 \tag{3.14}$$

and

$$a(u, v - u) - (f, v - u) - \langle \overline{\mu}, v - \psi \rangle \geq 0 \ \forall v \in H_0^1(\Omega). \tag{3.15}$$

Choosing $v = u + \varphi$ with $\varphi \in H_0^1(\Omega)$, by virtue of (3.14), we get

$$\langle \mathcal{L}u - f, \varphi \rangle - \langle \overline{\mu}, \varphi \rangle = 0 \tag{3.16}$$

and hence

$$\mathcal{L}u - f - \overline{\mu} = 0. \tag{3.17}$$

Then, from (3.14) and (3.17), we get

$$\overline{\mu} = \mathcal{L}u - f \tag{3.18}$$

and

$$\langle \overline{\mu}, u - \psi \rangle = 0. \tag{3.19}$$

From (3.18) and (3.19) the importance of $\overline{\mu}$ clearly follows.

4. A continuous model of transportation.

Following the suggestions of S. Dafermos (see [6]), we express the equilibrium obtained using the user's optimization approach, namely when the structure (1.1) is fulfilled, by means of a variational inequality.

Let Ω be a simply connected bounded domain in $\mathbb{R}^2$ of generic point $x = (x_1, x_2)$, with Lipschitz boundary $\partial\Omega$. The unknown flow at each point $x \in \Omega$ is described by a vector field $u(x)$, whose components $u_1(x)$, $u_2(x)$ represent the traffic density through a neighbourhood of x in the directions of the increasing axes x_1 and x_2.

It will be

$$u_1(x) \geq 0$$

$$u_2(x) \geq 0 \tag{4.1}$$

and $u_1(x)$, $u_2(x)$ have non negative fixed traces $\varphi_1(x)$ and $\varphi_2(x)$ on $\partial\Omega$ (or on a part of $\partial\Omega$) respectively; we suppose that $u(x) \in H^1(\Omega, \mathbb{R}^2)$ and $\varphi(x) = (\varphi_1(x), \varphi_2(x) \in H^{\frac{1}{2}}(\partial\Omega, \mathbb{R}^2)$. If we associate to each point $x \in \Omega$ a scalar field $t(x) \in L^2(\Omega)$, that represents the density of the flow originating or terminating at x, we can write the flow conservation law in the form:

$$\int\int_{\Omega'} t\,dx + \oint_{\partial\Omega'} u\,n\,ds = 0 \tag{4.2}$$

where n is the normal at the point $x \in \partial\Omega'$ and Ω' any subdomain of Ω with smooth boundary $\partial\Omega$.

If we write (4.2) for $\Omega' = \Omega$ we obtain:

$$\int\int_{\Omega} t\,dx_1\,dx_2 + \oint_{\partial\Omega} [\varphi_1(x)\,n_1 + \varphi_2(x)\,n_2]\,ds = 0 \tag{4.3}$$

which expresses the connection between $t(x)$ and the traces $\varphi_1(x)$, $\varphi_2(x)$. Moreover from (4.2), since $u \in H^1(\Omega, \mathbb{R}^2)$, we get:

$$\operatorname{div} u + t(x) = \frac{\partial u_1}{\partial x_1} + \frac{\partial u_2}{\partial x_2} + t(x) = 0. \tag{4.4}$$

Now we consider the "personal cost" c that will be a vector whose components $c_1(x, u(x))$, $c_2(x, u(x))$ represent the travel cost along the axes x_1 and x_2 respectively. For the case when we are not interested in the user's personal travel cost, but rather in an overall travel cost spent in the network, we can consider the "total cost" which can be expressed in the form:

$$c_T(x, u(x)) = c(x, u(x))u(x) = c_1(x, u(x))u_1(x) + c_2(x, u(x))u_2(x) \tag{4.5}$$

and the "global total cost"

$$F(u) = \int_\Omega [c_1(x, u(x)) u_1(x) + c_2(x, u(x)) u_2(x)] \, dx. \qquad (4.6)$$

We are concerned with the "user optimizing" approach in order to state an appropriate definition of equilibrium when the set of feasible flows is given by

$$\mathbb{K} = \{u(x) \in H^1(\Omega, \mathbb{R}^2) \; : u_1(x), \, u_2(x) \geq 0, \; u_1(x)|_{\partial\Omega} = \varphi_1(x), \qquad (4.7)$$
$$u_2(x)|_{\partial\Omega} = \varphi_2(x), \; \mathrm{div}\, u + t(x) = 0\}.$$

Definition 3 $u(x) \in \mathbb{K}$ *is an equilibrium distribution flow if there exists a potential* $\mu \in H^1(\Omega)$ *such that*

$$\left(c_i(x, u(x)) - \frac{\partial\mu}{\partial x_i}\right) u_i(x) = 0, \quad i = 1, 2 \quad a. \; e. \; in \; \Omega$$
$$c_i(x, u(x)) - \frac{\partial\mu}{\partial x_i}\right) \leq 0, \quad i = 1, 2 \quad a. \; e. \; in \; \Omega \qquad (4.8)$$

μ *measures the cost occurred when a network user travels from the point* x *to the boundary* $\partial\Omega$ *using the cheapest possible path.*

If we suppose that the cost function $c(x, y)$ is a Carathéodory function satisfying the growth condition:

$$\|c(x, u)\| \leq \alpha(x) + \|u\|_{\mathbb{R}^2} \qquad (4.9)$$

with $\alpha(x) \in L^2(\Omega)$, we can prove the following:

Theorem 4 $u \in \mathbb{K}$ *is an equilibrium distribution if and only if*

$$\int_\Omega c(x, u(x))(v(x) - u(x)) \, dx \geq 0 \quad \forall v \in \mathbb{K}. \qquad (4.10)$$

Proof. Let us prove that an equilibrium condition according to Definition 3 fulfills the Variational Inequality (4.10).

In fact, setting:

$$\Omega_1^+ = \left\{x \in \Omega : c_1(x, u(x)) > \frac{\partial\mu}{\partial x_1}\right\},$$

$$\Omega_2^+ = \left\{x \in \Omega : c_2(x, u(x)) > \frac{\partial\mu}{\partial x_2}\right\},$$

we have:

$$\int_\Omega c(x, u(x))(v(x) - u(x))\, dx = \int_{\Omega_1^+} c_1(x, u(x))(v_1(x) - u_1(x))\, dx +$$

$$+ \int_{\Omega\setminus\Omega_1^+} c_1(x, u(x))(v_1(x) - u_1(x))\, dx + \int_{\Omega_2^+} c_2(x, u(x))(v_2(x) - u_2(x))\, dx +$$

$$+ \int_{\Omega\setminus\Omega_2^+} c_2(x, u(x))(v_2(x) - u_2(x))\, dx = \int_{\Omega_1^+} c_1(x, u(x)) v_1(x)\, dx +$$

$$+ \int_{\Omega\setminus\Omega_1^+} \frac{\partial\mu}{\partial x_1}(v_1(x) - u_1(x))\, dx + \int_{\Omega_2^+} c_2(x, u(x)) v_2(x)\, dx +$$

$$+ \int_{\Omega\setminus\Omega_2^+} \frac{\partial\mu}{\partial x_2}(v_2(x) - u_2(x))\, dx \ge \int_{\Omega_1^+} \frac{\partial\mu}{\partial x_1}(v_1(x) - u_1(x))\, dx +$$

$$+ \int_{\Omega\setminus\Omega_1^+} \frac{\partial\mu}{\partial x_1}(v_1(x) - u_1(x))\, dx + \int_{\Omega_2^+} \frac{\partial\mu}{\partial x_2}(v_2(x) - u_2(x))\, dx +$$

$$+ \int_{\Omega\setminus\Omega_2^+} \frac{\partial\mu}{\partial x_2}(v_2(x) - u_2(x))\, dx =$$

$$= \int_\Omega \frac{\partial\mu}{\partial x_1}(v_1(x) - u_1(x))\, dx + \int_\Omega \frac{\partial\mu}{\partial x_2}(v_2(x) - u_2(x))\, dx = 0$$

because

$$\int_\Omega \frac{\partial\mu}{\partial x_1}(v_1(x) - u_1(x))\, dx + \int_\Omega \frac{\partial\mu}{\partial x_2}(v_2(x) - u_2(x))\, dx =$$

$$= \int_{\partial\Omega} \mu\, n_1 (v_1(x) - u_1(x))\, dx + \int_{\partial\Omega} \mu\, n_2 (v_2(x) - u_2(x))\, dx + \quad (4.11)$$

$$- \int_\Omega \mu\left(\frac{\partial v_1}{\partial x_1} + \frac{\partial v_2}{\partial x_2} - \frac{\partial u_1}{\partial x_1} - \frac{\partial u_2}{\partial x_2}\right) dx = 0.$$

The integral (4.11) is zero because v_i and u_i, $i = 1, 2$, have the same trace on $\partial\Omega$ and $u, v \in \mathbb{K}$ have the same divergence.

Conversely, let us prove that if $u \in \mathbb{K}$ is a solution to (4.10), then u fulfills Definition 3.

To this end, let us consider the function

$$\psi(v) = \int_\Omega c(x, u(x))(v(x) - u(x))\, dx \quad v \in \mathbb{K} \tag{4.12}$$

where $u \in \mathbb{K}$ is the solution to the Variational Inequality (4.10) and, as in Section n. 2, let us consider the following results.

Lemma 7 *The problem*

$$\min_{v\in\mathbb{K}} \psi(v) \tag{4.13}$$

is equivalent to the problem

$$\min_{v\in H^1_\varphi(\Omega,\mathbb{R}^2)}\ \sup_{(\mu,\lambda_1,\lambda_2)\in C^*}\ \left\{\psi(v)+\int_\Omega \mu(x)\Big(\frac{\partial v_1}{\partial x_1}+\frac{\partial v_2}{\partial x_2}+t(x)\Big)\,dx+\right.$$

$$\left.-\int_\Omega \lambda_1(x)v_1(x)\,dx-\lambda_2(x)v_2(x)\,dx\right\} \tag{4.14}$$

where

$$C^* = \{(\mu,\lambda_1,\lambda_2):\mu\in H^1(\Omega),\ \lambda_1(x),\ \lambda_2(x)\in L^2(\Omega),$$

$$\lambda_1(x)\geq 0,\ \lambda_2(x)\geq 0\ a.e.\ in\ \Omega\} \tag{4.15}$$

and

$$H^1_\varphi(\Omega,\mathbb{R}^2)=\{u\in H^1(\Omega,\mathbb{R}^2):u_1(x)|_{\partial\Omega}=\varphi_1(x),\ u_2(x)|_{\partial\Omega}=\varphi_2(x)\}.$$

Let us consider the dual problems

$$\max_{(\mu,\lambda_1,\lambda_2)\in C^*}\ \inf_{v\in H^1_\varphi(\Omega,\mathbb{R}^2)}\ \left[\psi(v)+\int_\Omega \mu(x)\Big(\frac{\partial v_1}{\partial x_1}+\frac{\partial v_2}{\partial x_2}+t(x)\Big)\,dx+\right.$$

$$\left.-\int_\Omega \lambda_1(x)v_1(x)\,dx-\int_\Omega \lambda_2(x)v_2(x)\,dx\right] \tag{4.16}$$

$$\max_{\Lambda\in\Delta}\Lambda \tag{4.17}$$

where

$$\Delta=\left\{\Lambda\in\mathbb{R}:\psi(v)+\int_\Omega \mu(x)\Big(\frac{\partial v_1}{\partial x_1}+\frac{\partial v_2}{\partial x_2}+t(x)\Big)\,dx+\right.$$

$$\left.-\int_\Omega \lambda_1(x)v_1(x)\,dx-\int_\Omega \lambda_2(x)v_2(x)\,dx\geq\Lambda\right. \tag{4.18}$$

$$\left.\forall v\in H^1_\varphi(\Omega),\ \forall\mu\in C\right\}.$$

We have the following results.

Lemma 8 $(\bar\mu,\bar\lambda_1,\bar\lambda_2)\in C$ *is a maximal solution to the dual problem (4.16) if and only if*

$$\bar\Lambda=\max_{(\mu,\lambda_1,\lambda_2)\in C^*}\ \inf_{v\in H^1_\varphi(\Omega,\mathbb{R}^2)}\ \left[\psi(v)+\int_\Omega \bar\mu(x)\Big(\frac{\partial v_1}{\partial x_1}+\frac{\partial v_2}{\partial x_2}+t(x)\Big)\,dx+\right.$$

$$\left.-\int_\Omega \bar\lambda_1(x)v_1(x)\,dx-\int_\Omega \bar\lambda_2(x)v_2(x)\,dx\right]$$

is a solution to (4.17)–(4.18).

Lemma 9 *If the primal problem (4.13) (or (4.14)) is solvable, then the dual problem (4.16) is also solvable and the extremal values of the two problems are equal.*

Now let us consider the Lagrangean function

$$L : H^1_\varphi(\Omega, \mathbb{R}^2) \times C^* \Rightarrow \mathbb{R}$$

defined by setting

$$L(v, \mu, \lambda_1, \lambda_2) = \psi(v) + \int_\Omega \mu(x) \left(\frac{\partial v_1}{\partial x_1} + \frac{\partial v_2}{\partial x_2} + t(x) \right) dx - \\ - \int_\Omega \lambda_1(x) v_1(x) \, dx - \int_\Omega \lambda_2(x) v_2(x) \, dx. \tag{4.19}$$

The following result holds true.

Theorem 5 *A point $(u, \overline{\mu}, \overline{\lambda}_1, \overline{\lambda}_2) \in H^1_\varphi(\Omega, \mathbb{R}^2) \times C$ is a saddle point of the Lagrangean function L if and only if u is a solution of the primal problem, $(\overline{\mu}, \overline{\lambda}_1, \overline{\lambda}_2)$ is a solution of the dual problem (4.16) and the extremal values of the two problems are equal.*

From Theorem 5 it follows that

$$\overline{\lambda}_1(x) \, u_1(x) = 0 \text{ a. e. in } \Omega$$
$$\overline{\lambda}_2(x) \, u_2(x) = 0 \text{ a. e. in } \Omega \tag{4.20}$$

and, taking into account that

$$\min_{u \in H^1_\varphi(\Omega, \mathbb{R}^2)} L(v, \overline{\mu}, \overline{\lambda}_1, \overline{\lambda}_2) = L(u, \overline{\mu}, \overline{\lambda}_1, \overline{\lambda}_2) \tag{4.21}$$

and

$$L(v, \overline{\mu}, \overline{\lambda}_1, \overline{\lambda}_2) = \psi(v) - \int_\Omega \left(\frac{\partial \overline{\mu}}{\partial x_1} v_1 + \frac{\partial \overline{\mu}}{\partial x_2} v_2 \right) dx + \\ + \int_\Omega t(x) \, \overline{\mu}(x) \, dx + \int_{\partial\Omega} \overline{\mu}(x)(\varphi_1(x) \, n_1 + \varphi_2(x) \, n_2) \, d\sigma - \\ - \int_\Omega \overline{\lambda}_1(x) v_1(x) \, dx - \int_\Omega \overline{\lambda}_2(x) v_2(x) \, dx, \tag{4.22}$$

we get:

$$c_1(x, u(x)) - \frac{\partial \overline{\mu}}{\partial x_1} - \overline{\lambda}_1(x) = 0 \text{ a. e. in } \Omega \tag{4.23}$$

$$c_2(x, u(x)) - \frac{\partial \overline{\mu}}{\partial x_2} - \overline{\lambda}_2(x) = 0 \text{ a. e. in } \Omega. \tag{4.24}$$

From (4.23) and (4.24), by virtue of (4.20), we get:

$$\left(c_i(x, u(x)) - \frac{\partial \overline{\mu}}{\partial x_i} \right) u_i(x) = 0 \quad i = 1, 2 \text{ a. e. in } \Omega$$

and hence the assertion follows.

Remark 7. In [8] the authors consider another kind of flow conservation law and, consequently, a different type of convexity. They suppose that Ω is a rectangular grid ($]0, a[\times]0, b[$) of vertexes $A \equiv (0, 0)$, $B \equiv (a, 0)$, $C \equiv (a, b)$, $D \equiv (0, b)$ and that two non negative functions $\varphi_1(x_2)$, $x_2 \in (0, b)$, $\varphi_2(x_1)$, $x_1 \in (0, a)$ are given. They represent the flows entering through the boundary AD and AB and the flow conservation law is written in the form

$$u_1(x) + u_2(x) = \varphi_1(x_2) + \varphi_2(x_1) \text{ a. e. in } \Omega. \tag{4.25}$$

The motivation of (4.25) derives from the discrete case: since in the continuum case each point $x \in \Omega$ can be considered at the same time as an origin and destination point, the sum of the flows connecting this O/D (Origin/Destination) pair must be equal to the demand at that point.

Remark 8. As in [12] instead of the space $H^1(\Omega)$, we can consider the space

$$L^2_{\text{div}}(\Omega) = \{ u \in L^2(\Omega, \mathbb{R}^2) : \text{ div } u \in L^2(\Omega) \}.$$

This is a well–known space used in fluid mechanics (see [14]).

Remark 9. In [12], [15], [16] the "system optimization approach" is considered and, from different points of view, a lot of results are given. It is worth mentioning that the special nature of the convex $\mathbb{K}$ presents further difficulties in order to ensure the existence of the equilibria.

References

[1] A. Auslender, *Noncoercive Optimization Problems*, Mathematics of Operation Research (1996).

[2] C. Baiocchi, G. Buttazzo, F. Gastaldi, F. Tomarelli, *General existence theorems for unilateral problems in continuum mechanics*, Arch. Rational Mech. Anal. 100 (1988), 149–189.

[3] C. Baiocchi, A. Capelo, *Variational and quasivariational inequalities: applications to free boundary problems*, J. Wiley and Sons, Chichester (1984).

[4] V. Barbu, *Optimal control of Variational Inequalities*, Research Notes in Mathematics 100 (1984).

[5] H. Brezis, *Problème Unilatéraux*, J. Math. Pures et Appl. 51 (1972), 1–168.

[6] S. Dafermos, *Continuum Modelling of Transportation Networks*, Transportation Res. 14 B (1980), 295–301.

[7] P. Daniele, *Lagrangean Function for Dynamic Variational Inequalities*, Rendiconti del Circolo Matematico di Palermo, Serie II, Suppl. 58 (1999), 101–119.

[8] P. Daniele – A. Maugeri, *Vector Variational Inequalities and a Continuum Modelling of traffic Equilibrium Problem*, In *Vector Variational Inequalities and Vector Equilibria*, F. Giannessi Ed., Kluwer Academic Publishers (2000), 97–111.

[9] G. Fichera, *Problemi elastostatici con vincoli unilaterali: il problema di Signorini con ambigue condizioni al contorno*, Atti Accad. Naz. Lincei Mem. Sez. I (8) 7 (1964), 71–140.

[10] G. Fichera, *Boundary value problems in elasticity with unilateral constraints*, Handbuch der Physik, IV a/2, Springer–Verlag, Berlin Heidelberg New York (1972), 347–389.

[11] G. Fichera, *Problemi unilaterali nella statica dei sistemi continui, problemi attuali di meccanica teorica e applicata*, Atti del Convegno Internazionale a ricordo di Modesto Ponetti, Torino (1977), 171–178.

[12] J. Gwinner, *On continuum Modelling of Large Dense Networks in Urban Road Traffic*, In *Mathematics in Transport Planning and Control* (J.D. Griffiths ed.), IMA Conference, Cardiff (1988).

[13] J.L. Lions, G. Stampacchia, *Variational Inequalities*, Comm. Pure Appl. Math. 20 (1967), 493–519.

[14] P.L. Lions, *Mathematical Topics in Fluid Mechanics*, Clarendon Press., Oxford, 1996.

[15] A. Maugeri, *New Classes of Variational Inequalities and Applications to Equilibrium Problems*, Methods of Operation Research 53 (1985), 129–131.

[16] A. Maugeri, *New classes of Variational Inequalities and Applications to Equilibrium Problems*, Rendiconti Accademia Nazionale delle Scienze detta dei XL 11 (1987), 277–284.

[17] A. Nagurney, *Network Economics - A Variational Inequality Approach*, Kluwer Academic Publishers, 1993.

[18] B.D. Reddy, F. Tomarelli, *The obstacle problem for an elastoplastic body*, Appl. Math. Optim. 21 (1990), 89–110.

[19] A. Signorini, *Questioni di elasticità non linearizzata e semilinearizzata*, Rend. Mat. 18 (1959), 95–139.

AXIOMATIZATION FOR APPROXIMATE SOLUTIONS IN OPTIMIZATION *

Henk Norde

Department of Econometrics and CentER,
Tilburg University, P.O. Box 90153,
5000 LE Tilburg, The Netherlands

Fioravante Patrone

Department of Mathematics,
Via Dodecaneso 35, University of Genoa,
16146 Genoa, Italy

Abstract   Approximate solutions to optimization problems are characterized by means of properties like consistency, non-emptiness, behaviour w.r.t. inclusion, invariance w.r.t. translation, multiplication.

Keywords: Approximate optimization, consistency, invariance properties, maximizing sequences, vector optimization.

1. Introduction.

This paper is a follow up of "Characterizing Properties of Approximate Solutions for Optimization Problems" [8], in which an "axiomatic" approach has been adopted to approximate solutions in optimization. This approach is quite common in social choice theory and game theory: to quote just the most outstanding examples, the "impossibility theorem" by Arrow ([1]), the Nash bargaining solution [7] and the Shap-

*The authors wish to thank A. Rustichini, P. Wakker and T. Zolezzi for helpful suggestions on earlier versions of this paper. The financial support of CNR-Italy, the GNAFA group of CNR-Italy and Tilburg University is gratefully acknowledged.

F. Giannessi et al (eds.),
Equilibrium Problems: Nonsmooth Optimization and Variational Inequality Models, 207–221.
© 2001 *Kluwer Academic Publishers.*

ley value [15] are all based on an axiomatic approach. That is: given a class of problems, one states desirable properties that the "solution" should satisfy and, then, one tries to identify which consequences can be derived. It may be well that the required properties are mutually incompatible (a well known example is the impossibility theorem by Arrow), or they can identify a particular interesting solution, that maybe can be described through some formula (as in the cases of the Nash bargaining solution and of the Shapley value). Clearly, in this case the formula gets a more sound foundation, and so it can be increased the willingliness to accept it as a good solution for the class of problems under scrutiny.

What is novel in [8] and in this paper is the exploitation of such an approach for the characterization of approximate solutions for optimization problems. We refer to the body of the paper for a thorough discussion of the axioms involved. Here we shall briefly discuss just a couple of properties (i.e.: "axioms"), to give a feeling of the approach.

One trivial property to be considered is "non-emptiness" (NEM). It is natural to require that the set of approximate solutions to an optimization problem is non-empty. Notice, however, that this is quite far from being an innocent assumption.

Two samples of the difficulties that can arise are the following:

- requiring NEM can have the effect that one rules out the possibility of considering as a (special) approximate solution the choice of exact maxima, if the class of problems under consideration contains problems without maximum

- referring to a field different from optimization, notice that the existence of ε-Nash equilibria, which are a natural candidate for "approximate" Nash equilibria, is not guaranteed in general (in zero-sum games, the existence of ε-Nash equilibria for every $\varepsilon > 0$ is equivalent to the existence of the value: see [16]

Another property that is interesting to analyze is the following. Consider two objective functions to be maximized, f_1 and f_2 (defined on the same set), with $\sup f_1 \leq \sup f_2$. A natural assumption is that in maximizing f_2 we allow for a "tolerance" which is not less than the tolerance allowed for f_1. This is another type of "axiom" that we shall consider. We shall name it as CCA, since it is a direct reminiscent of the Chernoff's Choice Axiom. It has also a close relationship with the well known and much debated Independence of Irrelevant Alternative axiom, which is an essential ingredient both in the "impossibility theorem" by Arrow and in the characterization of the Nash bargaining solution, that were quoted above.

The paper is organized as follows: in section 2 the setting is introduced, together with the main "axiom" which incorporates the "approximation consistency" condition that we shall stress in all of the paper: it will be seen which kind of simplification is provided by such assumption. In the remaining sections, the set of requirements on the approximate solutions will be enriched: in section 3 will be described the main structural axioms (translation and multiplication invariance, CCA). In section 4 the focus will be on the CCA axiom. This section contains results which were already proved in [8], in a slightly more restricted setting. In section 5 one of these results is extended to the context of vector optimization problems. In section 6, we shall consider still the axiomatic approach to approximate solutions, but now the idea will be that of associating to an optimization problem a set of sequences (that should play the role of "maximizing" sequences). This approach is especially relevant if one is interested in Tikhonov well-posedness (see, e.g., [3]). We shall investigate which is the effect of requiring that properties like translation and multiplication invariance have on the set of admissible "approximating" sequences. These invariance requirements are to be considered crucial when the objective functions are von Neumann - Morgenstern utility functions (whenever standard assumptions of decision making under risk are assumed, the functions u and $\alpha u + \beta$ represent the same preferences of the decision maker, provided of course that $\alpha > 0$). The "positive" result that we shall get (we characterize the usual maximizing sequences by means of quite natural axioms), should be contrasted with the "negative" results described in section 4, where we impose translation and multiplication invariance to sets of approximate solutions instead of sets of "approximating" sequences.

We emphasize the novelty of this approach to the subject, and that in our opinion there are many open questions that deserve attention. Particularly promising is the extension of this kind of considerations to vector optimization problems: what is given here is just a first step in that direction. The wide number of variants considered in vector optimization are likely to reflect also into this approach. The results that we get for sequences say that in the setting of von Neumann - Morgenstern utility theory results that guarantee the existence of ε-Nash equilibria for every $\varepsilon > 0$ (e.g.: [16]) do have a meaning. The same can be said, in optimization, about Tikhonov well-posedness ([3]; see [10] for remarks about the invariance of this property). We want to add, however, some warning about sequences. First, for practical reasons one has eventually to give one solution: so, even if sequences can be considered interesting for theoretical reasons, they do not solve the problem of finding an approximate solution for an optimization problem. Secondly, sometimes one is interested in solving (approximately) problems which are an approximate version of the "true" one: in this case, it may be important to consider asymptotically minimizing sequences (in the terminology of [3]). Considering these "solutions" falls outside the scope of our present investigation. Actually, we have completely skipped any reference to "continuity" properties of our solution maps. In our opinion, this is also a topic that deserves to be thoroughly studied.

Notation Throughout this paper we denote the set $\mathbb{R} \cup \{+\infty\}$ by $\mathbb{R}^*$ and the set $\mathbb{R} \cup \{-\infty, +\infty\}$ by $\overline{\mathbb{R}}$.

2. Optimization problems.

An *optimization problem* is a pair (A, u) where A is a non-empty set of alternatives and u is a real-valued function with domain A. Let $\mathcal{P}$ be a non-empty collection of optimization problems. A *solution* β on $\mathcal{P}$ is a map which assigns to every optimization problem $(A, u) \in \mathcal{P}$ a subset of A.

Example 1.1 For the following examples no special restriction is imposed upon $\mathcal{P}$.

a) The solution β_{tot} is defined by

$$\beta_{\text{tot}}(A, u) := A.$$

b) The solution $\beta_{\max}$ is defined by

$$\beta_{\max}(A, u) := \{a \in A : u(a) \geq u(a') \text{ for every } a' \in A\}.$$

c) For $\varepsilon > 0$ the solution β_ε is defined by

$$\beta_\varepsilon(A, u) := \{a \in A : u(a) \geq u(a') - \varepsilon \text{ for every } a' \in A\}.$$

d) For $k \in \mathbb{R}$ the solution β^k is defined by

$$\beta^k(A, u) := \{a \in A : u(a) \geq k\}.$$

e) For $\varepsilon > 0, k \in \mathbb{R}$ the solution $\beta_{\varepsilon,k}$ is defined by

$$\beta_{\varepsilon,k}(A, u) := \begin{cases} \beta_{\max}(A, u) & \text{if } \beta_{\max}(A, u) \neq \emptyset \\ \beta_\varepsilon(A, u) & \text{if } \beta_{\max}(A, u) = \emptyset \text{ and} \\ & \beta_\varepsilon(A, u) \neq \emptyset \\ \beta^k(A, u) & \text{otherwise} \end{cases}$$

Notice that $\beta_{\max}(A, u)$, $\beta_\varepsilon(A, u)$ and $\beta^k(A, u)$ can be empty; on the contrary, $\beta_{\text{tot}}(A, u) \neq \emptyset$ and $\beta_{\varepsilon,k}(A, u) \neq \emptyset$ for every $(A, u) \in \mathcal{P}$.

Two optimization problems (A, u) and (B, v) are *sup-equivalent* if

$$\sup_{x \in A} u(x) = \sup_{x \in B} v(x).$$

A solution β on $\mathcal{P}$ is *approximation consistent* if for every pair of sup-equivalent problems $(A, u), (B, v) \in \mathcal{P}$ the following statement is true:

if $b \in \beta(B, v)$ and $a \in A$ is such that $u(a) \geq v(b)$ then $a \in \beta(A, u)$.

So, if a solution β is approximation consistent, selection by β of an alternative $b \in B$ in some problem $(B, v) \in \mathcal{P}$, induces selection by β of all 'non-worse' alternatives in sup-equivalent problems. Note that many numerical methods for optimization problems "stop" when some level of accuracy is reached. Since real numbers are stored by computers using floating point representation, this level of accuracy strongly depends on the supremum of the optimization problem under consideration. Therefore, such methods can be seen as examples of approximation consistent solutions. Clearly, the solutions a)-d) in example 1.1 are approximation consistent. For approximation consistent solutions we have the following proposition, the proof of which is provided in [8]:

Proposition 1.1 *Let β be an approximation consistent solution on $\mathcal{P}$ and let (A, u), $(B, v) \in \mathcal{P}$ be such that $u(A) = v(B)$. Then there is a subset T of $u(A)(= v(B))$ such that $\beta(A, u) = u^{-1}(T)$ and $\beta(B, v) = v^{-1}(T)$.*

¿From this proposition it follows that, under approximation consistency, the set of approximate solutions only depends on the range $u(A)$ of u. This is clearly a restrictive condition (notice, however, that it is a very common one), that provides also a strong simplification of the setting, since we can identify from now on an optimization problem (A, u) with $u(A)$, the range of u, which is a subset of $\mathbb{R}$.

3. Axioms.

Let $\mathcal{S}$ be a non-empty collection of non-empty subsets of $\mathbb{R}$. A *solution σ on $\mathcal{S}$* is a map which assigns to every $S \in \mathcal{S}$ a subset $\sigma(S)$ of S.

A solution σ on $\mathcal{S}$ satisfies (AC) (*approximation consistency*) if for every $S_1, S_2 \in \mathcal{S}$ with $\sup S_1 = \sup S_2$ the following statement is true:

if $s_2 \in \sigma(S_2)$ and $s_1 \in S_1$ is such that $s_1 \geq s_2$ then $s_1 \in \sigma(S_1)$.

Proposition 2.1 gives reasons for the introduction of (AC): if a solution β on $\mathcal{P}$ is approximation consistent, as defined in the previous section, then the induced solution σ on the family S of ranges $u(A)$ $((A, u) \in \mathcal{P})$, satisfies (AC). Conversely, given a solution σ on S that satisfies (AC), this induces an approximation consistent solution β for every $\mathcal{P}$ with ranges in S.

The (AC) condition is not entirely satisfactory: one expects that $\sigma(S)$ can be described as $\{s \in S : s \geq \gamma\}$ or $\{s \in S : s > \gamma\}$ for some γ (depending on S). However, we can have strict or weak inequality, depending on the value of $\sup S$, as can be seen in the next example.

Example 1.2 Let S be the collection of all non-empty subsets of $\mathbb{R}$. The solution σ_{mix} on S, defined by

$$\sigma_{\mathrm{mix}}(S) := \begin{cases} \{s \in S : s \geq \sup S - 1\} & \text{if } \sup S \leq 0 \\ \{s \in S : s > \sup S - 1\} & \text{if } \sup S \in (0, +\infty) \\ \{s \in S : s \geq 22\} & \text{if } \sup S = +\infty \end{cases},$$

satisfies (AC).

One could think that adding the requirement that $\sigma(S)$ is a closed subset of S eliminates this kind of bad behaviour for the approximate solutions. Notice, however, that this addition does not "force" the parentheses to be closed, as shown by the following example.

Example 1.3 Consider $S = \{\{0, 1\}\} \cup \{[\alpha, 1] : \alpha \in (0, 1)\}$ and let σ be the solution on S, defined by $\sigma(\{0, 1\}) := \{1\}$ and $\sigma([\alpha, 1]) := [\alpha, 1]$ for every $\alpha \in (0, 1)$. The solution σ satisfies (AC) and $\sigma(S)$ is a closed subset of S for every $S \in S$, but there is no $\gamma \in \mathbb{R}$ such that $\sigma(S) = \{s \in S : s \geq \gamma\}$ for every $S \in S$.

So, we shall introduce the following axiom, an appropriate strenghtening of (AC).

A solution σ on S satisfies (SAC) (*strong approximation consistency*) if for every $S, S_1, S_2, \ldots \in S$ with $\sup S = \sup S_i$ for every $i \in \mathbb{N}$ the following statement is true:

if $s_i \in \sigma(S_i)$ for every $i \in \mathbb{N}$ and $s \in S$ is such that $s \geq \liminf_{i \to \infty} s_i$ then $s \in \sigma(S)$.

One easily verifies that (SAC) induces (AC). Moreover, (SAC) implies that $\sigma(S)$ is a closed subset of S for every $S \in S$. In fact, if S is a collection of intervals, then σ satisfies (SAC) if and only if σ satisfies (AC) and $\sigma(S)$ is a closed subset of S for every $S \in S$.

Notice that there is the possibility of following a different route: stick to (AC), but assume that the class $\mathcal{S}$ is a class of subsets which is *rich enough*. This approach is pursued in [8].

In the sequel we also make use of the following axioms.

A solution σ on $\mathcal{S}$ satisfies (NEM) (*non-emptiness*) if for every $S \in \mathcal{S}$ we have

$$\sigma(S) \neq \emptyset.$$

The collection $\mathcal{S}$ is *closed under translation* (CL+) if for every $S \in \mathcal{S}$ and $t \in \mathbb{R}$ we have $t + S := \{t + s : s \in S\} \in \mathcal{S}$. A solution σ on $\mathcal{S}$, obeying (CL+), satisfies (TI) (*translation invariance*) if for every $S \in \mathcal{S}$ and $t \in \mathbb{R}$ we have

$$\sigma(t + S) = t + \sigma(S).$$

The collection $\mathcal{S}$ is *closed under multiplication* (CL*) if for every $S \in \mathcal{S}$ and $\lambda > 0$ we have $\lambda S := \{\lambda s : s \in S\} \in \mathcal{S}$. A solution σ on $\mathcal{S}$, obeying (CL*), satisfies (MI) (*multiplication invariance*) if for every $S \in \mathcal{S}$ and $\lambda > 0$ we have

$$\sigma(\lambda S) = \lambda \sigma(S).$$

A solution σ on $\mathcal{S}$ satisfies (CCA) (*Chernoff's choice axiom*) if for every $S, T \in \mathcal{S}$ with $S \subseteq T$ one has

$$\sigma(T) \cap S \subseteq \sigma(S).$$

So, if σ satisfies (CCA), selection by σ of an element $s \in T$, implies selection by σ of s in any subset S of T with $s \in S$. This condition is a rephrasing of the Choice Axiom used by [2]. It turns out to be a weaker condition than similar assumptions used in the context of bargaining theory by [5] and [13], under the name Independence of Irrelevant Alternatives.

One can construct examples of solutions satisfying different groups of these axioms. A fairly large collection of such samples is provided in [8]: here we shall provide only one solution that will turn out to be of particular interest in relationship with the (CCA).

Example 1.4 Assume that $\mathcal{S}$ is the collection of all non-empty subsets of $\mathbb{R}$.

The solution $\hat{\sigma}_{\varepsilon,k}$ (where $\varepsilon > 0, k \in \mathbb{R}$), defined by

$$\hat{\sigma}_{\varepsilon,k}(S) := \begin{cases} \sigma_\varepsilon(S) & \text{if } \sup S \leq k + \varepsilon \\ \sigma^k(S) & \text{if } \sup S > k + \varepsilon \end{cases},$$

satisfies (SAC), (NEM) and (CCA). Notice that $\hat{\sigma}_{\varepsilon,k}(S) = \sigma_\varepsilon(S) \cup \sigma^k(S)$, where $\sigma_\varepsilon(S)$ and $\sigma^k(S)$ are derived in a straightforward way from the β_ε and β^k introduced in Example 2.1.

4. Characterizations of solutions.

Let S be a collection of non-empty subsets of $\mathbb{R}$. We write $S = \cup_{k \in \mathbb{R}^*} S_k$ where $S_k := \{S \in S : \sup S = k\}$ for every $k \in \mathbb{R}^*$. For a function $a : \mathbb{R}^* \to \overline{\mathbb{R}}$ we define the solution σ_a on S by

$$\sigma_a(S) := \{s \in S : s \geq a(\sup(S))\}.$$

So, σ_a selects, for every $S \in S_k$, the elements $s \in S$ with $s \geq a(k)$. Clearly, σ_a satisfies (SAC). The following proposition (an adaptation of a proposition in [8] to this context) shows that the solutions σ_a are completely characterized by (SAC).

Proposition 4.2 *Let S be a collection of non-empty subsets of $\mathbb{R}$ and let σ be a solution on S. The solution σ satisfies (SAC) if and only if $\sigma = \sigma_a$ for some function a.*

Proof We only prove the only-if-part. So, assume that σ satisfies (SAC). First we define the function $a : \mathbb{R}^* \to \overline{\mathbb{R}}$. Let $k \in \mathbb{R}^*$. Define

$$a(k) := \begin{cases} \text{arbitrarily} & \text{if } S_k = \emptyset \\ \inf(\cup_{S \in S_k} \sigma(S)) & \text{otherwise} \end{cases} \tag{4.1}$$

(with the convention that $\inf \emptyset = +\infty$). Now we have to prove that $\sigma = \sigma_a$, i.e. we have to prove that $\sigma(S) = \sigma_a(S)$ for every $S \in S$. So, let $S \in S$ and let $k := \sup S$ (which trivially induces that $S_k \neq \emptyset$). For every $s \in \sigma(S)$ we have $s \geq a(k)$ by definition of $a(k)$. Therefore $\sigma(S) \subseteq \sigma_a(S)$. Note that the converse inclusion $\sigma_a(S) \subseteq \sigma(S)$ is trivial when $\sigma_a(S) = \emptyset$. So, assume $\sigma_a(S) \neq \emptyset$ and let $s \in \sigma_a(S)$. Then $s \geq a(k)$ which implies $a(k) \neq +\infty$. Therefore, $\cup_{S \in S_k} \sigma(S) \neq \emptyset$, and hence, by definition of $a(k)$, there is a sequence $S_1, S_2, \ldots \in S_k$ and, for every $i \in \mathbb{N}$, an $s_i \in \sigma(S_i)$ such that $a(k) = \lim_{i \to \infty} s_i$. By (SAC) we get $s \in \sigma(S)$. Therefore $\sigma_a(S) \subseteq \sigma(S)$, which finishes the proof. ■

If we impose some feasibility condition upon the function a we get solutions which are characterized by (SAC) and (NEM).

Proposition 4.3 *Let S be a collection of non-empty subsets of $\mathbb{R}$ and let σ be a solution on S. The solution σ satisfies (SAC) and (NEM) if and only if $\sigma = \sigma_a$ for some function $a : \mathbb{R}^* \to \overline{\mathbb{R}}$ satisfying $a(k) \leq k$ for every $k \in \mathbb{R}^*$, with strict inequality for every $k \in \mathbb{R}^*$ for which there is an $S \in S_k$ with $k \notin S$.*

We omit the proof, which is similar to the proof of the corresponding proposition in [8]. The main result which takes into account (TI) and (MI) is the following "impossibility" result:

Proposition 4.4 *Let S be a collection of non-empty subsets of $I\!R$, which satisfies (CL+) and (CL*). Suppose, moreover, that there is at least one upper bounded $S \in S$ which has no maximum. Let σ be a solution on S. The solution σ satisfies (SAC), (NEM), (TI) and (MI) if and only if $\sigma = \sigma_a$, where $a : I\!R^* \to \overline{I\!R}$ is defined by $a(k) := -\infty$ for every $k \in I\!R^*$ (i.e. $\sigma_a = \sigma_{tot}$).*

Proof Again we only prove the only-if-part. Suppose σ satisfies (SAC), (NEM), (TI) and (MI). By (SAC) we may conclude that $\sigma = \sigma_a$ where $a : I\!R^* \to \overline{I\!R}$ is defined by (4.1). Let $S \in S$ be un upper bounded set without maximum and let $k := \sup S$. Since σ satisfies (NEM) there is an $s \in \sigma(S)$. Clearly $s < k$. For every $n \in I\!N$ we have, by (CL+) and (TI),

$$-(1 - n^{-1})k + S \in S$$

and

$$-(1 - n^{-1})k + s \in \sigma(-(1 - n^{-1})k + S).$$

Moreover, by (CL*) and (MI), we get

$$-(n - 1)k + nS = n(-(1 - n^{-1})k + S) \in S$$

and

$$-(n - 1)k + ns = n(-(1 - n^{-1})k + s) \in \sigma(-(n - 1)k + nS).$$

Since $\sup(-(n - 1)k + nS) = k$ and $\lim_{n\to\infty}(-(n - 1)k + ns) = -\infty$ we get $a(k) = -\infty$. By (TI) we infer that $a(l) = -\infty$ for every $l \in I\!R$. If $S_{+\infty} \neq \emptyset$ a similar argument can be used in order to show that $a(+\infty) = -\infty$. ∎

We refer to [8] for variants of this negative result. In particular, if it is assumed that all of the sets in S has a maximum, then one gets that $\sigma = \sigma_{tot}$ or $\sigma = \sigma_{max}$. In [8] is also investigated the effect of the restrictions that (TI) or, separately, (MI) impose on the solution.

In order to get a characterization which takes (CCA) into account, we have to impose some condition on how the class S behaves under appropriate transformations, similar to the (CL+) and (CL*) introduced in the previous section. Since (CCA) deals with inclusions, an appropriate condition upon S would be either to be *closed under taking subsets* (CL⊂) or to be *closed under taking supersets* (CL⊃). The following example shows that a nice characterization with (CCA), is not possible if S satisfies (CL⊂).

Example 4.5 Consider the class S of all non-empty and upper bounded subsets S which satisfy the condition that there exists a $t \in [0, 1)$ such that $S \subseteq t + \mathbf{Z}$. Define the solution σ_a by

$$a(k) := \left\{ \begin{array}{ll} k - 22 & \text{if } k \in \mathbf{Z} \\ k - 37 & \text{otherwise} \end{array} \right. .$$

Clearly σ_a satisfies (SAC) and (NEM). It also satisfies (CCA): this is due to the fact that for $S, T \in S$ with $S \subseteq T$ both are contained in the same $t + \mathbf{Z}$ for some $t \in [0, 1)$. In fact, one can prove that any σ_a, with a feasible function a which is non-decreasing on $t + \mathbf{Z}$ for every $t \in [0, 1)$, satisfies (SAC), (NEM) and (CCA).

The reason for strange examples as above lies in the fact that the collection S is too poor: S can be partitioned into several subcollections such that sets belonging to different subcollections are not related by inclusion. These problems do not occur if S satisfies (CL⊃).

Proposition 4.5 *Let S be a collection of non-empty subsets of $\mathbb{R}$ which satisfies (CL⊃) and let σ be a solution on S. The solution σ satisfies (SAC), (NEM) and (CCA) if and only if $\sigma = \sigma_a$ for some non-decreasing function $a : \mathbb{R}^* \to \overline{\mathbb{R}}$ satisfying $a(k) \leq k$ for every $k \in \mathbb{R}^*$, with strict inequality for every $k \in \mathbb{R}^*$ for which there is an $S \in S_k$ with $k \notin S$.*

We shall omit the proof, which is a straightforward adaptation of the proof for a similar result given in [8] Of course, an example of such a solution is given when one considers $a(k) = k - \varepsilon$, for some given $\varepsilon > 0$. This corresponds to ε-maximizers (and leads to ε-Nash equilibria for strategic games: see e.g. [14] or [16]). An example of a solution, satisfying the requirements above, is given by $\hat{\sigma}_{\varepsilon, k}$, described in example 1.4. Such kind of approximate solutions were used in the context of strategic games by [4], [6] and [9]. We refer to [11] for an axiomatic approach (derived from [12]) that links approximate solutions in optimization to approximate solutions for strategic games.

The proposition above is not completely satisfactory, due to the fact that (CL⊃) is a very strong requirement on the class S. However, example 4.5 showed that it is not easy to get rid of it. Another approach is that one asks for some strengthening of (SAC), instead of looking for too special classes S.

A solution σ on S satisfies (SMAC) (*strong monotonic approximation consistency*) if for every $S, S_1, S_2, \ldots \in S$ with $\sup S \leq \sup S_i$ for every $i \in \mathbb{N}$ the following statement is true:

if $s_i \in \sigma(S_i)$ for every $i \in \mathbb{N}$ and $s \in S$ is such that $s \geq \liminf_{i \to \infty} s_i$ then $s \in \sigma(S)$.

It is obvious from the definitions that (SMAC) implies (SAC). Conversely, if $S_\infty = \emptyset$, we have that (SAC) and (TI) imply (SMAC). An interesting result in which (SMAC) plays an essential role is the following.

Proposition 4.6 *Let S be a collection of non-empty subsets of $\mathbb{R}$ and let σ be a solution on S. The solution σ satisfies (SMAC) and (NEM) if and only if $\sigma = \sigma_a$ for some non-decreasing function $a : \mathbb{R}^* \to \overline{\mathbb{R}}$ satisfying $a(k) \leq k$ for every $k \in \mathbb{R}^*$, with strict inequality for every $k \in \mathbb{R}^*$ for which there is an $S \in S_k$ with $k \notin S$.*

Proof First we prove the if-part: suppose $\sigma = \sigma_a$ for some a as mentioned above. By proposition 4.3 we infer that σ satisfies ((SAC) and) (NEM). Suppose $S, S_1, S_2, \ldots \in S$ with $\sup S \leq \sup S_i$ for every $i \in \mathbb{N}$ and let $s \in S$, $s_i \in \sigma(S_i)$ for every $i \in \mathbb{N}$, be such that $s \geq \liminf_{i \to \infty} s_i$. Since $s_i \geq a(\sup S_i) \geq a(\sup S)$ for every $i \in \mathbb{N}$, we get $s \geq a(\sup S)$. Hence $s \in \sigma(S)$. So, σ satisfies (SMAC).

For the proof of the only-if-part suppose that σ satisfies (SMAC) and (NEM). Since (SMAC) induces (SAC) we get $\sigma = \sigma_a$, where a is defined by (4.1). Notice however, that a needs not be non-decreasing. Now define $\bar{a} : \mathbb{R}^* \to \overline{\mathbb{R}}$ by

$$\bar{a}(k) = \inf\{a(l) : l \text{ is such that } S_l \neq \emptyset \text{ and } l \geq k\}$$

(with the usual convention that $\inf \emptyset = +\infty$). Clearly, $\bar{a}$ is non decreasing. We prove that $\sigma_a = \sigma_{\bar{a}}$. So, let $S \in S$ and $k := \sup S$ (which induces $S_k \neq \emptyset$). Since $\bar{a}(k) \leq a(k)$ we have $\sigma_a(S) \subseteq \sigma_{\bar{a}}(S)$. Now let $s \in \sigma_{\bar{a}}(S)$. By definition of $\bar{a}(k)$ there is a sequence $l_1, l_2, l_3, \ldots$ with $S_{l_i} \neq \emptyset$ and $l_i \geq k$ for every $i \in \mathbb{N}$ and $\bar{a}(k) = \lim_{i \to \infty} a(l_i)$. For every $i \in \mathbb{N}$ there is, by definition of $a(l_i)$, an $S_i \in S_{l_i}$ and $s_i \in \sigma(S_i)$ such that $\lim_{i \to \infty} a(l_i) = \lim_{i \to \infty} s_i$. Since $s \geq \bar{a}(k) = \lim_{i \to \infty} a(l_i) = \lim_{i \to \infty} s_i$ we get by (SMAC): $s \in \sigma(S) = \sigma_a(S)$. So, $\sigma_{\bar{a}}(S) \subseteq \sigma_a(S)$. ∎

5. Vector optimization.

In [8] it is shown that there exist no non-trivial approximate solutions satisfying (NEM), (TI), (MI), and (CCA), provided that the domain of problems is rich enough. In this section we will show that a similar result holds for vector optimization problems. Formally, a vector optimization problem is a pair (A, u), where $u : A \to \mathbb{R}^m$ is a vector-valued function. Once again we will identify the problem (A, u) with its range $u(A)$, which is a subset of $I\!R^m$. The following terminology will be needed.

A problem $T \subseteq I\!R^m$ is *bounded* if there exist an $a \in I\!R^m_{++}$ and a $b \in I\!R$ with $T \subseteq \{x \in I\!R^m : (x, a) < b\}$. Here $(\cdot, \cdot)$ denotes the standard inner product in $I\!R^m$.

A collection $\mathcal{T}$ of subsets of $I\!R^m$ is *complete* if for all $a \in I\!R^m_{++}$ and $b \in I\!R$ the open half-space $\{x \in I\!R^m : (x, a) < b\}$ is an element of $\mathcal{T}$.

The collection $\mathcal{T}$ is *closed under translation* (CL+) if for every $T \in \mathcal{T}$ and $t \in I\!R^m$ we have $t + T := \{t + s : s \in T\} \in \mathcal{T}$.

The collection $\mathcal{T}$ is *closed under multiplication* (CL*) if for every $T \in \mathcal{T}$ and $\lambda = (\lambda_1, \ldots, \lambda_m) \in I\!R^m_{++}$ we have $\lambda T := \{(\lambda_1 t_1, \ldots, \lambda_m t_m) : (t_1, \ldots, t_m) \in T\} \in \mathcal{T}$.

The axioms (NEM), (TI), (MI), and (CCA) for solutions on $\mathcal{T}$, obeying (CL+) and (CL*), are defined in an obvious way.

Theorem 1 *Let $\mathcal{T}$ be a complete collection of non-empty bounded subsets of $I\!R^m$, which satisfies (CL+) and (CL*). Let τ be a solution on $\mathcal{T}$. The solution τ satisfies (NEM), (TI), (MI) and (CCA) if and only if $\tau = \tau_{tot}$ (i.e. $\tau(T) = T$ for every $T \in \mathcal{T}$).*

Proof Again, we only prove the only-if-part. So, let τ be a solution satisfying (NEM), (TI), (MI), and (CCA). We have to prove that $\tau(T) = T$ for every $T \in \mathcal{T}$. First we prove this property for the open half-space $T = \{x \in I\!R^m : (x, a) < 0\}$, where $a \in I\!R^m_{++}$. By (NEM) we know that there is a $t^* \in \tau(T)$. Now let $t \in T$. In order to show that $t \in \tau(T)$ define $\lambda := \frac{(t,a)}{(t^*,a)} > 0$, $\bar{\lambda} := (\lambda, \ldots, \lambda) \in I\!R^m_{++}$ and $u := t - \lambda t^* \in I\!R^m$. Clearly, $\bar{\lambda} T = T$ and since $< u, a > = 0$ we also have $u + T = T$. By (TI) and (MI) we infer that $t = u + \lambda t^* \in \tau(u + \bar{\lambda} T) = \tau(T)$. So, $\tau(T) = T$. By (TI) we may conclude that $\tau(T) = T$ for every open half-space T. For a $T \in \mathcal{T}$ which is not necessary an open half-space there is an open half-space $T' \in \mathcal{T}$ with $T \subseteq T'$. Since $\tau(T') = T'$ and τ satisfies (CCA) we get $\tau(T) = T$. This finishes the proof.   ∎

6. Approximation with sequences.

It has been proved that there was no solution, besides the trivial σ_{tot}, satisfying (SAC), (NEM), (TI) and (MI), in case $\mathcal{S}$ contains at least one upper bounded set without maximum. In this section we will get a positive result by considering generalized solutions. In order to do so we need some definitions.

Let S be a non-empty collection of non-empty subsets of $\mathbb{R}$. A *generalized solution* Σ on S is a map which assigns to every $S \in S$ a subset $\Sigma(S)$ of S_{inc}^N, the collection of non-decreasing sequences in S. A sequence $\underline{s} = s_1, s_2, \ldots$ in S_{inc}^N can be interpreted as a sequence of approximate optimal elements, where the degree of approximation gets better when indices are increasing.

For generalized solutions we use the following axioms.

A solution Σ on S satisfies (AC) (*approximation consistency*) if for every $S_1, S_2 \in S$ with $\sup S_1 = \sup S_2$ the following statement is true:

if $\underline{s_2} \in \Sigma(S_2)$ and $\underline{s_1} \in S_1$ is such that $\lim \underline{s_1} \geq \lim \underline{s_2}$ then $\underline{s_1} \in \Sigma(S_1)$.

Notice that this condition can be seen as a direct generalization of condition (SAC) that was introduced in section 3 for the solutions σ.

The definitions of (NEM), (TI) and (MI) for generalized solutions are obvious.

Example 4.6 The following two generalized solutions, defined on a collection S obeying (CL+) and (CL*), all satisfy (AC), (NEM), (TI) and (MI).

    a) Define Σ_{opt} by $\Sigma_{opt}(S) := \{\underline{s} \in S_{\text{inc}}^N : \lim \underline{s} = \sup S\}$.

    b) Define Σ_{tot} by $\Sigma_{tot}(S) := S_{\text{inc}}^N$.

The following proposition shows that all generalized solutions, satisfying (AC), (NEM), (TI) and (MI), are mixtures of the solutions in example 4.6.

Proposition 4.7 *Let S be a collection of non-empty subsets of $\mathbb{R}$ and let Σ be a generalized solution on S. The solution Σ satisfies (AC), (NEM), (TI) and (MI) if and only if Σ coincides with one of the solutions Σ_{opt} or Σ_{tot} on the collection of upper bounded subsets in S and if Σ coincides with one of these two solutions (but not necessarily the same) on $S_{+\infty}$.*

Proof The if-part of the proof is left to the reader. For the only-if-part assume that Σ satisfies (AC), (NEM), (TI) and (MI). We only prove that Σ coincides with one of the solutions Σ_{opt} or Σ_{tot} on the collection of upper bounded subsets. The proof for $S_{+\infty}$ is similar. We distinguish between two cases.

Case 1: $\Sigma(S) \subseteq \Sigma_{opt}(S)$ for every upper bounded $S \in S$.

Let $S \in S$ be upper bounded. By (NEM), there is a sequence $\underline{s} \in \Sigma(S)$. By (AC), comparison of every sequence $\underline{s}' \in \Sigma_{opt}(S)$ with $\underline{s}$, implies $\Sigma_{opt}(S) \subseteq \Sigma(S)$. Therefore, $\Sigma(S) = \Sigma_{opt}(S)$.

Case 2: there is an upper bounded $S' \in \mathcal{S}$ with $\Sigma(S') \not\subseteq \Sigma_{opt}(S')$.
Let $S' \in \mathcal{S}$ and $\underline{s}' \in \Sigma(S')$ be such that $\underline{s}' \notin \Sigma_{opt}(S')$, let $k' := \sup S'$
and $k'' := \lim \underline{s}'$. Since $\underline{s}' \notin \Sigma_{opt}(S')$ we have $k'' < k'$. Moreover, we
have for every $n \in I\!\!N$ by (CL+) and (TI),

$$-(1 - n^{-1})k' + S' \in \mathcal{S}$$

and

$$-(1 - n^{-1})k' + \underline{s}' \in \Sigma(-(1 - n^{-1})k' + S').$$

Moreover, by (CL*) and (MI), we get

$$-(n - 1)k' + nS' = n(-(1 - n^{-1})k' + S') \in \mathcal{S}$$

and

$$-(n - 1)k' + n\underline{s}' = n(-(1 - n^{-1})k' + \underline{s}') \in \Sigma(-(n - 1)k' + nS').$$

Now let $\underline{s} \in \Sigma_{tot}(S')$ and let $l := \lim \underline{s}$. Choose $n \in I\!\!N$ such that
$-(n - 1)k' + nk'' < l$. By (AC), comparison of the sequences $\underline{s}$ and
$-(n - 1)k' + n\underline{s}'$ yields $\underline{s} \in \Sigma(S')$. Therefore, $\Sigma(S') = \Sigma_{tot}(S')$. By (TI)
one infers that $\Sigma(S) = \Sigma_{tot}(S)$ for every upper bounded $S \in \mathcal{S}$.   ∎

References

[1] Arrow K. J. (1951), Social Choice and Individual Values. Wiley, New York.

[2] Chernoff H. (1954), Rational selection of decision functions. Econometrica **22**: 422-443.

[3] Dontchev A. and Zolezzi T. (1993), Well-posed Optimization Problems. Lecture Notes in Mathematics, **1543**, Springer, Berlin.

[4] Jurg P. and Tijs S. (1993), On the determinateness of semi-infinite bimatrix games. Internat. J. Game Theory **21**: 361-369.

[5] Kaneko M. (1980), An extension of the Nash bargaining problem and the Nash social welfare function. Theory and Decision **12**, 135-148.

[6] Lucchetti R., Patrone F., Tijs S. (1986), Determinateness of two-person games. Bollettino U.M.I. **6**: 907-924.

[7] Nash, J. F. Jr. (1950), The Bargaining Problem. Econometrica **18**: 155-162.

[8] Norde H., Patrone F. and Tijs S. (1999), Characterizing Properties of Approximate Solutions to Optimization Problems, Mimeo. To appear in Mathematical Social Sciences.

[9] Norde H. and Potters J. (1997), On the determinateness of $m \times \infty$-bimatrix games. Mathematics of Operations Research **22**: 631-638.

[10] Patrone F. (1987), Well-Posedness as an Ordinal Property. Rivista di Matematica pura ed applicata **1**: 95-104.

[11] Patrone F., Pieri G., Tijs S. and Torre A. (1998), On Consistent Solutions for Strategic Games. Internat. J. Game Theory, **27**: 191-200.

[12] Peleg B. and Tijs S. (1996), The Consistency Principle for Games in Strategic Form. Internat. J. Game Theory **25**: 13-34.

[13] Peters H. (1992), Axiomatic bargaining game theory. Kluwer Academic Publishers, Dordrecht.

[14] Radner R. (1980), Collusive behavior in non-cooperative epsilon equilibria of oligopolies with long but finite lives. Journal of Economic Theory, **22**: 121-157.

[15] Shapley L. S. (1953), A Value for n-Person Games. In: Contributions to the Theory of Games II, (eds.: Kuhn H. W. and Tucker A. W.). Annals of Math. Studies, **28**, Princeton University Press, Princeton (NJ): 307-317.

[16] Tijs S. (1981), Nash equilibria for noncooperative n-person games in normal form. SIAM Review **23**: 225-237.

NECESSARY AND SUFFICIENT CONDITIONS OF WARDROP TYPE FOR VECTORIAL TRAFFIC EQUILIBRIA

Werner Oettli

Universität Mannheim

Fakultät für Mathematik und Informatik

D-68131 Mannheim, Germany

e–mail:oettli@math.uni-mannheim.de

Abstract   We consider analogues of Wardrop's condition for transportation equilibria in the case when the flows and costs are vector-valued. No scalarization is involved.

Keywords: Traffic Equilibrium Problem, Vectorial Variational Inequality, Vectorial Wardrop Equilibria.

1. Introduction.

Wardrop's condition is a necessary and sufficient condition for characterizing variational equilibrium flows in certain transportation networks. Due to its decomposed form it is better suited for many practical purposes than the original, though equivalent, equilibrium definition [4],[5]. In this note we consider conditions of Wardrop type in the case when the flows and costs are vector-valued. It turns out that the Wardrop condition now splits up into a necessary and a sufficient variant, and it is only under rather stringent assumptions that the two coincide.

In view of this ambiguity the question arises on how one should in this situation define a Wardrop equilibrium (in the scalar case it is simply a solution of the Wardrop condition). Many authors consider Wardrop equilibria as the notion of primary interest, the variational equilibrium

F. Giannessi et al (eds.),

Equilibrium Problems: Nonsmooth Optimization and Variational Inequality Models, 223–229.

© *2001 Kluwer Academic Publishers.*

formulation being only used for ad hoc purposes. But rather the variational equilibrium concept, which remains invariant if one passes to the vectorial setting, should be the basic notion, and the definition of a Wardrop equilibrium must be adapted in each case.

We shall not employ scalarization. For characterization of vectorial transport equilibria, based on scalarization, we refer the reader to [1].

2. The scalar case.

Let us start by describing the network geometry, and by recalling the relevant results for the scalar case, which serve as model for the vectorial case. We follow the presentation given in [2] or [3].

In the transport network one has a set $\mathcal{W}$ of origin-destination pairs, and a set $\mathcal{R}$ of routes. Each route $r \in \mathcal{R}$ links exactly one origin-destination pair $w \in \mathcal{W}$. The set of all $r \in \mathcal{R}$ which link a given $w \in \mathcal{W}$ is denoted by $\mathcal{R}(w)$. For each $r \in \mathcal{R}$, $F_r \in \mathbb{R}$ denotes the flow in route r. We combine the route-flows into a vector $F \in \mathbb{R}^{\mathcal{R}}$, and call F simply a flow. A *feasible* flow has to satisfy capacity restrictions $\lambda_r \leq F_r \leq \mu_r$ for all $r \in \mathcal{R}$, and demand requirements $\sum_{r \in \mathcal{R}(w)} F_r = \rho_w$ for all $w \in \mathcal{W}$, where $\lambda_r \leq \mu_r$ and $\rho_w \geq 0$ are given in $\mathbb{R}$. Thus the set of all feasible flows is given by

$$K := \{F \in \mathbb{R}^{\mathcal{R}} \mid \lambda_r \leq F_r \leq \mu_r \ \forall r \in \mathcal{R},$$

$$\sum_{r \in \mathcal{R}(w)} F_r = \rho_w \ \forall w \in \mathcal{W}\}, \tag{1}$$

supposed to be nonempty. Furthermore, for all $r \in \mathcal{R}$ we are given a cost function $C_r : K \to \mathbb{R}$. The number $C_r(F)$ gives the marginal cost of sending one additional unit of flow through route r, when the flow F is already present. We combine the functions C_r ($r \in \mathcal{R}$) into a mapping $C : K \to \mathbb{R}^{\mathcal{R}}$. The crucial *definition* is then the following: A flow $H \in \mathbb{R}^{\mathcal{R}}$ is called an *equilibrium flow* iff

$$H \in K, \text{and} \sum_{r \in \mathcal{R}} C_r(H) \cdot (F_r - H_r) \geq 0 \quad \forall F \in K. \tag{2}$$

The following characterization is the basis for what follows.

Theorem 1.1 *Let $H \in K$ and $\overline{C} \in \mathbb{R}^{\mathcal{R}}$ be arbitrary. Then the following are equivalent:*

(i) $\sum_{r \in \mathcal{R}} \overline{C}_r \cdot (F_r - H_r) \geq 0 \quad \forall F \in K;$ \hfill (3)

(ii) *for every* $w \in \mathcal{W}$ *and all* $q, s \in \mathcal{R}(w)$ *there holds:*
$$\overline{C}_q < \overline{C}_s \Rightarrow H_q = \mu_q \text{ or } H_s = \lambda_s; \tag{4}$$

(iii) *for every* $w \in \mathcal{W}$ *there exists* $\gamma_w \in \mathbb{R}$ *such that, for all* $r \in \mathcal{R}(w)$,
$$\overline{C}_r < \gamma_w \Rightarrow H_r = \mu_r, \quad \overline{C}_r > \gamma_w \Rightarrow H_r = \lambda_r. \tag{5}$$

A proof of Theorem 1 can be found in [2] or [3], but it is also a special case of Theorem 2 below.

Choosing in particular $\overline{C} := C(H)$, (i) becomes the requirement that $H \in K$ is an equilibrium flow; (ii) and (iii) are then equivalent forms of Wardrop's condition. Roughly speaking, (ii) or (iii) says that expensive routes are operated at their lower capacity, whereas cheap routes are operated at their upper capacity.

3. The vectorial case.

The geometry of the network remains the same as introduced above. In order to extend Theorem 1 to the vectorial case, we assume that we are given three topological vector spaces X, Y, Z. Then $X^{\mathcal{R}}$ is the space of flows, Z is the space of costs. The elements of Y are linear mappings from X into Z. The value of $C \in Y$, applied to $F \in X$, is denoted by $C[F] \in Z$.

In X, Y, Z there are given convex cones P_X, P_Y, P_Z respectively (recall that P is a cone iff $\lambda P \subseteq P$ for all real $\lambda \geq 0$). These cones are nonempty and proper. They induce a vectorial ordering on the underlying space. We assume that int $P_X \neq \emptyset$, and that P_Y is pointed, i.e., $P_Y \cap (-P_Y) = \{0\}$. We postulate the following:

$$\text{if } F \in P_X \text{ and } C \in P_Y, \text{ then } C[F] \in P_Z; \tag{6}$$

$$\text{if } F \in - \text{int } P_X \text{ and } C \in P_Y \setminus \{0\}, \text{ then } C[F] \notin P_Z. \tag{7}$$

Now the set of feasible flows is given by

$$K := \{F \in X^{\mathcal{R}} \mid F_r \in \lambda_r + P_X \quad \forall r \in \mathcal{R}, \quad F_r \in \mu_r - P_X$$

$$\forall r \in \mathcal{R}, \quad \sum_{r \in \mathcal{R}(w)} F_r = \rho_w \quad \forall w \in \mathcal{W}\}, \tag{8}$$

where λ_r, μ_r, ρ_w are given elements of X. Finally, for each $r \in \mathcal{R}$ we are given a cost-function $C_r(\cdot) : K \to Y$. Thus for every $H \in K$ and $F_r \in X$, $C_r(H)[F_r]$ is in Z. The essential definition is the following: A feasible flow $H \in K$ is an *equilibrium flow* iff

$$\sum_{r \in \mathcal{R}} C_r(H)[F_r - H_r] \in P_Z \quad \forall F \in K. \tag{9}$$

4. Results.

For arbitrary $\overline{C} = (\overline{C}_r)_{r \in \mathcal{R}} \in Y^{\mathcal{R}}$ we want to give conditions for $H \in K$ to be a solution of

$$\sum_{r \in \mathcal{R}} \overline{C}_r[F_r - H_r] \in P_Z \quad \forall F \in K. \tag{10}$$

Proposition 4.1 *The following condition is necessary for $H \in K$ being a solution of (10):*
for all $w \in W$, all $q, s \in \mathcal{R}(w)$, there holds

$$\overline{C}_q - \overline{C}_s \in -(P_Y \setminus \{0\}) \Rightarrow$$

$$\Rightarrow H_q - \mu_q \notin - \operatorname{int} P_X \text{ or } H_s - \lambda_s \notin \operatorname{int} P_X. \tag{11}$$

Proof. Assume the condition is not satisfied. So there exist $w \in W$ and $q, s \in \mathcal{R}(w)$ such that

$$\overline{C}_q - \overline{C}_s \in -(P_Y \setminus \{0\}), \quad H_q - \mu_q \in - \operatorname{int} P_X, \quad H_s - \lambda_s \in \operatorname{int} P_X.$$

Then there exists $\delta \in \operatorname{int} P_X$ such that

$$H_q - \mu_q + \delta \in -P_X,$$

$$H_s - \lambda_s - \delta \in P_X.$$

Set $F_q := H_q + \delta$, $F_s := H_s - \delta$, $F_r := H_r$ for $r \neq q, s$. Then $F \in K$, and

$$\sum_{r \in \mathcal{R}} \overline{C}_r[F_r - H_r] = \overline{C}_q[F_q - H_q] + \overline{C}_s[F_s - H_s]$$

$$= (\overline{C}_q - \overline{C}_s)[\delta] \notin P_Z,$$

from (7) . So H is not a solution of (10).

Proposition 4.2 *The following condition is sufficient for $H \in K$ being a solution of (10):*
for every $w \in W$ there exists $\gamma_w \in Y$ such that, for all $r \in \mathcal{R}(w)$:

$$\overline{C}_r - \gamma_w \in P_Y \cup (-P_Y); \tag{12}$$

$$\overline{C}_r - \gamma_w \in -(P_Y \setminus \{0\}) \Rightarrow H_r = \mu_r; \tag{13}$$

$$\overline{C}_r - \gamma_w \in (P_Y \setminus \{0\}) \Rightarrow H_r = \lambda_r. \tag{14}$$

Proof. Let $F \in K$ be arbitrary. Because of (12) it suffices to consider the three cases $\overline{C}_r - \gamma_w \in P_Y \setminus \{0\}$, $\overline{C}_r - \gamma_w \in -(P_Y \setminus \{0\})$, $\overline{C}_r - \gamma_w = 0$.

If $\overline{C}_r - \gamma_w \in P_Y \setminus \{0\}$, then from (14), $H_r = \lambda_r$, hence $F_r - H_r = F_r - \lambda_r \in P_X$, since $F \in K$. Then from (6),

$$(\overline{C}_r - \gamma_w)[F_r - H_r] \in P_Z.$$

The same holds if $\overline{C}_r - \gamma_w \in -(P_Y \setminus \{0\})$. And if $\overline{C}_r - \gamma_w = 0$, then

$$(\overline{C}_r - \gamma_w)[F_r - H_r] = 0.$$

Since $P_Z + P_Z \subseteq P_Z$ and $0 \in P_Z$, it follows that

$$\begin{aligned}
\sum_{r \in \mathcal{R}} \overline{C}_r [F_r - H_r] &\in \sum_{w \in \mathcal{W}} \sum_{r \in \mathcal{R}(w)} \gamma_w [F_r - H_r] + P_Z \\
&= \sum_{w \in \mathcal{W}} \gamma_w [\rho_w - \rho_w] + P_Z \\
&= P_Z.
\end{aligned}$$

So H is solution of (10).

It is clear that the sufficient condition implies the necessary condition. For the reverse inclusion we need additional assumptions. Let $\succeq$ denote the vectorial ordering induced on Y by P_Y ($y_1 \succeq y_2 : \iff y_1 - y_2 \in P_Y$). In the remaining parts we assume that for every finite subset $A \subseteq Y$ there exists $\sup A$ (a least upper bound) and $\inf A$ (a greatest lower bound) with regard to $\succeq$. Then we have

Proposition 4.3 *Assume that*

$$P_Y \cup (-P_Y) = Y, \tag{15}$$

$$P_X = \operatorname{int} P_X \cup \{0\}. \tag{16}$$

Then the necessary and the sufficient condition for (10) are equivalent.

Proof. Assume that (7) holds. We have to verify (12)–(14). From (15), condition (12) is automatically fulfilled. In order to verify (13) and (14), we fix $w \in \mathcal{W}$, and set

$$A := \{\overline{C}_q \mid q \in \mathcal{R}(w), \quad H_q \neq \mu_q\},$$

$$B := \{\overline{C}_s \mid s \in \mathcal{R}(w), \quad H_s \neq \lambda_s\}.$$

If $\overline{C}_q \in A$ and $\overline{C}_s \in B$, then it follows from $H \in K$ and (16) that $H_q - \mu_q \in -\operatorname{int} P_X$, $H_s - \lambda_s \in \operatorname{int} P_X$, hence from (7) that $\overline{C}_q - \overline{C}_s \notin -(P_Y \setminus \{0\})$, hence from (15) that $\overline{C}_q - \overline{C}_s \in P_Y$. Thus we have obtained that $\overline{C}_q \succeq \overline{C}_s$ for all $\overline{C}_q \in A$, $\overline{C}_s \in B$. By hypothesis $\inf A$ and $\sup B$ with regard to $\succeq$ exist, and then $\inf A \succeq \sup B$. Thus there exists $\gamma_w \in Y$ such that $\inf A \succeq \gamma_w \succeq \sup B$. Then $\overline{C}_q \succeq \gamma_w$ for all $\overline{C}_q \in A$, and $\gamma_w \succeq \overline{C}_s$ for all $\overline{C}_s \in B$. Hence for every $r \in \mathcal{R}(w)$: If $\overline{C}_r \not\succeq \gamma_w$, then $\overline{C}_r \notin A$, hence $H_r = \mu_r$. And if $\gamma_w \not\succeq \overline{C}_r$, then $\overline{C}_r \notin B$, hence

$H_r = \lambda_r$. Now, if $\overline{C}_r - \gamma_w \in -(P_Y \setminus \{0\})$, then since P_Y is pointed, $\overline{C}_r - \gamma_w \notin P_Y$, i.e., $\overline{C}_r \not\geq \gamma_w$, and then, as already shown, $H_r = \mu_r$. Likewise if $\overline{C}_r - \gamma_w \in (P_Y \setminus \{0\})$, then $H_r = \lambda_r$. Thus (4.13) and (4.14) are satisfied.

Combining Propositions 4.1, 4.2, 4.3 we obtain

Theorem 4.2 *Assume that (15), (16) hold. Then for arbitrary $H \in K$ and $\overline{C} \in Y^{\mathcal{R}}$, the following are equivalent:*

(i) $\displaystyle\sum_{r \in \mathcal{R}} \overline{C}_r[F_r - H_r] \in P_Z \quad \forall F \in K;$

(ii) *for all $w \in \mathcal{W}$, all $q, s \in \mathcal{R}(w)$ there holds*
$\overline{C}_q - \overline{C}_s \in -(P_Y \setminus \{0\}) \Rightarrow H_q = \mu_q$ *or* $H_s = \lambda_s;$

(iii) *for every $w \in \mathcal{W}$ there exists $\gamma_w \in Y$ such that, for all $r \in \mathcal{R}(w)$,*
$\overline{C}_r - \gamma_w \in -(P_Y \setminus \{0\}) \Rightarrow H_r = \mu_r,$
$\overline{C}_r - \gamma_w \in (P_Y \setminus \{0\}) \Rightarrow H_r = \lambda_r.$

In this situation with $\overline{C} := C(H)$, any solution of (ii) or (iii) would be termed a Wardrop equilibrium.

The case of additional constraints on the flows, say $F \in D$ where $D \subseteq X^{\mathcal{R}}$ is convex, can be handled in the same way. We only have to replace for $H \in K \cap D$ the cost mapping $C(H)$ by $\overline{C} := C(H) + S$, where $S = (S_r)_{r \in \mathcal{R}}$ is such that

$$\sum_{r \in \mathcal{R}} S_r[F_r - H_r] \in -P_Z \text{ for all } F \in D.$$

For details see [2] or [3].

References

[1] G. Y. Chen, C. J. Goh, X. Q. Yang, *Vector network equilibrium problems and nonlinear scalarization methods*, Mathematical Methods of Operations Research 49 (1999), 239–253.

[2] P. Daniele, A. Maugeri, W. Oettli, *Time-dependent traffic equilibria*, Journal of Optimization Theory and Applications (to appear).

[3] A. Maugeri, W. Oettli, D. Schläger, *A flexible form of Wardrop's principle for traffic equilibria with side constraints*, Rendiconti del Circolo Matematico di Palermo, Serie II, Supplemento 48 (1997), 185–193.

[4] A. Nagurney, *Network Economics, A Variational Inequality Approach*. Kluwer, Dordrecht, 1993.

[5] J. G. Wardrop, *Some theoretical aspects of road traffic research*, Proceedings of the Institute of Civil Engineers, Part II, (1952), 325–378.

APPROXIMATE SOLUTIONS AND TIKHONOV WELL–POSEDNESS FOR NASH EQUILIBRIA

L. Pusillo Chicco

Department of Mathematics
University of Genova
Via Dodecaneso 35, 16146 Genova, Italy

Abstract   This contribution is in the framework of non-cooperative games with focus on the main solution concept: the Nash equilibrium (NE). The properties of Tikhonov well-posedness for Nash equilibria (briefly $T-$wp for NE) will be analized with a particular attention to its generalization: Tikhonov well-posedness in value (T^v-wp) which is an ordinal property. Metric characterizations of T^v-wp will be discussed and known results, which give existence and uniqueness of NE in oligopoly model, will be proved to guarantee $T-$wp too.

Keywords: Games, well posedness, Nash equilibria.

1. Introduction.

Since most of this contribution is in the context of games (in particular, non-cooperative games), it is perhaps appropriate to spend a few words about Game Theory.

Game Theory was born as an appendix to political economy, thanks to the French economist Augustin Cournot (1838) who wanted to solve the oligopoly problem and as a branch of mathematics when Zermelo (1913) began the formal study of chess-game. Game theory became an autonomous discipline between economy and mathematics in 1944 after the publication of "Theory of Games and Economic Behaviour" of John von Neumann and Oskar Morgenstern [19]. During the 50s and 60s the contributions of Game Theory were above all in mathematics.

F. Giannessi et al (eds.),
Equilibrium Problems: Nonsmooth Optimization and Variational Inequality Models, 231–246.
© 2001 *Kluwer Academic Publishers.*

In 1968 Harsanyi published the foundamental work about incomplete information games [11] and in 1975 Selten began the study on refinement of Nash equilibria [26].

Game Theory is now a foundamental tool for Economics, but it is applied also to Political Sciences, Philosophy, Computer Sciences, Engineering and Evolutive Biology. For more details the interested reader may consult the books [4], [8], [18], [20].

This contribution will be in the framework of non-cooperative games with focus on the main solution concepts which is Nash equilibrium. The properties of Tikhonov-well posedness $(T-\mathrm{wp})$ for Nash-equilibria (NE) will be analysed.

Given a game $G = (X, Y, f, g)$ where X and Y are topological spaces, we shall say that G is Tikhonov well-posed $(T-\mathrm{wp})$ if there is a unique NE $(\overline{x}, \overline{y})$ and every $a-\mathrm{NE}$ (x_n, y_n) converges to $(\overline{x}, \overline{y})$ where a sequence $(x_n, y_n) \in X \times Y$ is an asymptotically Nash equilibrium $(a-\mathrm{NE})$ if

$$\sup_{x \in X} f(x, y_n) - f(x_n, y_n) \to 0, \quad \sup_{y \in Y} g(x_n, y) - g(x_n, y_n) \to 0$$

Classically, $T-\mathrm{wp}$ was introduced for minimum problems ([28]) and then it was generalized to NE using the asymptotically Nash-equilibria, but a plain extension of the concept faces some problems: the uniqueness requirement of NE can be considered as too demanding, since even in the elementary context of finite games, there is not a unique equilibrium, generically there is an odd number of NE cfr. [30]). Taking into account these problems, a path could be to consider $T-\mathrm{wp}$ in generalized sense [7] where the uniqueness of Nash is no more required but it is necessary to extract from every asymptotically Nash-equilibria $(a-\mathrm{NE})$ a subsequence converging to a NE.

Basic definitions and notations will be given in §2. In this section the problem of ordinality of such notions will also be considered.

In §3 a new approach of $T-\mathrm{wp}$ is described: Tikhonov-well posedness in value $(T^v-\mathrm{wp})$ We have looked in [16] for some "intermediate" property between $T-\mathrm{wp}$ and *generalized* $T-\mathrm{wp}$, taking into account the "value". So we have defined the Tikhonov well-posedness in value as explained in § 4. $T^v-\mathrm{wp}$ is very important because it is an ordinal property (i.e. it is preserved under strictly increasing transformations of payoffs). In §5 the geometric characterization of $T^v-\mathrm{wp}$ is considered; for this result we have introduced a performance function so that the problem of NE can be reduced to a minimum problem and hence a metric characterization is possible (using the nice geometric characterization due to Furi-Vignoli for $T-\mathrm{wp}$ in optimization).

In [17] we have given metric characterizations of T^v−wp for games as Furi and Vignoli did for T−wp for minimum problem [7]. We have shown that if Ω_0 (the set of Nash equilibria) is compact and not empty and if there are no-repeated NE, T^v−wp for a game as a metric cheracterization is possible.

If Ω_0 is not compact, this is impossible; so we have provided also many examples from which one sees that the approaches used to characterize T^v−wp are very different from those used for T−wp.

The §6 is devoted to applications of T−wp regarding to oligopoly models; finally in §7 open problems and works in progress are briefly indicated.

2. T−wp for Nash equilibria.

Tikhonov well−posedness was introduced in minimum problems in [28]. A minimization problem is said to be T−wp if:
  i) there is a unique minimum point
  ii) every minimizing sequence converges to the minimum point.

(For a survey of this subject see [6] and [15]).

The concept of T−wp was generalized to other problems as saddle point problems [3], variational inequalities [13], [14], [24] and to Nash equilibria [2], [21].

To define T−wp for NE one needs the notion of "appproximate solution" for NE, which replaces the minimizing sequence for a minimum problem.

Definition 2.1. A game G of two players is a quadruple (X, Y, f, g) where X, Y are not empty sets and they are the sets of strategies, $f, g :$ $X \times Y \to \mathbf{R}$ are the payoff functions of the two players (this is a game in strategic form)

Definition 2.2. Given a game $G = (X, Y, f, g)$ a Nash equilibrium (NE) for G is a pair $(\overline{x}, \overline{y}) \in X \times Y$ s.t.

$$f(\overline{x}, \overline{y}) \geq f(x, \overline{y}) \quad \forall x \in X$$

$$g(\overline{x}, \overline{y}) \geq g(\overline{x}, y) \quad \forall y \in Y$$

Definition 2.3. Given a game $G = (X, Y, f, g)$ we shall say that a sequence $(x_n, y_n) \in X \times Y$ is an asymptotically Nash equilibrium $(a-\text{NE})$

if

$$\sup_{x\in X} f(x, y_n) - f(x_n, y_n) \rightarrow 0$$

$$\sup_{y\in Y} g(x_n, y) - g(x_n, y_n) \rightarrow 0$$

This definition was introduced in [3] for zero sum games. For the extension to the general case see [2], [21].

Note that existence of an a–NE is not guaranteed (for example, in zero sum games, existence of a–NE is equivalent to the fact that the game has value).

Definition 2.4. Given a game $G = (X, Y, f, g)$ where X and Y are topological spaces, we shall say that G is Tikhonov well-posed (T–wp) if there is a unique NE $(\bar{x}, \bar{y})$ and every a–NE (x_n, y_n) converges to $(\bar{x}, \bar{y})$.

Example 2.5 The game $G = (\mathbf{R}, \mathbf{R}, f, -f)$, where $f(x, y) = xy$, is T–wp.

Example 2.6 (the battle of sexes).
   Let $G = (X, Y, f, g)$ be the game where $X = Y = \{-1, 1\}$
   $f(x, y) = 3xy + x + y + 3$, $g(x, y) = 3xy - x - y + 3$. This game is not T–wp because there are two NE.

The definition of a game $G = (X, Y, f, g)$ has been given in Def. 2.1, but remind that f, g are usually interpreted as the utility functions of the two players (I, II): having this in mind, a game can be seen as a quadruple $\Gamma = (X, Y, \preceq_I, \preceq_{II})$ where $\preceq_I$, $\preceq_{II}$ are total preorders on $X \times Y$ [20] .

Definition 2.7. A relation $\preceq$ on a not empty set Z is said to be a total preorder if it is transitive and total. A function $h : Z \longrightarrow \mathbf{R}$ is said to represent $\preceq$ if

$$h(z_1) \leq h(z_2) \quad \Longleftrightarrow \quad z_1 \preceq z_2$$

Definition 2.8. Given a game $\Gamma = (X, Y, \preceq_I, \preceq_{II})$, we shall say that $G = (X, Y, f, g)$ represents Γ (or, equivalently, that G induces Γ) if:
   $f, g : X \times Y \longrightarrow \mathbf{R}$
   $\forall (x, y), (\bar{x}, \bar{y}) \in X \times Y,$
      $(x, y) \preceq_I (\bar{x}, \bar{y}) \iff f(x, y) \leq f(\bar{x}, \bar{y})$
      $(x, y) \preceq_{II} (\bar{x}, \bar{y}) \iff g(x, y) \leq g(\bar{x}, \bar{y}).$

So we say that f, g are ordinal utility functions representing respectively $\preceq_I$ and $\preceq_{II}$.

Has Tikhonov well posedness an ordinal character? That is, given two games G and G_o, both representing a given game Γ, T–wp for G is equivalent to T–wp for G_o?

The answer is positive for antagonistic games, while in general it is negative as shown by the following example (see [15]).

Example 2.9. Let $G = (\mathbf{R}, \mathbf{R}, \mathrm{f}, \mathrm{g})$, where $f(x, y) = g(x, y) = xy - x^2$. Then, G is T–wp. However, taking $f_o = f$ and $g_o = \arctan(g)$, then $G_o = (X, Y, f_o, g_o)$ is not T–wp.

In the paper [23] the authors show that the "positive" results about the ordinal character of T–wp for *antagonistic games* cannot be extended to *unilaterally competitive games* even if they are the "closest" to *antagonistic* among a lot of games considered.

We consider in fact the following example [23]:

Example 2.10. Consider $G = (\mathbf{N}, \mathbf{N}, f, g)$, where f is defined as:

$$f(i, j) = \begin{cases} \alpha \cdot (1/2^i) & \text{if } j<i, \\ 0 & \text{if } j=i, \\ \alpha \cdot (2^{(j-i)} - 1) & \text{if } j>i. \end{cases}$$

and where $g(i, j) = f(j, i)$. Let $\mathbf{N} \times \mathbf{N}$ be endowed with the discrete topology. Then G is T–wp. However, considering $f_o = \arctan(f)$ and $g_o(i, j) = f_o(j, i)$, then
$G_o = (\mathbf{N}, \mathbf{N}, \mathrm{f_o}, \mathrm{g_o})$ is not T–wp.

3. A new approach to Tikhonov well-posedness for Nash equilibria.

To overcome the "non-ordinality" character of T–wp and, at the same time, the fact that uniqueness requirement for NE is a very severe one, we have introduced an alternative definition of T–wp for NE, an "intermediate" property between T–wp and *generalized* T–wp: Tikhonov well posedness in value (T^v–wp).

It has to be noticed that *generalized* T–wp requires only the existence (not uniqueness) of Nash equilibrium and that every a–NE has a subsequence converging to a NE.

However if we consider mixed extensions of finite games, they generically have a finite number of NE, and this property is not taken in due account by *generalized $T-$wp*.

It is remarked that in minimum problems and in saddle point problems, there is a unique value of the involved function, which is associated to a solution, but this is not true for Nash equilibria (see Example 2.6).

So, in the paper [16] our attention has been focused on this problem and the following more general definition of well-posedness has been given :

Definition 3.1. Given a game $G = (X, Y, f, g)$ where X and Y are topological spaces, we shall say that G is Tikhonov well-posed in value $(T^v-\text{wp})$ if
   – there is at least one NE;
   – every $a^v NE$ (x_n, y_n) converges to a NE
where we shall say that a sequence $(x_n, y_n) \in X \times Y$ is an asymptotically Nash equilibrium in value $(a^v-\text{NE})$ if it is an $a-$NE and there exists a NE $(\bar{x}, \bar{y})$ s.t.

$$f(x_n, y_n) \to f(\bar{x}, \bar{y}), \quad g(x_n, y_n) \to g(\bar{x}, \bar{y})$$

(that is it converges in "value").

Remark 3.2. It is easy to show that

$$T-\text{wp} \implies T^v-\text{wp}.$$

Example 2.6 is a game not $T-$wp but it is T^v-wp.
   Example 2.10 is a game T^v-wp and $T-$wp.

If a game G has at least two different Nash equilibria, they cannot have the same payoffs, in fact:

Proposition 3.3. Assume that the game $G = (X, Y, f, g)$ is T^v-wp and that X, Y are Hausdorff topological spaces. Then:
A) $\forall (\lambda, \mu) \in \mathbf{R}^2$ there is at most one NE $(\bar{x}, \bar{y})$ s.t. $f(\bar{x}, \bar{y}) = \lambda$, $g(\bar{x}, \bar{y}) = \mu$ (briefly: G has no repeated NE);
B) Every a^v-NE (x_n, y_n) has a unique NE $(\bar{x}, \bar{y})$ associated to it in value and (x_n, y_n) converges to $(\bar{x}, \bar{y})$.

Proof. To get (A), we assume that $(\bar{x}, \bar{y})$ and $(\tilde{x}, \tilde{y})$ are two different NE with the same payoffs. Then the sequence which alternates $(\bar{x}, \bar{y})$ and $(\tilde{x}, \tilde{y})$ is an a^v-NE which cannot converge, and this is absurd.

To get (B), the fact that every a^v-NE (x_n, y_n) has a unique NE $(\bar{x}, \bar{y})$ associated to it in value follows immediately from (A). To prove that $(x_n, y_n) \longrightarrow (\bar{x}, \bar{y})$, let us suppose that $(x_n, y_n) \longrightarrow (\tilde{x}, \tilde{y}) \neq (\bar{x}, \bar{y}), (\tilde{x}, \tilde{y})$ NE. Then the sequence which alternates (x_n, y_n) to $(\tilde{x}, \tilde{y})$ should be an a^v-NE which doesn't converge, and this is absurd.   □

So T^v−wp is a reasonable step between T−wp and *generalized* T−wp taking into consideration the condition of T−wp namely the convergence of "approximate solutions" not only the convergence of a subsequence.

Proposition 3.4. Let $G = (X, Y, f, -f)$ be a zero-sum game and X, Y Hausdorff topological spaces. Then G is T^v-wp if and only if G is T−wp.

4. Ordinality of T^v−wp.

In the example 2.9, $G(\mathbf{N}, \mathbf{N}, \mathrm{f}, \mathrm{g})$ is T^v−wp and if we consider $G_o(\mathbf{N}, \mathbf{N}, \arctan \mathrm{f}, \arctan \mathrm{g})$, this is T^v−wp too.

So a question is natural: has T^v−wp the ordinal property that we are looking for?

In this paragraph we try to give an answer to this question.

Definition 4.1. Given a not empty set Z and $h_1, h_2 : Z \longrightarrow \mathbf{R}$,
h_2 is an admissible transformation of h_1 (briefly, $h_1 \top h_2$), if there is $\eta : K \longrightarrow \mathbf{R}$ s.t.
A) K is an interval containing $h_1(Z)$;
B) η is strictly increasing ;
C) $h_2 = \eta \circ h_1$.
So, if $h_1 \top h_2$, both of them represent the same preorder on Z.

Definition 4.2. $h_1, h_2 : Z \longrightarrow \mathbf{R}$ are said equivalent (briefly $h_1 \sim h_2$) if $h_1 \top h_2$ and $h_2 \top h_1$.

Proposition 4.3.   If Z is a connected topological space and $h_1, h_2 : Z \longrightarrow \mathbf{R}$ are continuous, then $h_1 \sim h_2$ if and only if h_1, h_2 represent the same preorder on Z
A proof can be found in [21].

Definition 4.4.   Given two games $G_1 = (X, Y, f_1, g_1)$ and $G_2 = (X, Y, f_2, g_2)$, G_2 is an admissible transformation of G_1 (briefly $G_1 \top G_2$) if $f_1 \top f_2$ and $g_1 \top g_2$. In a similar way, G_1 and G_2 are called equivalent ($G_1 \sim G_2$) if $f_1 \sim f_2$ and $g_1 \sim g_2$.

Lemma 4.5. Let $G_1 = (X, Y, f_1, g_1)$ and $G_2 = (X, Y, f_2, g_2)$ be two games s.t. $G_1 \top G_2$. If (x_n, y_n) is an a^v-NE associated in value to the NE $(\bar{x}, \bar{y})$ for G_2, then it is the same for G_1 too.

Proof. By Lemmata 4.6, 4.7 of [16] $\exists a, b, c, \in f_1(X \times Y), \bar{n} \in \mathbf{N}$ s.t.:

$$a \leq f_1(x_n, y_n) \leq b \quad \forall n \in \mathbf{N}$$

$$f_1(x, y_n) \leq c \quad \forall n > \bar{n}, \quad \forall x \in X.$$

First it is proved that (x_n, y_n) is an a-NE for G_1.
It is known that $\eta^{-1}\big|_{\eta([a,d])}$ is uniformly continuous, where $d = \max\{b, c\}$ (see Lemma 4.8 of [16]).

Given $\varepsilon > 0$, let δ be corresponding to ε by the uniform continuity of $\eta^{-1}\big|_{\eta([a,d])}$.

Since (x_n, y_n) is an a-NE for G_2, $\exists n_1 \in \mathbf{N}$ s.t.

$$f_2(x, y_n) \leq f_2(x_n, y_n) + \delta \quad \forall x \in X \quad \forall n > n_1.$$

Now, for $x \in X$ s.t.

$$f_2(x, y_n) \leq f_2(x_n, y_n)$$

applying η^{-1} of Definition 4.1, it results

(1) $$f_1(x, y_n) \leq f_1(x_n, y_n) < f_1(x_n, y_n) + \varepsilon.$$

On the contrary, for $x \in X$ s.t.

$$f_2(x_n, y_n) \leq f_2(x, y_n) \leq f_2(x_n, y_n) + \delta$$

it result

(2) $$f_1(x, y_n) \leq f_1(x_n, y_n) + \varepsilon$$

by the uniform continuity of $\eta^{-1}\big|_{\eta([a,d])}$.

From (1) and (2) it follows that

$$\sup_{x \in X} f_1(x, y_n) \leq f_1(x_n, y_n) + \varepsilon \quad \forall n > n_1$$

So, (x_n, y_n) is an a-NE for G_1.

Since $f_2(x_n, y_n) \longrightarrow f_2(\bar{x}, \bar{y})$ and $\eta^{-1}\big|_{\eta([a,d])}$ is uniformly continuous, then
$f_1(x_n, y_n) \longrightarrow f_1(\bar{x}, \bar{y})$ and this completes the proof.   □

From Lemma 4.5 we obtain an interesting result:

Corollary 4.6. *Let G_1, G_2 be two games s.t. $G_1 \sim G_2$. Then G_1 is T^v−wp if and only if G_2 is T^v−wp, and the same holds for generalized T^v−wp.*

This result is important because it permits us to say that T^v−wp has the ordinal property; in fact the Definition 4.5 of equivalent games is the same thing as ordinality for example if one of the following two conditions is valid:

i) G_1, G_2 are finite games, or mixed extensions of finite games;

ii) X, Y are connected topological spaces and the utility functions are continuous.

(For more details about T^v−wp see [16]).

5. Metric characterization of T^v−wp.

In [17] it is discussed and given metric characterizations of T^v−wp for NE analogous to that given by Furi and Vignoli in [7] for T−wp for minimum problems.

Let us see some definitions:

Definition 5.1. Given $\epsilon \geq 0$ we denote by

(i) Ω_ϵ the set of $\epsilon - NE$, i.e.:

$$\Omega_\epsilon = \{(\overline{x}, \overline{y}) \in X \times Y : \sup_{x \in X} f(x, \overline{y}) - f(\overline{x}, \overline{y}) \leq \epsilon,$$

$$\sup_{y \in Y} g(\overline{x}, y) - g(\overline{x}, \overline{y}) \leq \epsilon\}$$

(Obviously Ω_0 is the set of Nash equilibria).

(ii) Ω_ϵ^v the set of $\epsilon-NE$ in value, i.e. the set of ϵ-equilibria whose distance in value from a NE is less than ϵ:

$$\Omega_\epsilon^v = \Omega_\epsilon \cap B_\epsilon$$

where

$$B_\epsilon = \cup_{(\overline{x}, \overline{y}) \in \Omega_0} B_\epsilon(\overline{x}, \overline{y})$$

$$= \cup_{(\overline{x}, \overline{y}) \in \Omega_0} \{(x, y) : |f(x, y) - f(\overline{x}, \overline{y})| < \epsilon, |g(x, y) - g(\overline{x}, \overline{y})| < \epsilon\}$$

Definition 5.2. Let (X, d) be a metric space, $A, B \subset X$.

The diameter of a set A is

$$\delta(A) = \begin{cases} \sup\{d(x,y) : x, y \in A\} & \text{if } A \neq \emptyset, \\ 0 & \text{if } A = \emptyset. \end{cases}$$

The Hausdorff distance between two sets $A, B \neq \emptyset$ is defined as

$$d_H(A, B) = \max\{\sup_{a \in A} d(a, B), \sup_{b \in B} d(b, A)\}.$$

Proposition 5.3. A game $G = (X, Y, f, g)$ is $T-$wp if and only if there is a NE $(\overline{x}, \overline{y})$ and $\delta(\Omega_\epsilon) \to 0$ if $\epsilon \to 0$

In a similar way to that made in [3] to introduce $T-$wp for saddle point problems, it has been defined a "performance" function to give a metric characterization of T^v-wp in the case Ω_0 compact and not empty.

It has been introduced a performance function:

$$\psi : X \times Y \to R \cup \{+\infty\}$$

in the following way:

$$\psi(x_0, y_0) = \sup_{x \in X}\{f(x, y_0) - f(x_0, y_0)\} + \sup_{y \in Y}\{g(x_0, y) - g(x_0, y_0)\} +$$

$$+ \inf_{(x,y) \in \Omega_0} \{|f(x, y) - f(x_0, y_0)| + |g(x, y) - g(x_0, y_0)|\}$$

(where we take inf $\emptyset = +\infty$)

Proposition 5.4. The following hold:
(i) $\psi(x, y) \geq 0$
(ii) $\psi(x, y) = 0 \Leftrightarrow (x, y) \in \Omega_0$
(iii) if (x_n, y_n) is a^v-NE then $\psi(x_n, y_n) \to 0$ if $n \to +\infty$
(iv) if Ω_0 is compact, $\Omega_0 \neq \emptyset$, f, g are continuous functions, $(x_n, y_n) \in X \times Y$ and $\psi(x_n, y_n) \to 0$, then (x_n, y_n) is a^v-NE.

We want to point out that both the hypotheses: "Ω_0 compact and non-empty" and "f, g continuous" are necessary for the validity of (iv) (Prop 5.4). This is shown by the following examples:

Example 5.5. Let $G = (X, Y, f, g)$ be a game with $X = Y = \mathbf{R}$ and $f(x, y) = y - (x - y)^2$, $g(x, y) = x - (x - y)^2$. Clearly $\Omega_0 = \{(x, y) \in X \times Y : x = y\}$ is not compact, the sequence (n, n) is not a^v-NE but $\psi(n, n) = 0$.

Example 5.6. Let $G = (X, Y, f, g)$, be a game with $X = Y = [0, 1]$ and

$$f(x, y) = g(x, y) = \{0, \quad x \neq yx, \quad x = y \neq 12, \quad x = y = 1$$

In this case $\Omega_0 = \{(x, y) \in [0, 1] \times [0, 1] : x = y\}$ is compact, the sequence $(1 - 1/n, 1 - 1/n)$ is not a^v–NE, but $\psi(1 - 1/n, 1 - 1/n) = 0$.

Then we have a metric characterization of T^v–wp by means of the following theorem.

Theorem 5.7. Let $G = (X, Y, f, g)$, with f, g continuous, $\Omega_0 \neq \emptyset$, Ω_0 compact.
   Then, G is T^v–wp if and only if $d_H(\Omega_\epsilon^v, \Omega_0) \to 0$ and G has no repeated Nash-equilibria.

Note that a Furi-Vignoli type characterization is not valid if Ω_0 is not compact even if we choose another function $\tilde{\psi}$ which verifies (i) e (ii) of Proposition 5.4, and if instead of the Hausdorff distance, another distance d in $\mathcal{P}(X \times Y)$ is chosen as the following example shows.

Example 5.8. Let $G = (X, Y, f, g)$, with $X = \{0\}$, $Y = \mathbf{N}$ and

$$f(0, n) = \{0, \quad n = 0, 1/n, \quad n \neq 0 \qquad g(0, n) = 0.$$

Then $\Omega_0 = X \times Y$, G is not *generalized* T^v–wp because the sequence $(0, n)$ is a^v- NE but it has no convergent subsequences. On the contrary, if we define

$$\Omega_\epsilon(\psi) = \{(x, y) \in X \times Y : \psi(x, y) \leq \epsilon\}$$

the following equality is true:
   $d(\Omega_\epsilon(\tilde{\psi}), \Omega_0) = 0$ for any choice of $\tilde{\psi}$ and of d on $\mathcal{P}(X \times Y)$ because $\Omega_\epsilon(\tilde{\psi})) = \Omega_0$.
   For more details about metric characterizations of T^v–wp, see [17]

6. An application: oligopoly models.

There are some results available, which guarantee the existence and the uniqueness of Nash equilibria ([9], [12], [25], [29]); can these theorems be converted into "well-posedness" theorems?
   In particular it is interesting to study this case for oligopoly models: that is some manifacturers (generically n) produce a single type of item

and they compete with each other when selling in a market where the price depends on the total amount produced.

The duopoly model is studied in this section: in this game the two firms are assumed as players, the different quantities of an homogeneus product are the players' strategies and the sets of strategies are $[0, +\infty)$; the utility functions are depending on prices and on the cost of production.

The model was first studied by Cournot [5] who anticipated Nash's definition of equilibrium over a century. So Cournot's paper is one of the classics of Game Theory and it is also one of cornerstones of the theory of industrial organization [10]

Some theorems of existence and uniqueness were shown by Rosen [25], Gabay-Moulin [9], Karamardjan [12], Szidarovszky -Yakowitz [27] and Watts [29], the last did not use cardinal utility functions but ordinal relations.

In particular the paper of Szidarovszky-Yakowitz is interesting because they showed existence and uniqueness of NE under hypotheses of regularity on the price function (p) and on the cost function (c) $(p$ decreasing and concave, c increasing and convex) without hypotheses of compacteness.

Theorem 6.1 (Existence and uniqueness)
   Let $G = (X, Y, f, g)$ be the duopoly game, $X = Y = [0, +\infty)$

$$f(x, y) = xp(x + y) - c_1(x)$$
$$g(x, y) = yp(x + y) - c_2(y)$$

$p, c_i \in C^2[0, +\infty)$ $p(a) = 0$
$p' < 0$, $p'' < 0$
$c_i' > 0$, $c_i'' > 0$
further
$c_1(0) = c_2(0) = 0$, $p(0) > 0$
then exists one and only one Nash equilibrium.

Proof see [27]

Lemma 6.2 Let $G = (X, Y, f, g)$ be a game, where $X = Y = [0, +\infty)$, f, g continuous functions, then G is $T-$wp if and only if there exists one and only one Nash equilibrium and there exists $\bar{\epsilon}$ s.t. Ω_ϵ is compact and non empty for any $\epsilon < \bar{\epsilon}$.

Proof: " $\Leftarrow$ " if G is supposed, by contradiction, not T−wp then $\exists\ (x_n, y_n)$ a−NE sequence s.t. $(x_n, y_n) \to +\infty$ then $\exists\ \epsilon$ s.t. eventually $(x_n, y_n) \in \Omega_\epsilon$ so Ω_ϵ is not compact.

" $\Rightarrow$ " we assume by contradiction that $\forall \epsilon, \Omega_\epsilon$ is not compact, then we can choose $(x_n, y_n) \in \Omega_{1/n}$ s.t. $|(x_n, y_n)| > n$ but (x_n, y_n) is a−NE and it is not convergent: this is impossible.   □

Proposition 6.3 In the hypotheses of Theorem 6.1, the duopoly game is T−wp.

Proof: from lemma 6.2, it is sufficient to show that Ω_ϵ is compact for all ϵ small enough. From definition 5.1 Ω_ϵ is closed then it is sufficient to show that it is bounded.

It shows that $\Omega_\epsilon \in [0, a] \times [0, a]$ where $\epsilon < \min(c_1(a), c_2(a))$, Ω_ϵ is not empty because $\exists!\ NE$ (see theorem 6.1).

It is assumed, by contradiction, that $\forall \epsilon < \min(c_1(a), c_2(a))$, $\Omega_\epsilon \not\subset [0, a] \times [0, a]$ that is $\exists\ (x_1, y_1) \in \Omega_\epsilon$ s.t. $x_1 > a$ or $y_1 > a$. Suppose $x_1 > a$; the following relations are valid:

$$f(0, y_1) - f(x_1, y_1) \leq \epsilon$$

and

$$-x_1 p(x_1 + y_1) + c_1(x_1) > c_1(a)$$

but these two disequalities are in contradiction. (Analogously if $y_1 > a$)   □

Concluding it is useful to spend still a few words about this problem: as the oligopoly problem has interesting links with economic studies then it is investigating if several hypotheses which assure existence and uniqueness of Nash equilibrium (as strongly positivity of the jacobian matrix of the utility functions, pseudo-concavity, coercitivity of derivatives etc.) guarantee the T−wp of NE too.

The result is positive for the quadratic case, the general case (also for n players) is still in progress.

7. Open problems.

Let us see some comments about possible extensions: we can consider a sequence of games, instead of a fixed one, the definition of a^v-NE can be conveniently adapted, using the idea of asymptotically minimizing sequences [6]. Given $G_n = (X, Y, f_n, g_n)$ and $G = (X, Y, f, g)$, we may require that (x_n, y_n) verifies:

$$\sup_{x \in X} f_n(x, y_n) - f_n(x_n, y_n) \longrightarrow 0, \quad \sup_{y \in Y} g_n(x_n, y) - g_n(x_n, y_n) \longrightarrow 0$$

and there exists a NE $(\bar{x}, \bar{y})$ s.t.

$$f_n(x_n, y_n) \longrightarrow f(\bar{x}, \bar{y}), \quad g_n(x_n, y_n) \longrightarrow g(\bar{x}, \bar{y})$$

Another problem that can be considered as "classical" in the context of well-posedness in optimization is the connection between Tikhonov well-posedness and Hadamard well-posedness ([15]). It would be interesting to explore this relationship in the context of NE: given an approximate equilibrium, how to build a game "close" to the original one for which it will become an equilibrium?

This problem suggests another one: which is the relationship between the well-posedness of direct problem and that one of inverse problem? This research could be very interesting for its connections with problems in Medicine and in particular in Radiology [1].

As it has showned in §6 a lot of results available, which guarantee the existence and the uniqueness of Nash-equilibria can be converted into "well-posedness" theorems: which are the connections with variational-inequalities and their well-posedness ([11] [12] [20])?

Another research subject consists to provide a definition of $T^v - \text{wp}$ which does not use the payoffs but only the preferences of the players; is it possible give a metric characterization of $T^v - \text{wp}$ for a game, based only on the players' preferences?

Works are in progress about this issue.

Acknowledgment. I am grateful to Prof. F. Patrone for his careful reading and suggestions about this manuscript.

References

[1] M. Bertero, P. Boccacci, *Introduction to inverse problems in imaging*, Institute of Physics Publishing (1998).

[2] E. Cavazzuti, *Cobwebs and something else*, in *Decision Processes in Economics* (G. Ricci ed.), Proc. Conf. Modena/Italy 1989, Springer, Berlin (1990), 34–43.

[3] E. Cavazzuti, J. Morgan, *Well-Posed Saddle Point Problems*, in *Optimization, theory and algorithms* (J. B. Hirriart-Urruty, W. Oettli and J. Stoer eds.), Proc. Conf. Confolant/France 1981, (1983), 61–76.

[4] G. Costa, P.A. Mori,*Introduzione alla Teoria dei Giochi*, Il Mulino, 1994.

[5] A. Cournot, *Mathematical principles of the theory of wealth*, english traslation by N. O. Bacon, New York: The Macmillan Company, 1927.

[6] A. Dontchev, T. Zolezzi,*Well-Posed Optimization Problems*, Springer, Berlin, 1993.

[7] M. Furi, A. Vignoli, *About well-posed optimization problems for functionals in metric spaces*, J. Optimization Theory Appl. 5 (1970), 225–229.

[8] D. Fudenberg, J. Tirole, *Game Theory*, MIT Press, Cambridge (Massachusetts), 1991.

[9] D. Gabay, H. Moulin, *On the uniqueness and stability of Nash-equilibria in non-cooperative games*, Applied Stochastic Control in Econometrics and Management Science (A.Bensoussan, et al..Eds),New York: North Holland.

[10] R. Gibbons, *A Primer in Game Theory*, Il Mulino, Bologna, Hemel Hempstead etc. Prentice Hall International, 1994.

[11] J.C. Harsanyi, *Games with incomplete information played by Bayesian players*, Management Science 14 (1967-68).

[12] S. Karamardian, *The nonlinear complementarity problem with applications*, J. Opt. Th. Appl. 4 (1969), 87–98 and 167–181.

[13] R. Lucchetti, F. Patrone, *A characterization of Tyhonov well-posedness for minimum problems, with applications to variational inequalities*, Numer. Funct. Analysis Optimiz. 3 (1981), 461–476.

[14] R. Lucchetti, F. Patrone, *Some properties of "well-posed" variational inequalities governed by linear operators*, Numer. Funct. Analysis Optimiz. 5 (1982-83), 349–361.

[15] R. Lucchetti, J. Revalski (eds.), *Recent Developments in Well-Posed Variational Problems*, Kluwer, Dordrecht, 1995.

[16] M. Margiocco, F. Patrone, L. Pusillo Chicco, *A New Approach to Tikhonov Well-Posedness for Nash Equilibria*, Optimization 40 (1997), 385–400.

[17] M. Margiocco, F. Patrone, L. Pusillo Chicco, *Metric characterizations of Tikhonov well-posedness in value*, J. Opt. Theory Appl. 100, n. 2, (1999), 377–387.

[18] R.B. Myerson, *Game Theory: Analysis of Conflict*, Harvard University Press, Cambridge (MA), 1991.

[19] J. von Neumann, O. Morgenstern, *Theory of Games and economic behavior*, Princeton University Press, 1944.

[20] M. Osborne, A. Rubistein, *Bargaining and markets*, Academic Press, London, 1990.

[21] F. Patrone, *Well-Posedness as an Ordinal Property*, Rivista di Matematica pura ed applicata 1 (1987), 95–104.

[22] F. Patrone F, *Well-posed minimum problems for preorders*, Rend. Sem. Mat. Univ. Padova 84 (1990), 109–121.

[23] F. Patrone, L. Pusillo Chicco, *Antagonism for two-person games: taxonomy and applications to Tikhonov well-posedness*, preprint.

[24] J.P. Revalski, *Variational inequalities with unique solution*, in *Mathematics and Education in Mathematics*, Proc. 14th Spring Confer. of the Union of Bulgarian Mathematicians, Sofia, 1985.

[25] J.B. Rosen, *Existence and uniqueness of equilibrium points for concave N-person games*, Econometrica 33 (1965), 520–534.

[26] R. Selten, *Reexamination of the perfectness concept for equilibrium points in extensive games*, Int. J. Game Theory 4 (1975), 25–55.

[27] F. Szidarovszky, S. Yakowitz, *A new proof of the existence and uniqueness of the Cournot equilibrium*, Int. Economic Rev. 18 (1977), 787–789.

[28] A.N. Tikhonov, *On the stability of the functional optimization problem*, USSR J. Comp. Math. Math. Phys. 6(1966), 631–634.

[29] A. Watts, *On the uniqueness of equilibrium in Cournot oligopoly and other games*, Games and Economic Behavior 13 (1996), 269–285.

[30] R. Wilson, *Computing equilibria of n-person games*, SIAM J. Appl. Math. 21 (1971), 80–87.

EQUILIBRIUM IN TIME DEPENDENT TRAFFIC NETWORKS WITH DELAY

Fabio Raciti

Dipartimento di Matematica,
Università di Catania, Italy and
Consorzio Ennese Universitario

e–mail:fraciti@dipmat.unict.it

Abstract   We propose a new model of time dependent traffic networks which include delay effects. Starting from the new definition of "retarded equilibrium " we cast the problem in the framework of variational inequalities and give sufficient conditions for the existence of retarded equilibria.

Keywords: Traffic Networks, Variational Inequalities, Delay.

1. Introduction.

The importance of *delay effects* is well known since long time in physics litterature, as well as in mathematical modelling. In electromagnetic theory, for instance, the fact that signals cannot be transmitted faster than light is embodied through the retarded potentials. In mathematical modelling, the importance of delay effects in population dynamics has led several authors to propose a rich variety of *delay differential equations* [1]. For a brief survey on this kind of differential equations the interested reader can refer to [3].

In a recent paper [2] time dependent traffic equilibria have been studied in the general framework of variational inequalities. In [2] the authors have done the implicit assumption that information through the network is transmitted at infinite speed. In real cases, however, the fact that signals (or more generaly, *information*) travel at finite speed cannot always be neglected. In this note we want to take into account the case in which information within the network is propagated with a finite

F. Giannessi et al (eds.),
Equilibrium Problems: Nonsmooth Optimization and Variational Inequality Models, 247–253.
© 2001 *Kluwer Academic Publishers.*

velocity. Thus, we are naturally led to introduce a delay h in the traffic conservation law. Moreover, we put forth a new definition of equilibrium which we call *retarded equilibrium*. Before introducing our problem, let us briefly review the time dependent model proposed in [2]. (For general references on the traffic assignment problem, and on the functional methods based on variational inequalities see also references in [2]).

A traffic network consists of a set W of origin-destination pairs and a set $\mathcal{R}$ of routes. The set of all $r \in \mathcal{R}$ which link a given $w \in W$ is denoted by $\mathcal{R}(w)$. The traffic network is considered at all times $t \in \tau$, where $\tau = [0, T]$. For each time $t \in \tau$ there is a route-flow vector $F(t) \in \Re^{\mathcal{R}}$. Feasible flows are flows which satisfy the time dependent capacity constraints and demands, i.e., which belong to the set:

$$K := \{F \in \mathcal{L} \mid \lambda(t) \leq F(t) \leq \mu(t) \ \phi F(t) = \rho(t) \text{ a.e. on } \tau\}$$

where $\lambda(t) \leq \mu(t)$ and $\rho(t)$ are given, and ϕ is the well known pair-route incidence matrix whose elements $(\phi)_{w,r}$ are set equal 1 if route r connects the pair w, 0 else. $F : \tau \mapsto \Re^{\mathcal{R}}$ is the flow trajectory over time. Flows trajectories are supposed to be elements of $\mathcal{L} := L^p(\tau, \Re^{\mathcal{R}})$ with $p > 1$. The dual space, $L^q(\tau, \Re^{\mathcal{R}})$ with $\frac{1}{p} + \frac{1}{q} = 1$, will be denoted by $\mathcal{L}^*$. A Mapping $C := K \mapsto \mathcal{L}^*$ is then given which assigns to each flow trajectory $F(t) \in K$ the cost trajectory $C[F(t)] \in \mathcal{L}^*$.

On $\mathcal{L}^* \times \mathcal{L}$ the canonical bilinear form is then introduced:

$$\langle\langle G, F \rangle\rangle := \int_\tau \langle G(t), F(t) \rangle dt \ \ G \in \mathcal{L}^*, F \in \mathcal{L}$$

According to [2] $H \in \mathcal{L}$ is an *equilibrium flow* iff:

$$H \in K \text{ and } \langle\langle C(H), F - H \rangle\rangle \geq 0 \ \forall F \in K$$

which is equivalent to the following: $H \in K$ and:

$$\forall w \in W, \forall q, s \in \mathcal{R}(w), \text{and a.e on } \tau \text{ there holds}:$$

$$C_q[H(t)] < C_s[H(t)] \implies H_q(t) = \mu_q(t) \text{ or } H_s(t) = \lambda_s(t)$$

¿From the user point of view, however, one can prefer the last condition to define an equilibrium flow and use the variational inequality as a characterization. Starting from this framework the authors state existence theorems for equilibria. Moreover, in [2], a procedure is given for the approximate calculation of equilibria.

The plan of this paper is the following: in section 2 we present our model, based on the definition of *retarded equilibrium*; moreover we characterize this new kind of equilibrium through a variational inequality. In section 3 we give sufficient conditions for the existence of retarded equilibria. Finally, in section 4, we apply our model to a simple traffic network.

2. The model.

In our model we shall suppose that the traffic demand at time t is satisfied after a delay $h > 0$. The new feasible set is then:

$$K_h := \{F \in \mathcal{L}_h \mid \lambda(t) \leq F(t) \leq \mu(t) \ \ a.e. \ t \in [0, T + h],$$

$$\phi F(t + h) = \rho(t) \ \text{a.e. on} \ [0, T] \ \}$$

where $\mathcal{L}_h := L^p([0, T + h], \Re^{\mathcal{R}})$ with $p > 1$. Let us notice that K_h is not a translation of K, because capacities constraints are not retarded. An operator C is then defined on K_h, which assigns to each flow trajectory the correspondent cost trajectory, $C[F(t)] \in \mathcal{L}_h^*$. In order to establish the functional form of C one must, of course, specify the physical hypothesis of the model. For a more detailed physical analysis, in the discrete time dependent case, we refer to [4]. Now we give our basic definition:

Def. 1 A flux $H \in \mathcal{L}_h$ is a *retarded equilibrium flux* iff:

$$(1) \begin{cases} H \in K_h \text{ and } \forall w \in W \ \forall q, s \in \mathcal{R}(w), \text{and a.e on } \tau \text{ there holds}: \\ C_q[H(t)] < C_s[H(t)] \implies H_q(t + h) = \mu_q(t + h) \\ \text{or } \ H_s(t + h) = \lambda_s(t + h) \end{cases}$$

In the following we shall prove that Def.1 is equivalent to:

$$H \in K_h \text{ and } \int_{[0, T]} \langle C[H(t)], F(t + h) - H(t + h)\rangle dt \geq 0 \ \ \forall F \in K_h \quad (2)$$

Let us prove first that $(2) \implies (1)$.

Suppose that (1) does not hold, then $\exists w \in W, q, s \in \mathcal{R}(w), E \subset [0, T], |E| > 0$:

$$C_q[H(t)] < C_s[H(t)], \ H_q(t + h) = \mu_q(t + h) \ \text{ and } \ H_s(t + h) = \lambda_s(t + h)$$

$\forall t \in E$ let us define:

$$\delta(t + h) = min\{\mu_q(t + h) - H_q(t + h), H_s(t + h) - \lambda_s(t + h)\}$$

Now we introduce a new flux $F \in K_h$:

$$F_q : \begin{cases} F_q(t + h) = H_q(t + h) + \delta(t + h) & \forall t \in E \\ F_q(t + h) = H_q(t + h) & \forall t \in [0, T] \backslash E \\ F_q(t) = H_q(t) & \forall t \in [0, h] \end{cases}$$

$$F_s : \begin{cases} F_s(t+h) = H_s(t+h) - \delta(t+h) & \forall t \in E \\ F_s(t+h) = H_s(t+h) & \forall t \in [0,T] \backslash E \\ F_s(t) = H_s(t) & \forall t \in [0,h] \end{cases}$$

$$F_r(t) = H_r(t) \ \forall r \neq q, s, \forall t \in [0, T+h].$$

If we evaluate (2) for this new F, which belongs to K_h, we get:

$$\int_{[0,T]} \langle C[H(t)], F(t+h) - H(t+h) \rangle =$$

$$= \int_E \delta(t+h)[C_q[H(t)] - C_s[H(t)]]dt < 0.$$

Now let us show that $(1) \Longrightarrow (2)$.
For each fixed $t \in [0,T] \backslash I$ $(|I| = 0)$ the following implication holds:

$$C_q[H(t)] < C_s[H(t)] \Longrightarrow H_q(t+h) = \mu_q(t+h) \ \text{ or } \ H_s(t+h) = \lambda_s(t+h)$$

For each fixed $w \in W$ we consider the following sets:

$$A := \{q \in \mathcal{R}(w) : H_q(t+h) < \mu_q(t+h)\}$$

$$B := \{s \in \mathcal{R}(w) : H_s(t+h) > \lambda_s(t+h)\}$$

It follows that $C_q[H(t)] \geq C_s[H(t)] \ \forall q \in A, \ \forall s \in B$. Thus, $\exists \gamma_{w,t} \in \Re$: $\inf_A C_q[H(t)] \geq \gamma_{w,t} \geq \sup_B C_s[H(t)]$.
Let now $F \in K_h$ and consider $F(t)$ for a fixed $t \in [0, T+h]$. Then, $\forall r \in \mathcal{R}(w)$ if $C_r[H(t)] < \gamma_{w,t}$ then $r \notin A$, therefore $H_r(t+h) = \mu_r(t+h)$ and $F_r(t+h) - H_r(t+h) \leq 0$, so that: $(C_r[H(t)] - \gamma_{w,t})(F_r(t+h) - H_r(t+h)) \geq 0$. Analogously, if $C_r[H(t)] > \gamma_{w,t}$ then $(C_r[H(t)] - \gamma_{w,t})(F_r(t+h) - H_r(t+h)) \geq 0$. Thus:

$$\sum_{r \in \mathcal{R}(w)} C_r[H(t)](F_r(t+h) - H_r(t+h)) >$$

$$> \gamma_{w,t} \sum_{r \in (\mathcal{R})} (F_r(t+h) - H_r(t+h)) = 0$$

Summing up $\forall w \in W$, the thesis follows by integration on $[0,T]$.

3. Existence of Equilibria.

We say that C satisfies condition (i) on K_h if $\forall H, F \in K_h$ the following implication holds:

$$\int_{[0,T]} \langle C[H(t)], F(t+h) - H(t+h) \rangle dt \geq 0 \Rightarrow$$

$$\Rightarrow \int_{[0,T]} \langle C[F(t)], H(t+h) - F(t+h) \rangle dt \leq 0$$

We then say that C satisfies condition (ii) if $\forall F \in K_h$ the function:

$$H \longrightarrow \int_{[0,T]} \langle C[H(t)], F(t+h) - H(t+h) \rangle dt$$

is weakly upper semicontinuous.

At last we say that C satisfies condition (iii) if, $\forall F, G \in K_h$, the function:

$$H \longrightarrow \int_{[0,T]} \langle C[H(t)], F(t+h) - G(t+h) \rangle dt$$

is upper semicontinuous on the segment $[F, G]$.

We may proof the following existence results:

Theorem 3.1:

Each of conditions (i) and (iii) or (ii) is sufficient for the existence of a solution to (2).

Proof:

Let us observe, first of all, that K_h is closed, convex and bounded, hence weakly compact.

If we put $t + h = s$ our problem is: find $H \in K_1$:

$$\int_{[h,T+h]} \langle C[H(s-h)], F(s) - H(s) \rangle ds \geq 0 \quad \forall F \in K_1 \qquad (3)$$

where:

$$K_1 := \{ F \in L^p([h, T+2h], \mathcal{R}^\mathcal{R}) \mid \lambda(s) \leq F(s) \leq \mu(s)$$

$$\text{a.e. } s \in [0, T + 2h],$$

$$\phi F(s) = \rho(s - h) \text{ a.e. on } [h, T + h] \}$$

Then, let us introduce a mapping C_h which satisfies the following condition:

$$C[H(s-h)] = C_h[H(s)] \quad \forall s \in [h, T+h]$$

then (3) reads:

$$H \in K_1 \text{ and } \int_{[h,T+h]} \langle C_h[H(s)], F(s) - H(s) \rangle ds \geq 0 \ \forall F \in K_1 \qquad (4)$$

With respect to K_1 and to C_h we can then apply corollary 5.1 of [2] and give sufficient conditions for the existence of a solution to (4). However we want to give conditions directly on C.

If C satisfies condition (i) on K_h $\forall H, F \in K_h$ the following implication holds:

$$\int_{[0,T]} \langle C[H(t)], F(t+h) - H(t+h) \rangle dt \geq 0 \Rightarrow$$

$$\Rightarrow \int_{[0,T]} \langle C[F(t)], H(t+h) - F(t+h) \rangle dt \leq 0$$

which implies that C_h is pseudomonotone on K_1.

If C satisfies condition (ii) $\forall F \in K_h$ the function:

$$H \longrightarrow \int_{[0,T]} \langle C[H(t)], F(t+h) - H(t+h) \rangle dt$$

is weakly upper semicontinuous. This condition implies that C_h is weakly hemicontinuous on K_1.

At last if C satisfies condition (iii) , $\forall F, G \in K_h$, the function:

$$H \longrightarrow \int_{[0,T]} \langle C[H(t)], F(t+h) - G(t+h) \rangle dt$$

is upper semicontinuous on the segment $[F, G]$. Thus, the operator C_h is hemicontinuous along segments. According to the corollary above mentioned we can now state that each of conditions (i) and (iii) or (ii) is sufficient for the existence of a solution to (3).

4. An example.

Let us consider a network with only one O-D pair $w_1 := w$ and suppose that there are 3 routes which connect w: R_1, R_2, R_3. For our purposes we are not interested in the detailed structure of the network (i.e. the link structure of each route). First of all we consider the case without delay, in order to compare the two cases. Let $\rho(t) = 2 + t$ and $[0, T] = [0, 1]$, $\lambda(t) = 0$ and $\mu(t) = +\infty$, so that:

$$K := \{F \in L^2([0,1], \Re^3) \mid 0 \leq F(t)$$

$$F_1(t) + F_2(t) + F_3(t) = 2 + t \ \text{ a.e. } [0,1]\}$$

Now we assign the cost operator:

$$C : \begin{cases} C_1[F(t)] = F_1(t) + 2F_2(t) & \forall t \in [0,1] \\ C_2[F(t)] = F_2(t) & \forall t \in [0,1] \\ C_3[F(t)] = 2F_2(t) + F_3(t) & \forall t \in [0,1] \end{cases}$$

With these data is easily seen that $H(t) = (0, 2+t, 0)$ is an equilibrium flux. Let us now introduce a delay $h = 1$ and consider the new feasible set:

$$K_2 := \{F \in L^2([0,2], \Re^3) \mid 0 \leq F(t), \quad \text{a.e. } t \in [0,2],$$

$$F_1(t+1) + F_2(t+1) + F_3(t+1) = t + 2 \ \text{ a.e. } t \in [0,1] \}$$

The functional form of the cost operator is the same as before. In this case one can verify that: $H(t) = (0, t+1, 0)$ is a retarded equilibrium. Let us notice, however, that we could change the values of the flux in $[0,1]$ and still obtain an equilibrium flux (provided, of course, it belongs to the feasible set). This fact is easily understood because our concept of *retarded equilibrium* requires that fluxes "reorganize" in order to reach the equilibrium *after* they "know" the cost .

References

[1] J.D. Murray, *Mathematical Biology*, Springer Verlag, 1993.

[2] P. Daniele, A. Maugeri and W. Oettli, *Time-Dependent Traffic Equilibria*, Jota 103, No. 3 December 1999. (See also references therein).

[3] J. Hale, *Functional Differential Equations*, Springer Verlag New York 1971.

[4] M.J. Smith,*A new Dynamic Traffic Model And the Existence and calculation of Dynamic User Equilibria on Congested Capacity-Constrained Road Networks*, Transportation Research 27B (1993), 49–63.

NEW RESULTS ON LOCAL MINIMA AND THEIR APPLICATIONS

Biagio Ricceri

Department of Mathematics

University of Catania

Viale A. Doria 6, 95125 Catania, Italy

e–mail:ricceri@dipmat.unict.it

Abstract   In this paper, I give an overview of the rapidly growing research work originated from my recent results on local minima of sequentially weakly lower semicontinuous functionals in reflexive Banach spaces. Particular emphasis is placed on the applications of the basic theory to differential equations.

Keywords: Local minima; sequential weak lower semicontinuity; variational methods; Dirichlet problem; Neumann problem; multiplicity of solutions.

The aim of this paper is to give an overview of the rapidly growing research work originated from [7] and [10].

The problem treated in those papers is essentially the following: given a topological space X, a real interval I, and a function $f : X \times I \to \mathbf{R}$, find some $\lambda \in I$ such that the function $f(\cdot, \lambda)$ has local minima.

The contribution to this problem offered in [7] was as follows:

Theorem 1 ([7], Theorem 1). - *Let X be a topological space; $I \subseteq \mathbf{R}$ an interval; $f : X \times I \to \mathbf{R}$ a given function. Assume that:*
(a) for each $x \in X$, the set

$$\{\lambda \in I : f(x, \lambda) \geq 0\}$$

is closed in I and the set

$$\{\lambda \in I : f(x, \lambda) > 0\}$$

F. Giannessi et al (eds.),

Equilibrium Problems: Nonsmooth Optimization and Variational Inequality Models, 255–268.

is non-empty, connected and open in I;
(b) for each $\lambda \in I$, the set

$$\{x \in X : f(x, \lambda) \leq 0\}$$

is non-empty, closed and sequentially compact;
(c) there is $\lambda_0 \in I$ such that the set

$$\{x \in X : f(x, \lambda_0) \leq 0\}$$

is connected.

  Under such assumptions, there exist $x^* \in X$, $\lambda^* \in I$, a sequence $\{\lambda_n\}$ in I converging to λ^* and a neighbourhood U of x^* such that $f(x^*, \lambda^*) = 0$ and $f(x, \lambda_n) > 0$ for all $n \in \mathbf{N}$, $x \in U$. So, in particular, x^* is a local minimum for $f(\cdot, \lambda^*)$.

  In [8], applying Theorem 1 jointly with Theorem 1 of [6], I obtained the following three critical points theorem:

Theorem 2 ([8], Theorem 3.1). - *Let X be a separable and reflexive real Banach space; $\Phi : X \to \mathbf{R}$ a continuously Gâteaux differentiable and convex functional whose Gâteaux derivative admits a continuous inverse on X^*; $\Psi : X \to \mathbf{R}$ a continuously Gâteaux differentiable functional whose Gâteaux derivative is compact. Assume that*

$$\lim_{\|x\| \to +\infty} (\Phi(x) + \lambda\Psi(x)) = +\infty$$

for all $\lambda \geq 0$, and that there exists a continuous concave function $h : [0, +\infty[\to \mathbf{R}$ such that

$$\sup_{\lambda \geq 0} \inf_{x \in X} (\Phi(x) + \lambda\Psi(x) + h(\lambda)) < \inf_{x \in X} \sup_{\lambda \geq 0} (\Phi(x) + \lambda\Psi(x) + h(\lambda)) . \quad (1)$$

Then, there exists $\lambda^* > 0$ such that the equation

$$\Phi'(x) + \lambda^*\Psi'(x) = 0$$

has at least three solutions in X.

  A natural way to get (1), with a linear h, is provided by the following

  PROPOSITION 1 ([8], Proposition 3.1). - *Let X be a non-empty set, and Φ, J two real functions on X. Assume that there are $r > 0$, $x_0, x_1 \in X$ such that*

$$\Phi(x_0) = J(x_0) = 0,$$

$$\Phi(x_1) > r,$$

$$\sup_{x \in \Phi^{-1}(]-\infty,r])} J(x) < r\frac{J(x_1)}{\Phi(x_1)}.$$

Then, for each ρ satisfying

$$\sup_{x \in \Phi^{-1}(]-\infty,r])} J(x) < \rho < r\frac{J(x_1)}{\Phi(x_1)}$$

one has

$$\sup_{\lambda \geq 0} \inf_{x \in X} (\Phi(x) + \lambda(\rho - J(x))) < \inf_{x \in X} \sup_{\lambda \geq 0} (\Phi(x) + \lambda(\rho - J(x))).$$

Again in [8] (Remark 5.2), I asked whether it may happen that, for a suitable continuous concave function h, (1) holds, while, for every $\rho \in \mathbf{R}$, one has

$$\sup_{\lambda \geq 0} \inf_{x \in X} (\Phi(x) + \lambda(\Psi(x) + \rho)) = \inf_{x \in X} \sup_{\lambda \geq 0} (\Phi(x) + \lambda(\Psi(x) + \rho)) .$$

A complete and deep answer to this question was given by G. Cordaro ([3]) who proved what follows:

Theorem 3 ([3], Theorem 1). - *Let E be a separable and reflexive real Banach space, X a weakly closed and unbounded subset of E, and Φ, Ψ two (non-constant) sequentially weakly lower semicontinuous functionals on X such that*

$$\lim_{x \in X, \|x\| \to +\infty} (\Phi(x) + \lambda\Psi(x)) = +\infty$$

for all $\lambda \geq 0$.

Then, the following assertions are equivalent:

(i) For every $\rho \in \mathbf{R}$, one has

$$\sup_{\lambda \geq 0} \inf_{x \in X} (\Phi(x) + \lambda(\Psi(x) + \rho)) = \inf_{x \in X} \sup_{\lambda \geq 0} (\Phi(x) + \lambda(\Psi(x) + \rho)) .$$

(ii) For every $\rho \in]\inf_X \Psi, \sup_X \Psi[$, one has

$$\sup_{x \in \Psi^{-1}(]\rho,+\infty[)} \frac{\Phi(x) - \inf_{\Psi^{-1}(]-\infty,\rho])} \Phi}{\rho - \Psi(x)} \leq$$

$$\leq \inf_{x \in \Psi^{-1}(]-\infty,\rho[)} \frac{\Phi(x) - \inf_{\Psi^{-1}(]-\infty,\rho])} \Phi}{\rho - \Psi(x)} .$$

(iii) For every concave function $h : [0, +\infty[\to \mathbf{R}$, *one has*

$$\sup_{\lambda \geq 0} \inf_{x \in X} (\Phi(x) + \lambda \Psi(x) + h(\lambda)) = \inf_{x \in X} \sup_{\lambda \geq 0} (\Phi(x) + \lambda \Psi(x) + h(\lambda)) .$$

A few time after I wrote [8], J. Saint Raymond, in response to some questions of mine, proved the following very deep result:

Theorem 4 ([14]). - *Let* X *be a compact metric space,* $I \subseteq \mathbf{R}$ *a compact interval, and* $f : X \times I \to \mathbf{R}$ *a function satisfying the following conditions:*
(α) for each $x \in X$, *the function* $f(x, \cdot)$ *is continuous and concave;*
(β) for each $\lambda \in I$, *the function* $f(\cdot, \lambda)$ *is lower semicontinuous;*
(γ) one has

$$a = \sup_{\lambda \in I} \inf_{x \in X} f(x, \lambda) < b = \inf_{x \in X} \sup_{\lambda \in I} f(x, \lambda) .$$

Then, for each $c \in]a, b[$, *there exist a strictly increasing continuous function* $\varphi :]a, c[\to I$ *and an upper semicontinuous multifunction* $F :]a, c[\to 2^X$, *with non-empty closed values, such that, for every* $r \in]a, c[$, *one has*

$$\inf_{x \in F(r)} f(x, \varphi(r)) = r$$

and each $x \in F(r)$ *satisfying* $f(x, \varphi(r)) = r$ *is a local, non-absolute minimum of the function* $f(\cdot, \varphi(r))$.

Using Theorem 4 jointly with two propositions established in [7], I then was able to improve Theorem 2 in the following way:

Theorem 5 ([9], Theorem 1). - *Let* X *be a separable and reflexive real Banach space;* $\Phi : X \to \mathbf{R}$ *a continuously Gâteaux differentiable and sequentially weakly lower semicontinuous functional whose Gâteaux derivative admits a continuous inverse on* X^*; $\Psi : X \to \mathbf{R}$ *a continuously Gâteaux differentiable functional whose Gâteaux derivative is compact;* $I \subseteq \mathbf{R}$ *an interval. Assume that*

$$\lim_{\|x\| \to +\infty} (\Phi(x) + \lambda \Psi(x)) = +\infty$$

for all $\lambda \in I$, *and that there exists a continuous concave function* $h : I \to \mathbf{R}$ *such that*

$$\sup_{\lambda \in I} \inf_{x \in X} (\Phi(x) + \lambda \Psi(x) + h(\lambda)) < \inf_{x \in X} \sup_{\lambda \in I} (\Phi(x) + \lambda \Psi(x) + h(\lambda)) .$$

Then, there exist an open interval $J \subseteq I$ and a positive real number ρ such that, for each $\lambda \in J$, the equation

$$\Phi'(x) + \lambda \Psi'(x) = 0$$

has at least three solutions in X whose norms are less than ρ.

Very recently, in [5], S. A. Marano and D. Motreanu established a non-smooth version of Theorem 5.

Here are two applications of Theorem 5.

Theorem 6 ([9], Theorem 4; [8], Theorem 1.1). - *Let $\Omega \subseteq \mathbf{R}^n$ be an open bounded set, with smooth boundary, and $f, g : \mathbf{R} \to \mathbf{R}$ two continuous functions, with $\sup_{\xi \in \mathbf{R}} \int_0^\xi f(t)dt > 0$. Assume that there are four positive constants a, q, s, γ, with $q < \frac{n+2}{n-2}$ (if $n > 2$), $s < 2$ and $\gamma > 2$, such that*

$$\max\{|f(\xi)|, |g(\xi)|\} \le a(1 + |\xi|^q) \ \forall \xi \in \mathbf{R},$$

$$\max \left\{ \int_0^\xi f(t)dt, \left| \int_0^\xi g(t)dt \right| \right\} \le a(1 + |\xi|^s) \ \forall \xi \in \mathbf{R}$$

and

$$\limsup_{\xi \to 0} \frac{\int_0^\xi f(t)dt}{|\xi|^\gamma} < +\infty \ .$$

Then, there exists $\delta > 0$ such that, for each $\mu \in [-\delta, \delta]$, there exist a positive real number ρ_μ and an open interval $J_\mu \subseteq [0, +\infty[$ such that, for each $\lambda \in J_\mu$, the problem

$$\begin{cases} -\Delta u = \lambda(f(u) + \mu g(u)) & \text{in } \Omega \\ u_{|\partial\Omega} = 0 \end{cases}$$

has at least three distinct weak solutions in $W_0^{1,2}(\Omega)$ whose norms are less than ρ_μ.

Recall that, if $f : \Omega \times \mathbf{R} \to \mathbf{R}$ is a Carathéodory function, a weak solution of the problem

$$\begin{cases} -\Delta u = f(x, u) & \text{in } \Omega \\ u_{|\partial\Omega} = 0 \end{cases}$$

is any $u \in W_0^{1,2}(\Omega)$ such that

$$\int_\Omega \nabla u(x) \nabla v(x)dx - \int_\Omega f(x, u(x))v(x)dx = 0$$

for all $v \in W_0^{1,2}(\Omega)$.

For another result related to Theorem 6 see [4].

Theorem 7 ([2], Theorem 2). - *Let $f : \mathbf{R} \to \mathbf{R}$ be a continuous function. Assume that there exist four positive constants c, d, a, s, with $c < \sqrt{2}d$ and $s < 2$, such that*

$$f(t) \geq 0$$

for all $t \in [-c, \max\{c, d\}]$,

$$4d^2 \int_0^c f(t)dt < c^2 \int_0^d f(t)dt \, ,$$

and

$$\int_0^\xi f(t)dt \leq a(1 + |\xi|^s)$$

for all $\xi \in \mathbf{R}$.

Then, there exist a positive real number ρ and an open interval $J \subseteq [0, +\infty[$ such that, for each $\lambda \in J$, the problem

$$\begin{cases} -u'' = \lambda f(u) & in \ [0, 1] \\ u(0) = u(1) = 0 \end{cases}$$

has at least three distinct classical solutions whose derivatives have norms in $L^2([0, 1])$ less than ρ.

Now, come to [10]. There, I established the following general variational principle:

Theorem 8 ([10], Theorem 2.1). - *Let X be a topological space, and let $\Phi, \Psi : X \to \mathbf{R}$ be two sequentially lower semicontinuous functions. Denote by I the set of all $\rho > \inf_X \Psi$ such that the set $\Psi^{-1}(] - \infty, \rho[)$ is contained in some sequentially compact subset of X. Assume that $I \neq \emptyset$. For each $\rho \in I$, denote by $\mathcal{F}_\rho$ the family of all sequentially compact subsets of X containing $\Psi^{-1}(] - \infty, \rho[)$, and put*

$$\alpha(\rho) = \sup_{K \in \mathcal{F}_\rho} \inf_K \Phi.$$

Then, for each $\rho \in I$ and each λ satisfying

$$\lambda > \inf_{x \in \Psi^{-1}(]-\infty,\rho[)} \frac{\Phi(x) - \alpha(\rho)}{\rho - \Psi(x)}$$

the restriction of the function $\Phi + \lambda\Psi$ *to* $\Psi^{-1}(]-\infty, \rho[)$ *has a global minimum.*

One of the most significant features of Theorem 8 is the possibility to get from it multiplicity results for local minima. Indeed, we have

Theorem 9 ([10], Theorem 2.2). - *Let the assumptions of Theorem 8 be satisfied. In addition, suppose*

$$\sup I = +\infty$$

and

$$\gamma < +\infty ,$$

where

$$\gamma = \liminf_{\rho \to +\infty} \inf_{x \in \Psi^{-1}(]-\infty, \rho[)} \frac{\Phi(x) - \alpha(\rho)}{\rho - \Psi(x)} .$$

Finally, denote by τ *the weakest topology on* X *for which* Ψ *is upper semicontinuous.*

Then, for each $\lambda > \gamma$, *the following alternative holds: either* $\Phi + \lambda\Psi$ *has a global minimum, or there exists a sequence* $\{x_n\}$ *of* τ-*local minima of* $\Phi + \lambda\Psi$ *such that*

$$\lim_{n \to \infty} \Psi(x_n) = +\infty .$$

Another multiplicity result is as follows:

Theorem 10 ([10], Theorem 2.3). - *Let the assumptions of Theorem 8 be satisfied. In addition, suppose*

$$\delta < +\infty ,$$

where

$$\delta = \liminf_{\rho \to \left(\inf_{X} \Psi\right)^{+}} \inf_{x \in \Psi^{-1}(]-\infty, \rho[)} \frac{\Phi(x) - \alpha(\rho)}{\rho - \Psi(x)} .$$

Finally, denote by τ *the weakest topology on* X *for which* Ψ *is upper semicontinuous.*

Then, for each $\lambda > \delta$, *there exists a sequence of* τ-*local minima of* $\Phi + \lambda\Psi$ *which converges to a global minimum of* Ψ.

The next result groups together the versions of Theorems 8, 9 and 10 which are directly applicable to differential equations.

Theorem 11 ([10], Theorem 2.5). - *Let X be a reflexive real Banach space, and let $\Phi, \Psi : X \to \mathbf{R}$ be two sequentially weakly lower semicontinuous and Gâteaux differentiable functionals. Assume also that Ψ is (strongly) continuous and satisfies $\lim_{\|x\| \to +\infty} \Psi(x) = +\infty$. For each $\rho > \inf_X \Psi$, put*

$$\varphi(\rho) = \inf_{x \in \Psi^{-1}(]-\infty,\rho[)} \frac{\Phi(x) - \inf_{\overline{(\Psi^{-1}(]-\infty,\rho[))}_w} \Phi}{\rho - \Psi(x)},$$

where $\overline{(\Psi^{-1}(] - \infty, \rho[))}_w$ is the closure of $\Psi^{-1}(] - \infty, \rho[)$ in the weak topology. Furthermore, set

$$\gamma = \liminf_{\rho \to +\infty} \varphi(\rho)$$

and

$$\delta = \liminf_{\rho \to \left(\inf_X \Psi\right)^+} \varphi(\rho) .$$

Then, the following conclusions hold:
(a)   For each $\rho > \inf_X \Psi$ and each $\lambda > \varphi(\rho)$, the functional $\Phi + \lambda\Psi$ has a critical point which lies in $\Psi^{-1}(] - \infty, \rho[)$.
(b)   If $\gamma < +\infty$, then, for each $\lambda > \gamma$, the following alternative holds: either $\Phi + \lambda\Psi$ has a global minimum, or there exists a sequence $\{x_n\}$ of critical points of $\Phi + \lambda\Psi$ such that $\lim_{n \to \infty} \Psi(x_n) = +\infty$.
(c)   If $\delta < +\infty$, then, for each $\lambda > \delta$, the following alternative holds: either there exists a global minimum of Ψ which is a local minimum of $\Phi + \lambda\Psi$, or there exists a sequence of pairwise distinct critical points of $\Phi + \mu\Psi$, with $\lim_{n \to \infty} \Psi(x_n) = \inf_X \Psi$, which weakly converges to a global minimum of Ψ.

Let's see now some specific applications of Theorem 11. In [11], applying it (part (a)), I obtained the following general existence theorem:

Theorem 12 ([11], Theorem 1). - *Let $\Omega \subset \mathbf{R}^n$ $(n \geq 3)$ be an open bounded set, with smooth boundary, let $\alpha, \beta \in L^{\frac{2n}{n+2}}(\Omega)$, and let $f, g, h, l : \Omega \times \mathbf{R} \to \mathbf{R}$ be four Carathéodory functions satisfying, in $\Omega \times \mathbf{R}$, the following conditions*

$$\max\{|f(x,\xi)|, |g(x,\xi)|\} \leq a(1 + |\xi|^q) ,$$

$$\int_0^\xi (g(x,t) + h(x,t))dt \leq a(1 + |\xi|^s) ,$$

$$\max\{|h(x,\xi)|, |l(x,\xi)|\} \le a(1 + |\xi|^{\frac{n+2}{n-2}}) ,$$

where a, q, s are three positive constants, with $s < 2$ and $q < \frac{n+2}{n-2}$. Finally, assume that, for each $x \in \Omega$, both the functions $h(x, \cdot)$ and $l(x, \cdot)$ are non-increasing. For each $u \in W_0^{1,2}(\Omega)$, put

$$\Phi(u) = -\int_\Omega \left(\int_0^{u(x)} (f(x,\xi) + l(x,\xi))d\xi + \beta(x)u(x) \right) dx$$

and

$$\Psi(u) = \int_\Omega |\nabla u(x)|^2 dx -$$

$$- 2\int_\Omega \left(\int_0^{u(x)} (g(x,\xi) + h(x,\xi))d\xi + \alpha(x)u(x) \right) dx .$$

Then, for each $r > \inf_{W_0^{1,2}(\Omega)} \Psi$ and each μ satisfying

$$\mu > \inf_{u \in \Psi^{-1}(]-\infty,r[)} \frac{\Phi(u) - \inf_{\overline{(\Psi^{-1}(]-\infty,r[))}_w} \Phi}{r - \Psi(u)} ,$$

where $\overline{(\Psi^{-1}(] - \infty, r[))}_w$ is the closure of $\Psi^{-1}(] - \infty, r[)$ in the weak topology of $W_0^{1,2}(\Omega)$, the problem

$$\begin{cases} -\Delta u = g(x,u) + h(x,u) + \alpha(x) + \frac{1}{2\mu}(f(x,u) + l(x,u) + \beta(x)) & in \ \Omega \\ \\ u_{|\partial\Omega} = 0 \end{cases}$$

has at least one weak solution that lies in $\Psi^{-1}(] - \infty, r[)$.

Concerning Theorem 12, note that, even when the existence of a solution is obvious, it gives an additional information about the location of it. A typical example is as follows:

Proposition 2 ([11], Proposition 1). - *Let $n = 3$, let λ_1 be the first eigenvalue of the problem*

$$\begin{cases} -\Delta u = \lambda u & in \ \Omega \\ \\ u_{|\partial\Omega} = 0 \end{cases}$$

and set

$$c = \left(\sup_{u \in W_0^{1,2}(\Omega) \setminus \{0\}} \frac{(\int_\Omega |u(x)|^4 dx)^{\frac{1}{4}}}{(\int_\Omega |\nabla u(x)|^2 dx)^{\frac{1}{2}}} \right)^4 .$$

Then, for each $\beta \in L^2(\Omega) \setminus \{0\}$ and each a satisfying

$$0 < a < \frac{2\lambda_1}{27c\|\beta\|_{L^2(\Omega)}^2}$$

the (unique) weak solution u_0 of the problem

$$\begin{cases} -\Delta u = -u^5 + \beta(x) & in\ \Omega \\ u_{|\partial\Omega} = 0 \end{cases}$$

satisfies the inequality

$$\int_\Omega |\nabla u_0(x)|^2 dx + \frac{1}{3}\int_\Omega |u_0(x)|^6 dx + \frac{a}{2}\int_\Omega |u_0(x)|^4 dx < \left(\frac{2\|\beta\|_{L^2(\Omega)}}{ac\lambda_1^{\frac{1}{2}}}\right)^{\frac{2}{3}}.$$

The following result, by Ambrosetti and Rabinowitz, is certainly one the most classical existence theorems for nonlinear elliptic equations.

Theorem 13 ([1], Theorem 3.10). - *Let $\Omega \subset \mathbf{R}^n$ be a bounded open set, with smooth boundary, and $f : \Omega \times \mathbf{R} \to \mathbf{R}$ is a Carathéodory function. Assume that:*
(i) there are two positive constants a, q, with $q < \frac{n+2}{n-2}$ if $n \geq 3$, such that

$$|f(x, \xi)| \leq a(1 + |\xi|^q)$$

for all $(x, \xi) \in \Omega \times \mathbf{R}$;
(ii) there are constants $r \geq 0$ and $c > 2$ such that

$$0 < c\int_0^\xi f(x, t)dt \leq \xi f(x, \xi)$$

for all $(x, \xi) \in \Omega \times \mathbf{R}$ with $|\xi| \geq r$;
(iii) one has

$$\lim_{\xi \to 0} \frac{f(x, \xi)}{\xi} = 0$$

uniformly with respect to x.
  Then, the problem

$$\begin{cases} -\Delta u = f(x, u) & in\ \Omega \\ u_{|\partial\Omega} = 0. \end{cases}$$

has a non zero weak solution.

Concerning Theorem 13, what can be said whether condition (iii) is removed at all ? In [13], I provided an answer to this question, again applying Theorem 11. The result is as follows:

Theorem 14 ([13], Theorem 4). - *Assume that all the assumptions of Theorem 13 are satisfied but condition (iii).*
   Then, for each $\rho > 0$ and each μ satisfying

$$\mu > \inf_{u \in B_\rho} \frac{\sup_{v \in B_\rho} \int_\Omega \left(\int_0^{v(x)} f(x, \xi) d\xi \right) dx - \int_\Omega \left(\int_0^{u(x)} f(x, \xi) d\xi \right) dx}{\rho - \int_\Omega |\nabla u(x)|^2 dx} \, ,$$

where

$$B_\rho = \left\{ u \in W_0^{1,2}(\Omega) : \int_\Omega |\nabla u(x)|^2 dx < \rho \right\} ,$$

the problem

$$\begin{cases} -\Delta u = \frac{1}{2\mu} f(x, u) & \text{in } \Omega \\ \\ u_{|\partial\Omega} = 0 \end{cases}$$

has at least two weak solutions one of which lies in B_ρ.

I conclude recalling two results on the Neumann problem obtained in [12] as applications of parts (b) and (c) of Theorem 11 respectively.

In the sequel, $\Omega \subset \mathbf{R}^k$ is a bounded open set, with boundary of class C^1, and $\lambda \in L^\infty(\Omega)$, with ess $\inf_\Omega \lambda > 0$. Let $p > 1$, and let $\varphi : \Omega \times \mathbf{R} \to \mathbf{R}$ be a Carathéodory function.

Recall that a weak solution of the Neumann problem

$$\begin{cases} -\text{div}(|\nabla u|^{p-2}\nabla u) + \lambda(x)|u|^{p-2}u = \varphi(x, u) & \text{in } \Omega \\ \\ \frac{\partial u}{\partial \nu} = 0 & \text{on } \partial\Omega , \end{cases}$$

where ν is the outer unit normal to $\partial\Omega$, is any $u \in W^{1,p}(\Omega)$ such that

$$\int_\Omega |\nabla u(x)|^{p-2}\nabla u(x)\nabla v(x)dx + \int_\Omega \lambda(x)|u(x)|^{p-2}u(x)v(x)dx -$$

$$- \int_\Omega \varphi(x, u(x))v(x)dx = 0$$

for all $v \in W^{1,p}(\Omega)$.

Theorem 15 ([12], Theorem 3). - *Assume $p > k$. Let $f : \mathbf{R} \to \mathbf{R}$ be a continuous function, and $\{a_n\}$, $\{b_n\}$ two sequences in $\mathbf{R}^+$ satisfying*

$$a_n < b_n \ \forall n \in \mathbf{N}, \ \lim_{n \to \infty} b_n = +\infty, \ \lim_{n \to \infty} \frac{a_n}{b_n} = 0 \ ,$$

$$\max \left\{ \sup_{\xi \in [a_n, b_n]} \int_{a_n}^{\xi} f(t)dt, \ \sup_{\xi \in [-b_n, -a_n]} \int_{-a_n}^{\xi} f(t)dt \right\} \leq 0 \ \forall n \in \mathbf{N} \ ,$$

and

$$\limsup_{|\xi| \to +\infty} \frac{\int_0^{\xi} f(t)dt}{|\xi|^p} = +\infty \ .$$

Then, for every $\alpha, \beta \in L^1(\Omega)$, with $\min\{\alpha(x), \beta(x)\} \geq 0$ a.e. in Ω and $\alpha \neq 0$, and for every continuous function $g : \mathbf{R} \to \mathbf{R}$ satisfying

$$\sup_{\xi \in \mathbf{R}} \int_0^{\xi} g(t)dt \leq 0$$

and

$$\liminf_{|\xi| \to +\infty} \frac{\int_0^{\xi} g(t)}{|\xi|^p} > -\infty \ ,$$

the problem

$$\begin{cases} -\mathrm{div}(|\nabla u|^{p-2}\nabla u) + \lambda(x)|u|^{p-2}u = \alpha(x)f(u) + \beta(x)g(u) & in \ \Omega \\ \\ \frac{\partial u}{\partial \nu} = 0 & on \ \partial\Omega \end{cases}$$

admits an unbounded sequence of weak solutions in $W^{1,p}(\Omega)$.

An explicit example of application of Theorem 15 is as follows:

Example 1 ([12], Example 2). - Let $p > k$. Then, for each $\eta \in L^1(\Omega)$ with ess $\inf_\Omega \eta > 0$, the problem

$$\begin{cases} -\mathrm{div}(|\nabla u|^{p-2}\nabla u) = \\ \quad = \eta(x) \left(\sum_{n=1}^{\infty} \left(\mathrm{dist}(u, \mathbf{R} \setminus [n!n, (n+1)!]) \right)^p - |u|^{p-2}u \right) & in \ \Omega \\ \\ \frac{\partial u}{\partial \nu} = 0 & on \ \partial\Omega \end{cases}$$

admits an unbounded sequence of weak solutions in $W^{1,p}(\Omega)$.

Theorem 16 ([12], Theorem 4). - *Assume $p > k$. Let $f : \mathbf{R} \to \mathbf{R}$ be a continuous function, and $\{a_n\}$, $\{b_n\}$ two sequences in $\mathbf{R}^+$ satisfying*

$$a_n < b_n \ \forall n \in \mathbf{N}, \ \lim_{n \to \infty} b_n = 0, \ \lim_{n \to \infty} \frac{a_n}{b_n} = 0 \ ,$$

$$\max\left\{\sup_{\xi\in[a_n,b_n]}\int_{a_n}^{\xi}f(t)dt,\ \sup_{\xi\in[-b_n,-a_n]}\int_{-a_n}^{\xi}f(t)dt\right\}\leq 0\ \forall n\in\mathbf{N}\ ,$$

and

$$\limsup_{\xi\to 0}\frac{\int_0^{\xi}f(t)dt}{|\xi|^p}=+\infty\ .$$

Then, for every $\alpha,\beta\in L^1(\Omega)$, with $\min\{\alpha(x),\beta(x)\}\geq 0$ a.e. in Ω and $\alpha\neq 0$, and for every continuous function $g:\mathbf{R}\to\mathbf{R}$ satisfying

$$\sup_{\xi\in\mathbf{R}}\int_0^{\xi}g(t)dt\leq 0$$

and

$$\liminf_{\xi\to 0}\frac{\int_0^{\xi}g(t)}{|\xi|^p}>-\infty\ ,$$

the problem

$$\begin{cases} -\mathrm{div}(|\nabla u|^{p-2}\nabla u)+\lambda(x)|u|^{p-2}u=\alpha(x)f(u)+\beta(x)g(u) & in\ \Omega \\ \frac{\partial u}{\partial\nu}=0 & on\ \partial\Omega \end{cases}$$

admits a sequence of non zero weak solutions which strongly converges to 0 in $W^{1,p}(\Omega)$.

I wish to stress that the weak solutions whose existence is ensured by Theorems 15 and 16 are actually local minima of the corresponding energy functional. This remark is important in connection with the regularity of such solutions.

References

[1] A. Ambrosetti and P. H. Rabinowitz, *Dual variational methods in critical point theory and applications*, J. Funct. Anal. 14 (1973), 349–381.

[2] G. Bonanno, *Existence of three solutions for a two point boundary value problem*, Appl. Math. Lett., to appear.

[3] G. Cordaro, *On a minimax problem of Ricceri*, J. Inequal. Appl., to appear.

[4] S. A. Marano and D. Motreanu, *Existence of two nontrivial solutions for a class of elliptic eigenvalue problems*, Arch. Math. (Basel), to appear.

[5] S. A. Marano and D. Motreanu, *On a three critical points theorem for non-differentiable functions and applications to nonlinear boundary value problems*, Nonlin. Anal., to appear.

[6] P. Pucci and J. Serrin, *A mountain pass theorem*, J. Differential Equations, 60 (1985), 142–149.

[7] B. Ricceri, *A new method for the study of nonlinear eigenvalue problems*, C. R. Acad. Sci. Paris, Série I, 328 (1999), 251–256.

[8] B. Ricceri, *Existence of three solutions for a class of elliptic eigenvalue problems*, Math. Comput. Modelling, special issue on "Advanced topics in nonlinear operator theory", to appear.

[9] B. Ricceri, *On a three critical points theorem*, Arch. Math. (Basel), to appear.

[10] B. Ricceri, *A general variational principle and some of its applications*, J. Comput. Appl. Math. 113 (2000), 401–410.

[11] B. Ricceri, *Existence and location of solutions to the Dirichlet problem for a class of nonlinear elliptic equations*, Appl. Math. Lett., to appear.

[12] B. Ricceri, *Infinitely many solutions of the Neumann problem for elliptic equations involving the p-Laplacian*, Bull. London Math. Soc., to appear.

[13] B. Ricceri, *On a classical existence theorem for nonlinear elliptic equations*, in "Experimental, constructive and nonlinear analysis", M. Théra ed., Canad. Math. Soc., to appear.

[14] J. Saint Raymond, *On a minimax theorem*, Arch. Math. (Basel), to appear.

AN OVERVIEW ON PROJECTION–TYPE METHODS FOR CONVEX LARGE–SCALE QUADRATIC PROGRAMS *

Valeria Ruggiero
Dipartimento di Matematica
Università di Ferrara
Via Machiavelli, 35
44100 – Ferrara, Italy
e–mail: rgv@dns.unife.it

Luca Zanni
Dipartimento di Matematica
Università di Modena e Reggio Emilia
Via Campi, 213/b
41100 Modena, Italy
e–mail: zanniluca@unimo.it

Abstract   A well-known approach for solving large and sparse linearly constrained quadratic programming (QP) problems is given by the splitting and projection methods. After a survey on these classical methods, we show that they can be unified in a general iterative scheme consisting in to solve a sequence of QP subproblems with the constraints of the original problem and an easily solvable Hessian matrix. A convergence theorem is given for this general scheme. In order to improve the numerical performance of these methods, we introduce two variants of a projection-type scheme that use a variable projection parameter at each step. The two variable projection methods differ in the strategy used to assure a sufficient decrease of the objective function at each iteration. We prove, under very general hypotheses, the convergence of these schemes and we propose two practical, nonexpensive and efficient updating rules for the

*This work was supported by MURST Project "Numerical Analysis: Methods and Mathematical Software" and by CNR Research Contribution N. 98.01029.CT01, Italy.

F. Giannessi et al (eds.),
Equilibrium Problems: Nonsmooth Optimization and Variational Inequality Models, 269–300.
© 2001 *Kluwer Academic Publishers.*

projection parameter. An extensive numerical experimentation shows
the effectiveness of the variable projection–type methods.

Keywords: Convex quadratic programs, large-scale problems, projection type methods, splitting methods.

1. Introduction.

Large scale quadratic programming (QP) problems are basic to many
applications of data analysis, such as image restoration [1], pattern recog-
nition based upon support vector machine technique [6] and constrained
bivariate interpolation [12]. Furthermore, efficient QP solvers are crucial
to the success of the class of symmetric cost approximation methods [37]
for large–scale nonlinear programming or variational inequality problems
when the auxiliary function is a convex quadratic function or is the gradient map of a convex quadratic function. See, for example, the sequential quadratic programming method or, for the variational inequalities,
the linear approximation methods [33] and the descent methods based
upon projective gap functions [43]. These methods are used in several
large–scale applications arising in, for example, discrete optimal control problems [19], [27], in planning problems under uncertainty [4], in
multicommodity network flow and traffic equilibrium problems [36].
In developing numerical algorithms for QP problems attention is largely
focused on three classes of methods: active–set methods, interior–point
and projection–type methods. Active–set methods are dominant in the
development of software for convex quadratic programs [26]. For example, the recent release Mark 18 of the NAG library introduces, for sparse
QP problems, the routine E04NKF [29] based on an active–set strategy.
(see also SNOPT [17] and MINOS [28] packages). Interior–point
methods, particularly efficient for solving linear programs, have been
extended in the last years to QP problems ([26], [30], [35], [42] and
references therein). Projection–type methods are developed within the
linear complementarity and the variational inequality problems, of which
the QP problems are a special case. This work is concerned with the
theoretical and numerical features of this last approach in the solution
of large–scale sparse QP problems. Indeed, the projection methods and
the splitting methods (that are special projection methods with the projection parameter equal to 1) allow any sparsity of the objective and
constraint matrices to be readily exploited, use little storage and are
well suited to implementation on parallel computers. Furthermore, these
methods involve the solution of a sequence of QP subproblems with easy

Hessian matrices; for their solution, we can make use of solvers well suited to the type of constraints required by the application in question. For example, in the case of a problem with bounds on the variables and only one equality constraint, the formulation of the QP problem as a sequence of separable QP subproblems enables specialized algorithms ([31], [34]) to be used for solving each subproblem.

In section 2, the main results relating to the classical projection–type and splitting methods for convex QP problems are summarized. In order to accelerate these methods, in sections 3 and 4 we propose two variants of an iterative scheme scaled gradient projection scheme that uses a variable projection parameter at each step. The two versions differ in the strategy that assures a sufficient decrease in the objective function $f(x)$ and, consequently, the convergence of the schemes.

The first variant, called Variable Projection Method (VPM), is introduced in [38] for strictly convex QP problems. Each iteration consists in a projection step with a variable projection parameter given by a rule based on heuristic considerations; when this step does not produce a sufficient decrease in $f(x)$, a limited line search for $f(x)$ along a feasible descent direction determined by the projection step is performed. In section 3, we prove the convergence of the VPM for convex QP problems. The second variant, called Adaptive Variable Projection Method (AVPM), can be viewed as a projection method with an adaptive choice of the projection parameter. As in the VPM, at each iteration an appropriate rule gives a tentative projection parameter. When the projection step with this value does not determine a sufficient decrease in $f(x)$, we use a search procedure on a projection arc [2], i.e., we reduce the projection parameter and perform a new projection step until a convenient decrease in $f(x)$ is obtained. In section 4, we describe the AVPM and we prove its convergence for convex QP problems.

In section 5, we propose two updating rules for varying the projection parameter in the previous methods and, in section 6, we discuss the solution of the inner subproblems.

Finally, in section 7, we report the results of numerical experiments involving the methods described in the previous sections. From these results, we can conclude that when, in the VPM and the AVPM, a convenient updating rule for the choice of the projection parameter is used, the corrective step along a descent direction or a projection arc arises in few iterations and these variable projection schemes have an efficient numerical behaviour.

In the sequel of the work, we denote by $\lambda_{min}(G)$ and $\lambda_{max}(G)$ the minimum and the maximum eigenvalues, respectively, of a symmetric matrix G. We indicate by $\|x\|$ the usual Euclidean norm of x and by $\|x\|_G$ the elliptic norm of x defined as $\sqrt{x^T G x}$ where G is a symmetric positive definite matrix. Furthermore, we denote by $\|B\|$ the norm of a matrix B induced by the vector Euclidean norm. Finally, we indicate by $[x]^+$ the orthogonal projection of a vector x onto a closed convex set K. i.e., $[x]^+ \arg\min_{y \in K} \|y - x\|$.

2. The projection and splitting methods.

We consider the following linearly constrained convex QP problem:

$$\begin{aligned} minimize \quad & f(x)\tfrac{1}{2}x^T G x + q^T x \\ subject\ to \quad & C x d, \qquad A x \geq b, \end{aligned} \tag{2.1}$$

where G is a symmetric positive semidefinite matrix of order n, C is an $m_e \times n$ matrix of full row rank ($m_e \leq n$) and A is an $m_i \times n$ matrix. We assume that the feasible region $K\{ex \mid Cxd; Ax \geq b\}$ is nonempty and that $f(x)$ is bounded from below on K; then, the set of optimal solutions of (2.1), denoted by K^*, is a nonempty polyhedral set [21], given by $K^*\{x \in K \mid E(x - x^*)0; q^T(x - x^*)0\}$, where x^* is a solution of (2.1) and E is a matrix such that $GE^T E$. Furthermore, $\nabla f(x)Gx + q$ is invariant over K^*. When G is positive definite, the existence of the solution of (2.1) is guaranteed and the solution is unique, i.e., $K^*\{x^*\}$. In the following we assume that the matrices G, C and A are large, sparse and without a particular structure.

The classical projection and splitting methods have a common iterative scheme, involving the generation, starting from an arbitrary vector $x^{(0)}$, of a sequence of vectors $\{x^{(k)}\}$, $k1,2,...$, that are the solution of the following QP subproblems:

$$\begin{aligned} minimize \quad & \tfrac{1}{2}x^T \tfrac{D}{\rho} x + (q + (G - \tfrac{D}{\rho})x^{(k-1)})^T x \\ subject\ to \quad & C x d, \qquad A x \geq b, \end{aligned} \tag{2.2}$$

where D is a prefixed symmetric positive definite matrix of order n and ρ is a positive parameter. In this way, $x^{(k)}$ is the projection of the vector $x^{(k-1)} - \rho D^{-1}(Gx^{(k-1)} + q)$ onto K with respect to the norm $\|\cdot\|_D$, i.e., $x^{(k)} \arg\min_{x \in K} \|x - (x^{(k-1)} - \rho D^{-1}(Gx^{(k-1)} + q))\|_D$. When $\rho 1$ in (2.2), we refer to the iterative scheme as a splitting method while, for $\rho \neq 1$, we have a projection method. scalar product $z^T D w$.

¿From a practical point of view, D must be an easily solvable matrix (diagonal or block diagonal matrix); the computational complexity per iteration thus consists essentially in computing the matrix–vector product $\left(G - \tfrac{D}{\rho}\right) x^{(k-1)}$ and solving the QP subproblem (2.2).

In literature, there exists a vast collection of results regarding the convergence of the projection and splitting methods (see for example [5]). The linear convergence (in the root sense [32]) of the splitting method is proved in [21] under the hypotheses that D is a positive definite matrix and $(D, G - D)$ is a P–regular splitting of G. In this case, since D may be also an asymmetric matrix, $x^{(k)}$ is the solution of a subproblem given by the following variational inequality:

$$(Dx^{(k)} + (G - D)x^{(k-1)} + q)^T(x - x^{(k)}) \geq 0 \quad \text{for any } x \in K. \quad (2.3)$$

¿From the computational point of view, when K does not present a special structure, this subproblem is generally more difficult than the QP subproblem (2.2). Possible exceptions are the symmetric monotone linear complementarity problems and the QP problems where $m_i 0$, $x^T G x > 0$ for any x such that $Cx0$ and $m_e \ll n$. In this last case, each subproblem can be formulated [10] as a small–scale linear system of the form

$$CD^{-1}C^T \lambda d + D^{-1}C((G - D)x^{(k-1)} + q) \quad (2.4)$$

and it can be solved by a direct method. Nevertheless, when the system (2.4) is of medium– or large–scale, the choice of a symmetric matrix D enables a more efficient conjugate gradient method to be used for its solution.

For strictly convex QP problems, under the additional hypothesis that D is symmetric, the splitting method has a geometrical convergence with rate of convergence equal to the spectral radius of the matrix $(I - D^{-\frac{1}{2}}GD^{-\frac{1}{2}})$ [14].

The convergence of the projection method with D symmetric positive definite can be immediately derived from the projection schemes for the variational inequality problems. In the case of strictly convex QP problems, by proceeding as in [8], the convergence of $\{x^{(k)}\}$ can be obtained by a contraction argument for $\rho < \frac{2\lambda_{min}(G)}{\lambda_{max}(GD^{-1}G)}$; the method has a geometrical convergence with rate

$$(1 - \frac{\rho}{\lambda_{max}(D)}(2\lambda_{min}(G) - \rho\lambda_{max}(GD^{-1}G)))^{\frac{1}{2}}.$$

For convex QP problems, the convergence of the projection method is proved under the sufficient condition $\rho < \frac{2\lambda_{min}(D)}{\lambda_{max}(G)}$ [25].

In the following theorem we summarize some of the above convergence results under a weaker condition for the projection method.

Theorem 2.1. Let the problem (2.1) be feasible with G symmetric positive semidefinite and $f(x)$ bounded from below on K. Let D be a symmetric positive definite matrix. Given an arbitrary vector $x^{(0)}$, the sequence $\{x^{(k)}\}$ generated by the iterative scheme (2.2) is well–defined and is linearly convergent (in the root sense) to a solution of (2.1), under one of the following conditions:

- $\rho 1$ and $(D, G - D)$ is a P–regular splitting of G;

- $0 < \rho < \dfrac{2}{\lambda_{max}(D^{-\frac{1}{2}} G D^{-\frac{1}{2}})}$.

Proof. For the uniqueness of the solution of the strictly convex QP subproblems (2.2), the sequence $\{x^{(k)}\}$ is well–defined.

Let x^* be a solution of (2.1). We consider the first order optimality conditions for the solution $x^{(k)}$ of the subproblem (2.2) and for the solution x^* of (2.1),

$$\left(\frac{D}{\rho}(x^{(k)} - x^{(k-1)}) + G x^{(k-1)} + q \right)^T (x - x^{(k)}) \geq 0 \qquad ,$$

$$\text{for any} \quad x \in K, \quad (2.5)$$

$$(G x^* + q)^T (x - x^*) \geq 0 \qquad ,$$

$$\text{for any} \quad x \in K. \quad (2.6)$$

Setting $x x^*$ in (2.5) and $x x^{(k)}$ in (2.6) and adding the two inequalities, we have:

$$(x^{(k)} - x^{(k-1)})^T D(x^* - x^{(k)}) + \rho(x^{(k-1)} - x^*)^T G(x^* - x^{(k)}) \geq 0. \quad (2.7)$$

Using Lemmas 2.1 and 2.2 of [25], we can write

$$\|x^* - x^{(k)}\|_D^2 \leq \|x^* - x^{(k-1)}\|_D^2 +$$

$$-(x^{(k)} - x^{(k-1)})^T D^{\frac{1}{2}} \left(I - \frac{\rho}{2} D^{-\frac{1}{2}} G D^{-\frac{1}{2}} \right) D^{\frac{1}{2}} (x^{(k)} - x^{(k-1)})$$

$$\leq \|x^* - x^{(k-1)}\|_D^2 - \gamma \|x^{(k)} - x^{(k-1)}\|_D^2 \quad (2.8)$$

where $\gamma(1 - \frac{\rho}{2} \lambda_{max}(D^{-\frac{1}{2}} G D^{-\frac{1}{2}}))$.

When $\rho 1$ and $(D, G - D)$ is a P–regular splitting of G, the eigenvalues of $D^{-1/2} G D^{-1/2}$ are nonnegative real values in the interval $[0, 2)$; on the other hand, for $0 < \rho < \dfrac{2}{\lambda_{max}(D^{-\frac{1}{2}} G D^{-\frac{1}{2}})}$, the matrix $(I - \frac{\rho}{2} D^{-\frac{1}{2}} G D^{-\frac{1}{2}})$ is positive definite. Then $\gamma > 0$. Thus, the inequality (2.8) enables the convergence of the sequence $\{x^{(k)}\}$ to a solution of (2.1) to be proved by the same routine argument used in Theorem 2.1 of [25].

Now, in order to prove the R–linear convergence (i.e. the linear convergence in the root sense) of $\{x^{(k)}\}$, we define $\psi(x) \min_{\overline{x} \in K^*} \|\overline{x} - x\|_D^2$.

From the Theorem 2.1 of [21], we can assume that, for some positive scalars β and δ, the following condition holds:

$$\min_{\overline{x} \in K^*} \|\overline{x} - x\| \leq \beta \|x - [x - (Gx + q)]^+\|, \tag{2.9}$$

for any x with $\|x - [x - (Gx + q)]^+\| \leq \delta$.
Then, from the definition of $\psi(x)$, it follows that:

$$\psi(x) \leq \|D\|\beta^2 \|x - [x - (Gx + q)]^+\|^2, \tag{2.10}$$

for any x with $\|x - [x - (Gx + q)]^+\| \leq \delta$.
Since the solution $x^{(k)}$ of the subproblem (2.2) can be considered as the orthogonal projection of an appropriate vector onto K, i.e.,

$$x^{(k)} \left[x^{(k)} - \frac{D}{\rho}(x^{(k)} - x^{(k-1)}) - (Gx^{(k-1)} + q) \right]^+ \tag{2.11}$$

from the nonexpansive property of the projection operator, we have

$$\|x^{(k-1)} - [x^{(k-1)} - (Gx^{(k-1)} + q)]^+\| \quad \|x^{(k-1)} - [x^{(k-1)} -$$
$$- (Gx^{(k-1)} + q)]^+ + -x^{(k)} + [x^{(k)} - (D/\rho)(x^{(k)} - x^{(k-1)}) -$$
$$-(Gx^{(k-1)} + q)]^+\| \leq \left(2 + \frac{\|D\|}{\rho}\right) \frac{1}{\sqrt{\lambda_{min}(D)}} \|x^{(k)} - x^{(k-1)}\|_D. \tag{2.12}$$

Since $\|x^{(k)} - x^{(k-1)}\|_D$ converges to 0 as $k \to \infty$, the sequence $\|x^{(k-1)} - [x^{(k-1)} - (Gx^{(k-1)} + q)]^+\|$ also converges to 0 as $k \to \infty$. Then, for k greater than a convenient positive integer $\overline{k}$, we have

$$\|x^{(k-1)} - [x^{(k-1)} - (Gx^{(k-1)} + q)]^+\| \leq \delta$$

and, from (2.10),

$$\psi(x^{(k-1)}) \leq \|D\|\beta^2 \left(2 + \frac{\|D\|}{\rho}\right)^2 \frac{1}{\lambda_{min}(D)} \|x^{(k)} - x^{(k-1)}\|_D^2. \tag{2.13}$$

If x^* is the element of K^* such that $\psi(x^{(k-1)})\|x^* - x^{(k-1)}\|_D^2$, we have, for $k > \overline{k}$:

$$\begin{aligned} \psi(x^{(k)}) &\leq \|x^{(k)} - x^*\|_D^2 \\ &\leq \psi(x^{(k-1)}) - \gamma\|x^{(k)} - x^{(k-1)}\|_D^2 \\ &\leq \psi(x^{(k-1)}) - \zeta\psi(x^{(k-1)}) \\ &\quad (1 - \zeta)\psi(x^{(k-1)}) \end{aligned} \tag{2.14}$$

where $\zeta \frac{\gamma \lambda_{min}(D)}{\beta^2 \|D\|(2+\frac{\|D\|}{\rho})^2}$ and the second and the third inequalities follow
from (2.8) and (2.13), respectively. Thus $\psi(x^{(k)})$ converges Q–linearly
to 0. Finally, from Lemma 3.1 of [21] and from the following inequality:

$$\psi(x^{(k-1)}) - \psi(x^{(k)}) \geq \gamma \|x^{(k)} - x^{(k-1)}\|_D^2 \quad \text{for} \quad k > \overline{k},$$

it follows that the sequence $\{x^{(k)}\}$ converges R–linearly to a solution of
(2.1).   ∎

¿From the viewpoint of numerical effectiveness, in the splitting methods the practical requirement to select an easily solvable matrix D is
very often in conflict with the choice of a splitting of G that implies fast
convergence (unless the splitting is suggested by the underlying application; see, for example, the quadratic program arising in the constrained
bivariate interpolation problem [13], [38]). On the other hand, the projection schemes allow any positive definite matrix D to be selected, but
the values of ρ that satisfy the sufficient convergence conditions are often
so small as to determine very slow convergence. Furthermore, in some
cases, it is possible to find values of ρ that are greater than the values
required by the sufficient convergence condition and that enable a solution to be determined in a lower number of iterations (see numerical
experiments in [39]).
There exist many strategies to accelerate the classical projection and
splitting methods. The recent modified projection–type methods ([18],
[40]) are gradient projection methods combined with a correction formula that makes these methods contractive. In particular, at each iteration, the solution $y^{(k)}$ of a particular QP subproblem with identity
matrix as Hessian matrix is computed,

$$y^{(k)}[x^{(k-1)} - (Gx^{(k-1)} + q)]^+$$

and then $y^{(k)}$ is modified by the following correction formula:

$$x^{(k)}x^{(k-1)} + \delta_{k-1}P^{-1}(I + G)(y^{(k)} - x^{(k-1)})$$

where $\delta_{k-1}\theta \frac{\|y^{(k)} - x^{(k-1)}\|_2^2}{\|P^{-\frac{1}{2}}(I+G)(y^{(k)} - x^{(k-1)})\|_2^2}$, $\theta \in (0,2)$ and P is a symmetric
positive definite matrix. The choice of the parameters P and θ is crucial
for the performance of this class of methods. The computational cost of
the correction formula is dependent on the choice of P but can generally
be considered equivalent to the cost of a matrix–vector product (see, for
example, the choice $PI + G$ proposed in [18]). A numerical evaluation
of this approach is reported in [39].

Another technique for accelerating the projection and splitting methods consists in varying the projection parameter at each iteration on the basis of a nonexpensive updating rule. In this case, in order to ensure the convergence, we need to introduce a strategy to guarantee a sufficient decrease in the objective function after each projecton step. We consider two strategies that give rise to two different methods: the VPM and the AVPM.

In the VPM, after each projection step, a limited exact line search along the feasible descent direction determined by the projection step is performed. In this way, when the projection step does not produce a sufficient decrease in $f(x)$,

the current stepsize is reduced.

In the AVPM, when the projection step related to the current projection parameter does not determine a sufficient decrease in $f(x)$, we reduce the projection parameter and we repeat the projection step until a convenient decrease in $f(x)$ is obtained. In this way, starting from the current projection parameter, we determine adaptively a good value of this parameter by a search procedure on a projection arc.

It is well known that the projection and the splitting methods can be viewed as scaled gradient projection methods "with constant stepsize" ([2, p. 207]; in the notation of [2] the projection parameter is called stepsize). In the same way, the VPM and the AVPM generalize the scaled gradient projection schemes with a "limited minimization rule" over the interval $(0,1]$ ([2, p. 205]) and with an "Armijo rule along a projection arc" ([2, p. 206]), respectively. The original idea of the VPM and the AVPM with respect to this last class of methods is the introduction of an updating rule for the choice of the variable projection parameter. When a convenient updating rule is used, the convergence rate is improved and the correction of the stepsize in the VPM or the adaptive procedure for the determination of the projection parameter in the AVPM is present only in few iterations.

3. The Variable Projection Method.

Given a symmetric positive definite matrix D, the Variable Projection Method for the solution of (2.1) is stated as follows:

1 Let $x^{(0)}$ be an arbitrary vector and ρ_1 be an arbitrary positive constant; let ϵ be a prefixed small tolerance and $k \leftarrow 1$;

2 Compute the unique solution $y^{(k)}$ of the subproblem

$$minimize \quad \tfrac{1}{2}x^T \tfrac{D}{\rho_k}x + (q + (G - \tfrac{D}{\rho_k})x^{(k-1)})^T x$$
$$subject\ to \quad Cxd, \quad Ax \geq b; \tag{3.1}$$

3 Set $d^{(k)}y^{(k)} - x^{(k-1)}$;

4 If $(Gx^{(k-1)} + q)^T d^{(k)} < 0$ and $k \neq 1$, compute the solution θ_k of the problem:

$$\min\{f(x^{(k-1)} + \theta d^{(k)}); \quad \theta \in (0,1]\} \tag{3.2}$$

else

$$\theta_k 1;$$

5 Compute

$$x^{(k)} x^{(k-1)} + \theta_k d^{(k)}; \tag{3.3}$$

6 Terminate if $x^{(k)}$ satisfies a stopping rule, otherwise update ρ_{k+1} by an appropriate rule; compute

$$\rho_{k+1} \begin{cases} \rho_k & \text{for } \|Gd^{(k)}\|^2 \leq \epsilon\|d^{(k)}\|^2 \\ \dfrac{d^{(k)T}Gd^{(k)}}{d^{(k)T}GD^{-1}Gd^{(k)}} & \text{otherwise} \end{cases} \tag{3.4}$$

then $k \leftarrow k + 1$ and go to step 2.

When the matrix D is such that the computation of the vector $Dx^{(k-1)}$ is not expensive, the computational complexity of the VPM is essentially the same as the classical projection and splitting methods.

Indeed, if we keep stored the vector $tGx^{(k-1)}$, the computation of θ_k for the additional correction step (3.3) requires to compute the matrix-vector product $\bar{t}Gy^{(k)}$ and some less expensive vector operations. Now, the vector $\bar{t}$ is also used for computing , with some vector operations, ρ_{k+1} and for updating the vector t by the following rule:

$$t \leftarrow t + \theta_k(\bar{t} - t).$$

We will prove the convergence of the VPM under the following assumptions:

AS1. G is a symmetric positive semidefinite matrix, K is a nonempty set and $f(x)$ is bounded from below on K;

AS2. the intersection between K and the level set $\Gamma(x^{(1)})\{x|f(x) \leq f(x^{(1)})\}$ is a compact set (this condition is guaranteed when G is positive definite);

AS3. D is a symmetric positive definite matrix and the sequence of scalars $\{\rho_k\}$ is bounded from below and above by positive constants: $0 < \rho_m \leq \rho_k \leq \rho_M$.

This last assumption on $\{\rho_k\}$ is very general; the choice of a particular bounded sequence $\{\rho_k\}$ (as that defined by (5.1)) is based only on its numerical effectiveness. The choice of a constant projection parameter is a special case of this.
The aim of the initial step of the scheme is to generate, from an arbitrary point $x^{(0)}$, a vector $x^{(1)} \in K$. When $\lambda_{min}(D)$ is easily computable, a convenient value for ρ_1 is $\rho_1 \lambda_{min}(D)/\left(\sum_{i1}^{n} \sum_{j1}^{n} g_{ij}^2\right)^{1/2}$ where g_{ij} is the ij-th element of G; this value satisfies the sufficient condition for the convergence of the projection method and guarantees that (2.8) holds with $k1$.
The following Lemmas state some basic assertions for the convergence of the VPM.

Lemma 3.1. If, for a given $k > 1$, $x^{(k-1)}$ is feasible to (2.1) and $x^{(k-1)} \neq y^{(k)}$, then the following assertions hold:

a. $y^{(k)}$ is well–defined and is a feasible point of (2.1); furthermore, $d^{(k)}$ is a descent
direction for the objective function $f(x)$ in $x^{(k-1)}$;

b.
$$0 < \theta_{min} \min\left\{1, \frac{\lambda_{min}(D)}{\rho_M \lambda_{max}(G)}\right\} \leq \theta_k \leq 1. \qquad (3.5)$$

Proof. (see [38, Lemma 2.1]).

Lemma 3.2. The sequence $\{x^{(k)}\}$ generated by the VPM is well–defined, $x^{(k)} \in K$ for all $k \geq 1$ and $\{f(x^{(k)})\}$ is a sequence monotonically nonincreasing and convergent as $k \to \infty$.

Proof. Since $x^{(1)} \in K$, from the assertions of Lemma 3.1 and the convexity of K, it follows that $\{x^{(k)}\}$ is a well–defined sequence of feasible points of (2.1) for $k \geq 1$.

Now, we observe that, for $k > 1$, the exact solution of the problem (3.2) is given by the following rule:

$$\theta_k \left\{ \begin{array}{ll} \min\left\{ -\frac{(Gx^{(k-1)}+q)^T d^{(k)}}{d^{(k)T} G d^{(k)}}, 1 \right\} & \text{if } d^{(k)T} G d^{(k)} \neq 0 \\ 1 & \text{otherwise.} \end{array} \right. \qquad (3.6)$$

Furthermore, we consider the first order optimality condition for the solution $y^{(k)}$ of the subproblem (3.1):

$$\left(\frac{D}{\rho_k}(y^{(k)} - x^{(k-1)}) + Gx^{(k-1)} + q \right)^T (x - y^{(k)}) \geq 0, \quad \text{for any } x \in K. \qquad (3.7)$$

If we set $xx^{(k-1)}$ in (3.7), we obtain:

$$(Gx^{(k-1)} + q)^T d^{(k)} \leq -d^{(k)T} \frac{D}{\rho_k} d^{(k)} \leq 0. \qquad (3.8)$$

The equality to zero holds when $x^{(k-1)}y^{(k)}$. In this case $x^{(k-1)} \in K^*$. ¿From (3.2), (3.3) and the following equality:

$$f(x^{(k-1)}+\theta d^{(k)}) - f(x^{(k-1)})\theta(Gx^{(k-1)}+q)^T d^{(k)} + \frac{1}{2}\theta^2 d^{(k)T} G d^{(k)}, \qquad (3.9)$$

we have

$$f(x^k) \leq f(x^{(k-1)}) + \theta(Gx^{(k-1)} + q)^T d^{(k)} + \frac{1}{2}\theta^2 d^{(k)T} G d^{(k)}, \qquad (3.10)$$

for any $\theta \in (0,1]$. If $d^{(k)T} G d^{(k)} 0$ or $-\frac{(Gx^{(k-1)}+q)^T d^{(k)}}{d^{(k)T} G d^{(k)}} \geq 1$, the solution of the problem (3.2) is obtained for $\theta_k 1$; from (3.10), (3.8) and (3.5), we have

$$\begin{aligned} f(x^k) &\leq f(x^{(k-1)}) + \frac{1}{2}(Gx^{(k-1)} + q)^T d^{(k)} \\ &\leq f(x^{(k-1)}) - \frac{\theta_{min}}{2\rho_M}\|d^{(k)}\|_D^2. \end{aligned} \qquad (3.11)$$

If $d^{(k)T} G d^{(k)} \neq 0$ and $-\frac{(Gx^{(k-1)}+q)^T d^{(k)}}{d^{(k)T} G d^{(k)}} < 1$, the solution of the problem (3.2) is obtained for $\theta_k - \frac{(Gx^{(k-1)}+q)^T d^{(k)}}{d^{(k)T} G d^{(k)}}$; from (3.10), (3.8) and (3.5), we have

$$\begin{aligned} f(x^k) &\leq f(x^{(k-1)}) - \frac{1}{2}\frac{((Gx^{(k-1)} + q)^T d^{(k)})^2}{d^{(k)T} G d^{(k)}} \\ & f(x^{(k-1)}) + \frac{\theta_k}{2}(Gx^{(k-1)} + q)^T d^{(k)} \\ &\leq f(x^{(k-1)}) - \frac{\theta_{min}}{2\rho_M}\|d^{(k)}\|_D^2. \end{aligned} \qquad (3.12)$$

Thus, from (3.11) and (3.12), we conclude that

$$f(x^{k-1}) - f(x^{(k)}) \geq \frac{\theta_{min}}{2\rho_M} \|d^{(k)}\|_D^2 \geq 0, \tag{3.13}$$

where the equality to zero holds when $x^{(k-1)}y^{(k)}x^{(k)} \in K^*$. Since $f(x)$ is bounded from below on K, the sequence $\{f(x^{(k)})\}$ is convergent as $k \to \infty$. $\blacksquare$

Lemma 3.3. The sequences $\{d^{(k)}\}$ and $\{x^{(k)} - x^{(k-1)}\}$ are convergent to the null vector as $k \to \infty$. Furthermore, each accumulation point of the sequence $\{x^{(k)}\}$ is a solution of (2.1) and $\min_{\overline{x} \in K^*} \|x^{(k)} - \overline{x}\|$ converges to 0 as $k \to \infty$.

Proof. The convergence of $\{d^{(k)}\}$ to the null vector follows immediately from (3.13) and from the convergence of the sequence $\{f(x^{(k)})\}$. From (3.3) and the boundedness of θ_k, we have that $\{x^{(k)} - x^{k-1}\} \to 0$ as $k \to \infty$.

Since $\{x^{(k)}\}$ is a sequence of points of the set $K \cap \Gamma(x^{(1)})$, $\{x^{(k)}\}$ is bounded and has at least one accumulation point $\tilde{x} \in K \cap \Gamma(x^{(1)})$. Now, we prove that each accumulation point is a solution of (2.1). Let $\{x^{(k_p)}\}$ be a subsequence of $\{x^{(k)}\}$ convergent to $\tilde{x}$ as $p \to \infty$. From (3.7) computed for kk_p, we have

$$\left(\frac{D}{\rho_{k_p}} (y^{(k_p)} - x^{(k_p-1)}) + Gx^{(k_p-1)} + q \right)^T (x - y^{(k_p)}) \geq 0,$$

$$\text{for any } x \in K. \tag{3.14}$$

Since $x^{(k_p-1)}x^{(k_p-1)} - x^{(k_p)} + x^{(k_p)}$ and $y^{(k_p)}y^{(k_p)} - x^{(k_p-1)} + x^{(k_p-1)}$, $\{x^{(k_p-1)}\}$ and $\{y^{(k_p)}\}$ converge to $\tilde{x}$ as $p \to \infty$. From (3.14), as $p \to \infty$ we have

$$(G\tilde{x} + q)^T (x - \tilde{x}) \geq 0, \quad \text{for any } x \in K.$$

Then, $\tilde{x} \in K^*$. Finally, from Theorem 14.1.4 of [32], it follows that $\min_{\overline{x} \in K^*} \|x^{(k)} - \overline{x}\|$ converges to the null vector as $k \to \infty$. $\blacksquare$

When K^* is a finite set (when, for example, G is positive definite), the sequence $\{x^{(k)}\}$ converges to an element of K^* as $k \to \infty$. In the following theorem we prove that this arises also for a nonfinite set K^*.

Theorem 3.1. The sequence $\{x^{(k)}\}$ generated by the VPM is convergent to a solution of (2.1).

Proof. At the beginning, we prove that the following inequality holds:

$$\|x^{(k)} - x^*\|_D^2 \leq \|x^{(k-1)} - x^*\|_D^2 + \Psi(f(x^{(k-1)}) - f(x^{(k)})) \qquad (3.15)$$

where $x^* \in K^*$ and $\Psi \frac{2}{\theta_{min}} \rho_M (1 + \rho_M \frac{\lambda_{max}(G)}{2\lambda_{min}(D)})$. First we observe that, from the first order optimality conditions for the problems (3.1) and (2.1), it is possible to write:

$$\left(\frac{D}{\rho_k}(y^{(k)} - x^{(k-1)}) + Gx^{(k-1)} + q \right)^T (x^* - y^{(k)}) \geq 0$$

$$(Gx^* + q)^T(y^{(k)} - x^*) \geq 0$$

and then

$$(x^* - y^{(k)})^T D(y^{(k)} - x^{(k-1)}) \geq \rho_k(x^{(k-1)} - x^*)^T G(y^{(k)} - x^*). \quad (3.16)$$

Now, from (3.3), we have

$$
\begin{aligned}
\|x^{(k)} - x^*\|_D^2 \quad & \|x^{(k-1)} - x^*\|_D^2 + \theta_k^2 \|y^{(k)} - x^{(k-1)}\|_D^2 \\
& +2\theta_k(x^{(k-1)} - x^*)^T D(y^{(k)} - x^{(k-1)}) \\
\leq \quad & \|x^{(k-1)} - x^*\|_D^2 + \theta_k^2 \|y^{(k)} - x^{(k-1)}\|_D^2 + \\
& -2\theta_k \|y^{(k)} - x^{(k-1)}\|_D^2 + \qquad\qquad (3.17) \\
& -2\theta_k \rho_k(x^{(k-1)} - x^*)^T G(y^{(k)} - x^*),
\end{aligned}
$$

where the last inequality is obtained from (3.16). By proceeding as in the proof of the Lemma 2.1 of [25], we have

$$
\begin{aligned}
2(x^{(k-1)} - x^*)^T G(x^* - y^{(k)}) \quad & 2(x^{(k-1)} - x^*)^T G(x^* - x^{(k-1)}) + \\
& +2(x^{(k-1)} - x^*)^T G(x^{(k-1)} - y^{(k)})^T \\
\leq \quad & \frac{1}{2}(x^{(k-1)} - y^{(k)})^T G(x^{(k-1)} - y^{(k)})^T \quad (3.18)
\end{aligned}
$$

Using (3.18) in (3.17) and neglecting the negative terms, we obtain

$$
\begin{aligned}
\| \quad x^{(k)} - x^*\|_D^2 \leq & \|x^{(k-1)} - x^*\|_D^2 + \\
& +\theta_k^2 \|y^{(k)} - x^{(k-1)}\|_D^2 + \frac{\theta_k \rho_k}{2}(x^{(k-1)} - y^{(k)})^T G(x^{(k-1)} - y^{(k)}) \\
\leq & \|x^{(k-1)} - x^*\|_D^2 + \left(1 + \rho_M \frac{\lambda_{max}(G)}{2\lambda_{min}(D)}\right) \|y^{(k)} - x^{(k-1)}\|_D^2.
\end{aligned}
$$

Finally, from (3.13), we obtain (3.15).

Now, we prove by contradiction that the sequence $\{x^{(k)}\}$ is convergent. We assume that there exist two subsequences $\{x^{(k_p)}\}$ and $\{x^{(h_p)}\}$ of

$\{x^{(k)}\}$ convergent to two different points of K^*, denoted by $\overline{x}$ and x^* respectively, such that $\|\overline{x} - x^*\|_D\sqrt{\tau}$, $\tau > 0$. Since $\{x^{(h_p)}\} \to x^*$ as $p \to \infty$, there exists a positive integer p_1 such that $\|x^{(h_p)} - x^*\|_D < \frac{\sqrt{\tau}}{2}$ for $p > p_1$. Consequently, for $p > p_1$, we can write

$$\frac{\sqrt{\tau}}{2} < \|x^* - \overline{x}\|_D - \|x^{(h_p)} - x^*\|_D \leq \|x^{(h_p)} - \overline{x}\|_D. \qquad (3.19)$$

Since $\{x^{(k_p)}\} \to \overline{x}$ as $p \to \infty$, for p greater than a convenient positive integer p_2, we have

$$\|x^{(k_p)} - \overline{x}\|_D^2 < \frac{\tau}{8}. \qquad (3.20)$$

Since $\{f(x^{(k)})\}$ is a Cauchy sequence, for any i, j greater than a convenient positive integer $\overline{k}$ and $j > i$, we have

$$f(x^{(i)}) - f(x^{(j)}) < \frac{\tau}{8\,\Psi}. \qquad (3.21)$$

Let p be an integer such that $p > \max(p_1, p_2)$ and $h_p > k_p > \overline{k}$. From (3.19) and (3.15) with kh_p and, successively, $kh_p - 1, ..., k_p + 1$, we have

$$
\begin{aligned}
\frac{\tau}{4} \;<\; & \|x^{(h_p)} - \overline{x}\|_D^2 \leq \|x^{(h_p-1)} - \overline{x}\|_D^2 + \Psi(f(x^{(h_p-1)}) - f(x^{(h_p)})) \\
\leq\; & \|x^{(h_p-2)} - \overline{x}\|_D^2 + \Psi(f(x^{(h_p-2)}) - f(x^{(h_p-1)}) + \\
& \qquad\qquad + f(x^{(h_p-1)}) - f(x^{(h_p)})) \\
\leq\; & \cdots \\
\leq\; & \|x^{(k_p)} - \overline{x}\|_D^2 + \Psi(f(x^{(k_p)}) - f(x^{(h_p)})) \\
<\; & \frac{\tau}{8} + \frac{\tau}{8}\frac{\tau}{4} \qquad\qquad\qquad\qquad\qquad\qquad\qquad (3.22)
\end{aligned}
$$

where the last inequality follows from (3.20) and (3.21). Thus a contradiction is established and the sequence $\{x^{(k)}\}$ is convergent to a solution of (2.1). $\blacksquare$

The rule (5.1) for updating ρ_{k+1} is a generalization of that proposed in [38] on the basis of heuristic and numerical considerations for strictly convex QP problems. In practice, ρ_{k+1} is an approximation of the value of ρ that minimizes an upper bound of the ratio $\frac{\|x^{(k+1)}-x^*\|_D^2}{\|x^{(k)}-x^*\|_D^2}$. Other different rules could be given (see, for example, section 4). Nevertheless, the rule (5.1) appears slightly dependent on the scaling of G. With the choice (5.1), the sequence $\{\rho_k\}$ is bounded from below and above:

$$\rho_m \min\left(\rho_1, \frac{\lambda_{min}(D)}{\lambda_{max}(G)}\right) \leq \rho_k \leq \rho_M \max\left(\rho_1, \frac{\lambda_{max}(G)\lambda_{max}(D)}{\epsilon}\right)$$

where the bounds are obtained by using the inequality

$$x^T G x \geq \frac{1}{\|G\|} \|G x\|^2.$$

4. The Adaptive Variable Projection Method.

Given a symmetric positive definite matrix D and two positive scalars $\beta \in (0,1)$ and $\eta > \frac{1}{2}$, the Adaptive Variable Projection Method for the solution of (2.1) is stated as follows:

1. Let $x^{(0)}$ be an arbitrary vector and $\bar{\rho}_1$ be an arbitrary positive constant; let ϵ be a prefixed small tolerance; $k \leftarrow 1$; $\rho_k \bar{\rho}_1$; $m_k 0$;

2. Compute the unique solution $x(\rho_k)$ of the subproblem

$$
\begin{aligned}
&minimize \quad \tfrac{1}{2} x^T \tfrac{D}{\rho_k} x + (q + (G - \tfrac{D}{\rho_k}) x^{(k-1)})^T x \\
&subject\ to \quad C x d, \qquad A x \geq b;
\end{aligned}
\tag{4.1}
$$

3. Set $d^{(k)} x(\rho_k) - x^{(k-1)}$;

4. If $-(G x^{(k-1)} + q)^T d^{(k)} < \eta d^{(k)T} G d^{(k)}$ and $k \neq 1$, then $m_k m_k + 1, \rho_k \beta^{m_k} \bar{\rho}_k$ and go to step 2, otherwise $x^{(k)} x(\rho_k)$;

5. Terminate if $x^{(k)}$ satisfies an appropriate stopping rule, otherwise update ρ_{k+1} by an appropriate rule; compute

$$
\rho_{k+1}
\begin{cases}
\rho_k & \text{for } \|G d^{(k)}\|^2 \leq \epsilon \|d^{(k)}\|^2 \\[2mm]
\dfrac{d^{(k)T} D d^{(k)}}{d^{(k)} \eta G d^{(k)}} & \text{otherwise}
\end{cases}
\tag{4.2}
$$

then $\bar{\rho}_{k+1} \rho_{k+1}$; $m_{k+1} 0$; $k \leftarrow k + 1$ and go to step 2.

The computational complexity of each iteration of the AVPM depends on the number $m_k + 1$ of inner subproblems that can be solved to find a convenient value for ρ_k. Since the test at step 4. requires the matrix–vector product $G x(\rho_k)$, each iteration of the AVPM is essentially equivalent to $m_k + 1$ iterations of the classical projection and splitting methods.

As in the iterative scheme of section 3, the initial step is used to generate a point $x^{(1)} \in K$. The value of $\bar{\rho}_1$ can be chosen as suggested in section 3 for ρ_1.

Now, we observe that the vector $x(\rho_k)$, satisfying the condition

$$-(Gx^{(k-1)} + q)^T(x(\rho_k) - x^{(k-1)}) \geq$$

$$\geq \eta(x(\rho_k) - x^{(k-1)})^T G(x(\rho_k) - x^{(k-1)}), \qquad (4.3)$$

can be found after a finite number of trials. Indeed, if we consider the first order optimality conditions of the subproblem (4.1) computed in $xx^{(k-1)}$, we have

$$(Gx^{(k-1)} + q)^T(x^{(k-1)} - x(\rho_k)) \geq$$

$$\geq \frac{1}{\rho_k}(x^{(k-1)} - x(\rho_k))^T D(x^{(k-1)} - x(\rho_k)). \qquad (4.4)$$

For $\rho_k \leq \frac{\lambda_{min}(D)}{\eta\lambda_{max}(G)}$, the condition (4.3) is verified. Thus, if m_k is the first nonnegative integer such that

$$\beta^{m_k}\overline{\rho}_k \leq \frac{\lambda_{min}(D)}{\eta\lambda_{max}(G)} \qquad (4.5)$$

for $\rho_k\beta^{m_k}\overline{\rho}_k$, the condition (4.3) is satisfied.
The inequality (4.5) states a very strong sufficient condition; from the viewpoint of numerical effectiveness, it would be convenient to work with projection parameters greater than the value obtained by (4.5) and such that $m_k > 0$ for few k. the inequalities (4.3) and (4.4) suggest the following estimate for ρ_k:

$$\rho_k\frac{d^{(k)T}Dd^{(k)}}{d^{(k)T}\eta Gd^{(k)}} \quad \text{if } d^{(k)T}Gd^{(k)} \neq 0 \qquad (4.6)$$

where $d^{(k)}x(\rho_k) - x^{(k-1)}$. Since $x(\rho_k)$ has to be computed, we use $x^{(k-2)}$ instead of $x(\rho_k)$, obtaining the rule (5.2).
Consequently, the sequence of positive scalars $\{\rho_k\}$ is bounded from below and above:

$$0 < \rho_m \min\left(\rho_1, \frac{\beta\lambda_{min}(D)}{\eta\lambda_{max}(G)}\right) \leq \rho_k \leq \rho_M$$

$$\max\left(\rho_1, \frac{\lambda_{min}(D)}{\eta\lambda_{max}(G)}, \frac{2\lambda_{max}(D)\lambda_{max}(G)}{\epsilon}\right)$$

Moreover, in place of (5.2), we can use other different rules, as, for example, (5.1). In general, an efficient updating rule must give values of the projection parameter greater than that obtained by (4.5) and such that $m_k > 0$ for few k.
Now we prove the convergence of the AVPM for convex QP problems under the general assumption that the sequence $\{\rho_k\}$ is bounded.

Theorem 4.1. Under the assumptions AS1 and AS3 of section 3, given an arbitrary point $x^{(0)}$, the sequence $\{x^{(k)}\}$ generated by the AVPM is well–defined and converges linearly (in the root sense) to a solution of (2.1).

Proof. Since the solution of the subproblem (4.1) is unique and, at each iteration, the steps 2.-3.-4. of the AVPM are repeated a finite number of times, $x^{(k)}$ is a well–defined element of K.

The R–linear convergence of the sequence $\{x^{(k)}\}$ follows from Theorem 3.1 of [21], if the following relations hold:

$$f(x^{(k-1)}) - f(x^{(k)}) \geq \alpha_1 \|x^{(k-1)} - x^{(k)}\|^2 \qquad (4.7)$$

$$\|x^{(k-1)} - [x^{(k-1)} - (Gx^{(k-1)} + q)]^+\|^2 \leq$$
$$\leq \alpha_2(f(x^{(k-1)}) - f(x^{(k)})) \qquad (4.8)$$

$$x^{(k)}[x^{(k-1)} - (Gx^{(k-1)} + q) + e^{(k)}]^+ \qquad (4.9)$$

where α_1 and α_2 are positive scalars and the sequence $\{e^{(k)}\}$ converges to the null vector as $k \to \infty$.

Now, we prove that (4.7), (4.8) and (4.9) hold.

Since $x^{(k)}$ is such that the condition $-(Gx^{(k-1)} + q)^T(x^{(k)} - x^{(k-1)}) \geq \eta(x^{(k)} - x^{(k-1)})^T G(x^{(k)} - x^{(k-1)})$ holds, we can write

$$\begin{aligned}
f(x^{(k)}) - f(x^{(k-1)}) \quad & (Gx^{(k-1)} + q)^T(x^{(k)} - x^{(k-1)}) \\
& + \frac{1}{2}(x^{(k)} - x^{(k-1)})^T G(x^{(k)} - x^{(k-1)}) \\
\leq \quad & (Gx^{(k-1)} + q)^T(x^{(k)} - x^{(k-1)}) \\
& - \frac{1}{2\eta}(Gx^{(k-1)} + q)^T(x^{(k)} - x^{(k-1)}) \\
\leq \quad & (1 - \frac{1}{2\eta})(Gx^{(k-1)} + q)^T(x^{(k)} - x^{(k-1)}) \\
\leq \quad & -\frac{1}{\rho_k}\left(1 - \frac{1}{2\eta}\right)\|x^{(k)} - x^{(k-1)}\|_D^2 \qquad (4.10)
\end{aligned}$$

where the last inequality follows from (4.4) with $x^{(k)}$ in place of $x(\rho_k)$. Thus, we have

$$f(x^{(k-1)}) - f(x^{(k)}) \geq \frac{1}{\rho_M}(1 - \frac{1}{2\eta})\lambda_{min}(D)\|x^{(k)} - x^{(k-1)}\|^2$$

and (4.7) holds with $\alpha_1\left(1 - \frac{1}{2\eta}\right)\frac{\lambda_{min}(D)}{\rho_M}$.

The inequality (4.8) follows from (2.12) (using ρ_k in place of ρ) with $\alpha_2\left(2 + \frac{\|D\|}{\rho_m}\right)^2 \frac{1}{\lambda_{min}(D)\alpha_1}$.

Finally, we observe that

$$x^{(k)}[x^{(k-1)} - (Gx^{(k-1)} + q) + e^{(k)}]^+$$

where $e^{(k)} \left(I - \frac{D}{\rho_k}\right)(x^{(k)} - x^{(k-1)})$ and

$$\left\|e^{(k)}\right\| \leq \max \left(\left|1 - \frac{\lambda_{min}(D)}{\rho_M}\right|, \left|1 - \frac{\lambda_{max}(D)}{\rho_m}\right|\right) \|x^{(k)} - x^{(k-1)}\|. \quad (4.11)$$

Since the sequence $\{f(x^{(k)})\}$ is monotonically nonincreasing and $f(x)$ is bounded from below on K, $\{f(x^{(k)})\}$ is convergent, as $k \to \infty$; from (4.7), it follows that $\{x^{(k)} - x^{(k-1)}\}$ converges to 0 as $k \to \infty$. Thus, from (4.11), also $\{e^{(k)}\} \to 0$ as $k \to \infty$ and then (4.9) holds. ∎

5. Updating Rules for the Projection Parameter.

The numerical behaviour of the VPM and of the AVPM depends on the choice of the sequence of projection parameters $\{\rho_k\}$. The only hypothesis required on $\{\rho_k\}$ for the convergence of these methods is that the sequence is bounded. Then, at the k-th iteration we can try to identify a "good" value for ρ_k on the basis of heuristic considerations from the last values of the quantities involved in the iterative scheme. just computed values of $x^{(k-1)}$, $y^{(k-1)}$, $f(x^{(k-1)})$,... In this way we can find some updating rules for varying the projection parameter in the VPM and in the AVPM. Of course, these rules can have different numerical effectiveness.

We may consider an updating rule for ρ_k to be numerically efficient when it enables an approximate solution to (2.1) to be obtained by a number of iterations of the VPM and the AVPM significantly lower than that of the classical projection and splitting methods. At the same time, the introduction of this rule must not increase the computational complexity of one iteration and must imply that the corrective step along a descent direction in the VPM and along a projection arc in the AVPM is present only in few iterations, i.e. $\theta_k 1$ and, above all, $m_k 0$ for almost all k.

In the following, we propose two updating rules.

The first is given by

$$\rho_k \begin{cases} \rho_{k-1} & \\ \quad \text{for} \;\; |Gd^{(k-1)}|^2 \leq \epsilon \|d^{(k-1)}\|^2 & \quad k = 2, 3, \ldots \quad (5.1) \\ \dfrac{d^{(k-1)T}Gd^{(k-1)}}{d^{(k-1)T}GD^{-1}Gd^{(k-1)}} & \quad \text{otherwise} \end{cases}$$

where $d^{(k-1)}y^{(k-1)} - x^{(k-2)}$ in the VPM and $d^{(k-1)}x^{(k-1)} - x^{(k-2)}$ in the AVPM. The positive scalar ϵ is a prefixed small tolerance. The rule is a generalization of that proposed in [38] for strictly convex QP problems and it is deduced from the following considerations. If $\{x^{(k)}\}$ converges to $x^* \in K^*$ as $k \to \infty$, rearranging the terms of (3.16), we obtain

$$\|x^* - y^{(k)}\|_D^2 \le (x^* - x^{(k-1)})^T (D - \rho_k G)(x^* - y^{(k)}).$$

This last inequality holds in the VPM; for the AVPM we can substitute $y^{(k)}$ with $x^{(k)}$. Now, by using the Cauchy–Schwarz inequality, we obtain

$$\|x^* - y^{(k)}\|_D^2 \le \| (D - \rho_k G)(x^* - x^{(k-1)})\|_{D^{-1}} \|x^* - y^{(k)}\|_D$$

and, after dividing through by $\|x^* - y^{(k)}\|_D$, squaring and expanding, we have

$$\|x^* - y^{(k)}\|_D^2 \le \tau_k \|x^* - x^{(k-1)}\|_D^2,$$

where

$$\tau_k \quad 1 - \frac{\rho_k}{\|x^* - x^{(k-1)}\|_D^2} \Big(2(x^* - x^{(k-1)})^T G(x^* - x^{(k-1)}) + $$
$$- \rho_k (x^* - x^{(k-1)})^T GD^{-1} G(x^* - x^{(k-1)}) \Big).$$

If $(x^* - x^{(k-1)})^T GD^{-1} G(x^* - x^{(k-1)}) \ne 0$, the quantity τ_k is minimized for

$$\rho_k \frac{(x^* - x^{(k-1)})^T G(x^* - x^{(k-1)})}{(x^* - x^{(k-1)})^T GD^{-1} G(x^* - x^{(k-1)})};$$

then, we can try to obtain a "good" value of ρ_k by using $x^{(k-2)}$ in place of the solution x^*. Now, we remember that in the VPM the following equality holds: $x^{(k-1)} - x^{(k-2)} \theta_{k-1} (y^{(k-1)} - x^{(k-2)}) \theta_{k-1} d^{(k-1)}$. In this way, since the vectors $Gy^{(k-1)}$ and $Gx^{(k-2)}$ are already computed, we may define the updating rule (5.1) without essentially increasing the computational complexity of each iteration of the VPM and the AVPM. In the context of the projection methods for variational inequalities, a similar value for the step size ρ_k has been deduced, again by heuristic arguments, in [11].

With the choice (5.1), in the VPM the sequence $\{\rho_k\}$ is bounded from below and above as follows:

$$\min \Big(\rho_1, \frac{\lambda_{min}(D)}{\lambda_{max}(G)} \Big) \le \rho_k \le \max(\rho_1, \frac{\lambda_{max}(G)\lambda_{max}(D)}{\epsilon}),$$

where the bounds are obtained by using the inequality

$$x^T Gx \ge \frac{1}{\|G\|} \|Gx\|^2.$$

In the AVPM, because of the adaptive procedure for determining ρ_k, the resulting sequence $\{\rho_k\}$ is bounded from below and above in the following way:

$$\min\left(\rho_1, \frac{\lambda_{min}(D)}{\lambda_{max}(G)}, \frac{\beta\lambda_{min}(D)}{\eta\lambda_{max}(G)}\right) \leq \rho_k \leq$$

$$\leq \max(\rho_1, \frac{\lambda_{min}(D)}{\eta\lambda_{max}(G)}, \frac{\lambda_{max}(G)\lambda_{max}(D)}{\epsilon}).$$

Another updating rule for ρ_k may be suggested by the inequalities (4.3) and (4.4) stated in the AVPM:

$$\rho_k \frac{d^{(k)T}Dd^{(k)}}{d^{(k)T}\eta Gd^{(k)}} \quad \text{if } d^{(k)T}Gd^{(k)} \neq 0,$$

where $d^{(k)}x(\rho_k) - x^{(k-1)}$. Since $x(\rho_k)$ is not available, we use $x^{(k-2)}$ instead of $x(\rho_k)$, obtaining the following rule for the projection parameter:

$$\rho_k \begin{cases} \rho_{k-1} \\ \quad \text{for } \|Gd^{(k-1)}\|^2 \leq \epsilon\|d^{(k-1)}\|^2 \\ \dfrac{d^{(k-1)T}Dd^{(k-1)}}{d^{(k-1)}\eta Gd^{(k-1)}} \quad \text{otherwise} \end{cases} \quad k = 2,3,\ldots \quad (5.2)$$

$d^{(k-1)}x^{(k-1)} - x^{(k-2)}$ and ϵ is a prefixed small tolerance. In the VPM we can use the rule (5.2) with $d^{(k-1)}y^{(k-1)} - x^{(k-2)}$ and $\eta 1$. Consequently, in the VPM the new sequence $\{\rho_k\}$ is bounded as in the previous case while, in the AVPM, the bounds for the new sequence $\{\rho_k\}$ are:

$$\min\left(\rho_1, \frac{\beta\lambda_{min}(D)}{\eta\lambda_{max}(G)}\right) \leq \rho_k \leq \max\left(\rho_1, \frac{\lambda_{min}(D)}{\eta\lambda_{max}(G)}, \frac{2\lambda_{max}(D)\lambda_{max}(G)}{\epsilon}\right).$$

With regard to the choice of ρ_1, we observe that, when $\lambda_{min}(D)$ is easily computable, a convenient value for ρ_1 is $\rho_1\lambda_{min}(D)/\left(\sum_{i1}^n \sum_{j1}^n g_{ij}^2\right)^{1/2}$ where g_{ij} is the ij-th element of G; this value satisfies the sufficient condition for the convergence of the projection method and guarantees that (2.8) holds when $k1$.

6. Solution of the Inner QP Subproblems.

In all the projection-type and the splitting methods, as well as in the VPM and the AVPM, it is required to solve a sequence of strictly convex QP subproblems, having the following form

$$\begin{aligned} & minimize \quad \tfrac{1}{2}x^T\Delta x + q^{(k-1)T}x \\ & subject\ to \quad Cxd, \qquad Ax \geq b, \end{aligned} \qquad (6.1)$$

where $q^{(k-1)}(G - \Delta)x^{(k-1)} + q$, $k1,2,...$, and Δ is an easily solvable matrix, for instance a diagonal or block diagonal matrix.

When the constraints have particular features, we can use specialized inner solvers. For example, for single constrained separable strictly convex QP problems with simple bounds on the variables see [31], [34] and references therein; for equality constrained strictly convex QP problems with simple bounds on the variables see the dual ascent methods in [20]. When the set of constraints does not present a particular structure, it is possible to formulate the subproblem (6.1) as a mixed Linear Complementarity Problem (LCP) [20]. Indeed, by using the Karush–Kuhn–Tucker optimality conditions, we can derive the solution $\bar{x}$ of (6.1) in terms of its corresponding Lagrange multipliers $\bar{\lambda}$ and $\bar{\mu}$:

$$\bar{x}\Delta^{-1}(-q^{(k-1)} + A^T\bar{\lambda} + C^T\bar{\mu})$$

where $\begin{pmatrix}\bar{\lambda} \\ \bar{\mu}\end{pmatrix}$ is the solution of the following mixed LCP:

$$\begin{pmatrix}u \\ 0\end{pmatrix}\begin{pmatrix}z_1^{(k-1)} \\ z_2^{(k-1)}\end{pmatrix} + M\begin{pmatrix}\lambda \\ \mu\end{pmatrix} \tag{6.2}$$

$$u \geq 0, \quad \lambda \geq 0, \quad u^T\lambda 0$$

where $z_1^{(k-1)} - b - A\Delta^{-1}q^{(k-1)}$, $z_2^{(k-1)} - d - C\Delta^{-1}q^{(k-1)}$ and the matrix

$$M\begin{pmatrix}A \\ C\end{pmatrix}\Delta^{-1}(A^T \quad C^T) \tag{6.3}$$

is a symmetric positive semidefinite matrix of order $\nu(m_i + m_e)$ with positive diagonal entries. The mixed LCP (6.2) is solvable because it arises from the strictly convex QP problem (4.1). In the case of equality constraints only, M is the symmetric positive definite matrix $CD^{-1}C^T$ and (6.2) is reduced to the following positive definite linear system:

$$M\mu - z_2^{(k-1)}. \tag{6.4}$$

Thus, we can determine the solution of (6.1) by solving an equivalent problem whose size is equal to the number of constraints.

When this number is small, we can solve the problem (6.2) or (6.4) by direct methods. In the case of large–scale sparse symmetric monotone LCPs, it appears convenient to use again a splitting iterative scheme, such as the classical Projected SOR scheme of Cryer [7] and the Projected Symmetric SOR scheme of Mangasarian [22].

On parallel computers, we can easily achieve an efficient implementation of the projection type and splitting methods, as well as of the VPM and the AVPM, by distributing the matrix–vector products and the vector operations on the available processors and by using a parallel inner solver for the LCPs. A simple parallel solver for LCP is obtained by considering a splitting of the matrix M such that each iteration can be decomposed in independent processes. In [16] we show the effectiveness of the this approach in the case of a splitting method for QP problems combined with the Parallel SOR [23], the Parallel Gradient Projection SOR [24] and the Overlapping Parallel SOR [15] for the solution of the large inner LCPs. When the inner subproblem (4.1) can be reformulated in the linear system (6.4), we can use as iterative solver the Preconditioned Conjugate Gradient Method, that is also well suited for implementation on parallel computers. The absence of a particular structure in the matrix M suggests to use as preconditioner the classical SSOR preconditioner [?] or the Arithmetic Mean preconditioner [?].

7. Computational Experiments.

In this section we report the results of a set of numerical experiments that show the computational behaviour of the methods described in the previous sections. All the experiments are carried out on a Digital Personal Workstation 500au, using the double precision (macheps $2.22 \cdot 10^{-16}$) and a set of programs written in Fortran 90.

In Table 1 we describe the test problems used in the numerical experimentation.

These test problems are randomly generated with assigned features; in particular, we prefix the sizes n, m_e, m_i, the number nac of inequality constraints that are active in the solution of (2.1), a solution x^* and the corresponding Lagrange multipliers, the level of sparsity (denoted by spars($\cdot$)), the spectral condition number (denoted by $K(\cdot)$), the rank, the euclidean norm and the distribution of singular values of the matrices G and $\begin{pmatrix} A \\ C \end{pmatrix}$. These matrices are generated following a technique similar to that introduced in [41], but using Givens rotations instead of Householder elementary transformations for obtaining the required level of sparsity. The values considered for several features of the test problems reflect those of the problems arising in many practical applications. Furthermore, even if these experiments concern test problems without structure, it is well known that the projection–type methods are suited to exploiting the structures of the Hessian matrix or the constraint matrices that often appear in the real problems.

$n\,5000$			
spars(G) 99.8		$\|G\|1$	$K(G)\,100$
spars($(A^T\,C^T)^T$) 99.9		$\|(A^T\,C^T)^T\|1$	$K((A^T\,C^T)^T)10$
name	m_e	m_i	nac	rank(G)
TP1	1000	1000	500	4900
TP2	2000	2000	1000	4900
TP3	3500	1000	500	4900
TP4	2000	2000	1000	4500
TP5	2000	2000	1000	4000
TP6	2000	2000	1000	3500

Table 1 – Features of the test problems of Tables 2–4

The aim of the first set of experiments is to evaluate the numerical behaviour of the VPM and the AVPM with respect to the class of the gradient projection methods. In particular, in this comparison, we consider the following schemes:

- the projection method with a limited minimization rule ([2, p. 205]). This scheme is equal to the VPM with DI and $\rho_k\rho$ for any k; it is denoted by FPM in Table 2.

- the projection method with an Armijo rule along a projection arc ([2, p. 206]); for the choice of the projection parameter we can use the following strategy: given a positive integer γ and a constant value $\rho > 0$, for the tentative projection parameter at the k-th iteration with k such that $mod(k - 1, \gamma)0$, we start the search procedure with the value ρ and, for the next $(\gamma - 1)$ iterations we use as projection parameter the value obtained by the search procedure in the previous iteration. In practice, the scheme is equal to the AVPM with DI and $\bar\rho_k\rho$ when $mod(k - 1, \gamma)0$ and $\bar\rho_{k+i+1}\rho_{k+i}$ for $i0, ..., \gamma - 2$. When $\gamma1$, $\bar\rho_k\rho$ for any $k \geq 1$. The above scheme is denoted by AFPM-γ in the following tables.

For all methods, the inner QP subproblems are solved by the projected SOR method of Cryer.

ρ	FPM			AFPM-10			AFPM-1		
	it	n_c	time	it	c_n	time	it	c_n	time
TP1									
1.4	557 (1747)	1	35.9	557 (1747)	0	34.7	557 (1747)	0	35.3
2	403 (1277)	290	26.0	401 (1290)	1	25.5	398 (1274)	2	25.7
3.3	398 (1272)	398	26.0	333 (1205)	36	23.7	329 (1789)	225	35.1
5.00	395 (1280)	395	25.6	263 (1042)	52	20.3	270 (1809)	289	35.0
7	399 (1312)	399	25.4	241 (1022)	60	19.2	249 (2157)	405	39.9
10.0	402 (1353)	402	25.7	202 (974)	67	17.4	211 (2160)	446	40.3
20	393 (1448)	393	25.3	163 (944)	75	15.9	220 (3006)	683	54.7
100	344 (2692)	340	24.7	92 (882)	65	11.4	239 (5978)	1340	95.4
200	622 (5123)	620	44.6	122 (1214)	96	15.1	330 (9692)	2177	152.0
1000	1866 (20475)	1864	138.5	146 (2071)	142	20.5	265 (13231)	2281	158.2
TP2									
1.4	517 (1617)	1	47.4	517 (1616)	0	47.5	517 (1616)	0	47.8
2	374 (1185)	72	36.3	373 (1204)	1	36.4	370 (1189)	2	36.7
3.3	347 (1116)	347	34.4	321 (1602)	40	38.2	339 (2525)	258	59.2
5.00	329 (1078)	327	32.9	272 (1882)	62	38.1	689 (7660)	864	152.0
7	333 (1154)	324	33.4	242 (1651)	64	35.1	238 (3380)	406	66.5
10.00	483 (2333)	342	50.0	212 (2094)	79	37.2	204 (2099)	433	57.2
20	305 (1797)	258	35.8	140 (1617)	61	28.5	203 (2838)	633	73.9
100	909 (5487)	885	96.4	108 (1575)	72	26.9	247 (6140)	1385	144.4
200	1951 (11850)	1928	201.8	250 (3754)	193	57.2	308 (9138)	2033	206.2
TP3									
1.00	688 (2450)	1	77.7	690 (2462)	1	80.1	688 (2459)	2	79.2
1.4	481 (1513)	3	54.6	482 (1523)	1	53.7	481 (1520)	1	55.6
2	356 (1126)	340	42.1	346 (1120)	1	41.0	343 (1104)	2	42.8
3.3	356 (1134)	356	42.3	289 (1050)	32	38.5	287 (1568)	198	56.0
5.0	227 (771)	223	30.4	228(907)	46	34.2	312 (2342)	408	76.8
7	197 (700)	184	27.1	206 (940)	59	34.2	239 (2139)	401	69.8
10.0	281 (983)	271	35.3	185 (1145)	66	35.1	191 (2010)	402	65.2
20	206 (1080)	171	31.6	145 (1129)	66	32.4	192 (2785)	594	84.8
100	467 (2927)	456	65.7	103 (1136)	71	30.2	228 (6403)	1281	168.3
200	1606 (9008)	1606	196.9	123 (1477)	97	36.2	292 (9837)	1928	248.1

Table 2

In the SOR method we use an empirical optimal value of the relaxation parameter ($\omega 1.5$) and an inner progressive termination rule, in the sense that the accuracy in the solution of each subproblem depends on the quality of the previous iterate $x^{(k-1)}$: the closer $x^{(k-1)}$ is to satisfying the external stopping criterion for the outer iteration, the more accurately the corresponding subproblem is solved. In this way, unnecessary inner iterations are avoided when $x^{(k-1)}$ is far from the solution.

In the experiments we use the following stopping rule:

$$\frac{\|x^{(k+1)} - x^{(k)}\|_2}{\|x^{(k+1)}\|_2} \leq 10^{-6}, \tag{7.1}$$

and, for all methods, we obtain that $|f(x_{it}) - f(x^*)|/|f(x^*)| \leq 10^{-11}$.

In Tables 2 and 3 we report the results of the FPM, the AFPM-1, the AFPM-10, the VPM and the AVPM for the test problems TP1, TP2, TP3; these last two methods are combined with the rules (5.1) and (5.2) for ρ_k. In the VPM and in the AVPM we put DI.

In the tables we denote by it and $time$ the number of iterations and the time in seconds to obtain the solution. The total number of inner iterations for the solution of the inner QP subproblems is reported in brackets. Furthermore, we denote by n_c the number of correction steps performed after the line search in the FPM and the VPM. In the AFPM-γ and in the AVPM, n_c is the number $\sum_{k1}^{it} m_k$ of the additional projection steps performed in the search procedure of the projection parameter. In these last methods, $\beta 0.5$ and $\eta 0.9$.

AVPM – rule (5.1)			AVPM – rule (5.2)			VPM – rule (5.1)			VPM – rule (5.2)		
it	n_c	time	it	n_c	time	it	n_c	time	it	n_c	time
TP1											
71 (391)	18	6.9	103 (756)	113	14.3	82 (403)	21	6.6	390 (1261)	195	24.8
TP2											
90 (1184)	17	19.9	85 (1555)	87	26.0	71 (850)	14	16.5	156 (938)	71	21.3
TP3											
80 (528)	12	18.9	70 (665)	59	22.5	86 (560)	13	19.3	342 (1144)	169	40.8

Table 3

¿From the Tables 2 and 3, we can make the following remarks:

- in the gradient projection methods (Table 2), as ρ increases, the number it of outer iterations goes down to a minimum value and then begins to increase again. On the contrary, as ρ increases, n_c constantly increases. The variation of it is more consistent in the AFPM-γ. Nevertheless, while the line search procedure and the corrective step of the FPM are nonexpensive, the additional projection steps in the AFPM-γ affect the computational complexity considerably. In particular, in AFPM-1 the adaptive search procedure is always performed for large values of ρ and it requires a considerable number of projection steps. Consequently, the AFPM-10 is generally more efficient than the FPM (because of the lower value of it) and it is more efficient than the AFPM-1 (because of the lower value of n_c). The highest performance of AFPM-10 is obtained for the minimum value of $(it + n_c)$.

- The VPM and the AVPM combined with the updating rule (5.1) have a similar numerical behaviour (Table 3). They are more efficient than the gradient projection methods: the number of outer iterations is lower and the corrective steps of the VPM or the additional projection steps of the AVPM arise in few iterations. Furthermore, the VPM does not require any prefixed scalar parameter.

- The updating rule (5.2) derived from heuristic considerations about the AVPM is nonconvenient when it is combined with the VPM. In the case of the AVPM as well, rule (5.2) appears less efficient than rule (5.1). Indeed, when we use rule (5.2), the total number of the iterations of the inner solver increases.

The above considerations about the numerical behaviour of the AFPM-10, the VPM and the AVPM combined with rule (5.1) are confirmed in Table 4, where we consider some test problems with a large number of null eigenvalues in the Hessian matrix G. Here, Z denotes the $n \times (n - (m_e + nac))$ matrix having as columns an orthonormal basis of the null space of the matrix of the equality constraints and of the inequality constraints active in the solution. Then, $Z^T G Z$ is the reduced Hessian matrix; when this matrix has a full column rank, problem (2.1) has a unique solution, otherwise, K^* is not a singleton. We may observe that, on these test problems, the effectiveness of the three methods appears independent on the rank of G.

Finally, Table 5 shows the results obtained by the VPM and the AVPM with rule (5.1) on some large–scale and very large–scale test problems of the CUTE library [3]. [] These results confirm that when, in the VPM and the AVPM, a convenient updating rule for the choice of the projection parameter is used, the corrective step along a descent direction or a projection arc arises in few iterations and the two schemes show an efficient numerical behaviour.

rank($Z^T G Z$)	VPM			AVPM			AFPM-10VPM		
	it	n_c	time	it	n_c	time	it	n_c	time
TP4									
2000	80 (494)	11	13.8	97 (601)	22	16.5	102 (1145)	71	21.7
1750	83 (521)	12	14.2	74 (523)	18	14.7	94 (856)	62	20.3
1500	89 (399)	16	14.4	79 (421)	13	14.7	109 (937)	69	23.7
TP5									
2000	86 (366)	18	11.8	81 (442)	30	13.6	92 (727)	65	18.4
1500	84 (831)	9	16.7	74 (857)	13	17.2	240(3118)	164	51.8
1000	98 (1023)	8	19.7	90 (995)	15	20.2	137 (2170)	93	36.5
TP6									
2000	79 (327)	15	15.9	82 (409)	25	18.5	99 (727)	64	25.0
1250	87 (358)	20	13.7	67 (376)	25	14.2	92 (727)	65	21.6
500	76 (544)	9	13.8	81 (665)	15	15.7	91 (793)	60	19.5

Table 4

n	AVPM – rule (5.1)			VPM – rule (5.1)		
	it	n_c	time	it	n_c	time
$m_e n/4$						
3000	310 (817)	1	5.6	308 (775)	3	5.0
5000	257 (460)	2	5.5	249 (425)	1	5.2
7000	247 (392)	1	7.8	264 (433)	1	8.4
9000	315 (882)	3	20.3	304 (784)	2	18.8
$m_e n/2$						
3000	114 (204)	2	17.5	116 (218)	6	19.4
7000	101 (200)	3	54.2	103 (190)	3	50.3
9000	133 (420)	2	148.6	135 (350)	1	127.1
11000	135 (245)	0	107.2	135 (245)	0	107.0
13000	148 (232)	3	117.8	142 (229)	2	112.3
17000	129 (206)	6	159.6	128 (218)	3	172.2

Table 5 – Test problem CVXQP from CUTE

References

[1] H.C. Andrews, B.R. Hunt, *Digital Image Restoration*, Prentice Hall, Englewood Press, New Jersey, 1977.

[2] D.P. Bertsekas, *Nonlinear Programming*, Athena Scientific, Belmont, MA, 1995.

[3] I. Bongartz, A.R.Conn, N. Gould, Ph.L. Toint, *CUTE: constrained and unconstrained testing environment*, ACM Transaction on Mathematical Software 1995, 21, 123–160.

[4] Y. Censor, S.A. Zenios, *Parallel Optimization: Theory, Algorithms and Applications*, Oxford University Press, New York, 1997.

[5] H.G. Chen, R.T. Rockafellar, *Convergence rates in forward–backward splitting*, SIAM Journal on Optimization 7 (1997), 421–444.

[6] C. Cortes, V.N. Vapnik, *Support vector network*, Machine Learning 20 (1995), 1–25.

[7] C.W. Cryer, *The solution of a quadratic programming problem using systematic overrelaxation*, SIAM Journal on Control 9 (1971), 385–392.

[8] S. Dafermos, *Traffic equilibrium and variational inequalities*, Transportation Science 14 (1980), 14:42–54.

[9] P.L. De Angelis, P.M. Pardalos, G. Toraldo, *Quadratic Programming with Box Constraints*, In *Developments in Global Optimization*, I.M. Bomze et al. eds.,A0A0A0 Kluwer Academic Publishers, Dordrecht, 1997.

[10] N. Dyn, J. Ferguson, *The numerical solution of equality constrained quadratic programming problems*, Mathematics of Computation 41 (1983), 165–170.

[11] C.S. Fisk, S. Nguyen, *Solution algorithms for network equilibrium models with asymmetric user costs*, Transportation Science 16 (1982), 361–381.

[12] E. Galligani, C^1 *surface interpolation with constraints*, Numerical Algorithms 5 (1993), 549–555.

[13] E. Galligani, V. Ruggiero, L. Zanni, *Splitting methods for quadratic optimization in data analysis*, International Journal of Computer Mathematics 63 (1997), 289–307.

[14] E. Galligani, V. Ruggiero, L. Zanni, *Splitting methods for constrained quadratic programs in data analysis*, Computers & Mathematics with Applications 32 (1996), 1–9.

[15] E. Galligani, V. Ruggiero, L. Zanni, *Splitting methods and parallel solution of constrained quadratic programs*, Proceedings of *Equilibrium Problems with Side Constraints. Lagrangean Theory and Duality II*, 1996 May 17 - 18; Scilla, Italy; Rendiconti del Circolo Matematico di Palermo, Ser. II, Suppl. 48 (1997), 121–136.

[16] E. Galligani, V. Ruggiero, L. Zanni, *Parallel solution of large scale quadratic programs*. Proceedings of *High Performance Algoritms and Software in Nonlinear Optimization*, June 4 - 6 1997, Italy; Kluwer Academic Publishers, Dordrecht, 1998.

[17] P.E. Gill, W. Murray, M.A. Saunders, *SNOPT: An SQP Algorithm for Large-scale Constrained Optimization*, Numerical Analysis Report 96-2. Departement of Mathematics, University of California, San Diego, 1996.

[18] B. He, *Solving a class of linear projection equations*, Numerische Mathematik 68 (1994), 71–80.

[19] F. Leibftitz, E.W. Sachs, *Numerical Solution of Parabolic State Constrained Control Problems Using SQP and Interior-Point-Methods*, In *Optimization: State of the Art*, W.W. Hager et al. eds., Kluwer Academic Publishers, Dordrecht, 1994.

[20] Y.Y. Lin, J.S. Pang, *Iterative methods for large convex quadratic programs: a survey*, SIAM Journal on Control and Optimization 25 (1987), 383–411.

[21] Z.Q. Luo, P. Tseng, *On the linear convergence of descent methods for convex essentially smooth minimization*, SIAM Journal on Control and Optimization 30 (1992), 408–425.

[22] O.L. Mangasarian, *Solution of symmetric linear complementarity problems by iterative methods*, Journal of Optimization Theory and Applications 22 (1977), 465–485.

[23] O.L. Mangasarian, R. De Leone, *Parallel successive overrelaxation methods for symmetric linear complementarity problems and Linear Programs*, Journal of Optimization Theory and Applications 54 (1987), 437–446.

[24] O.L. Mangasarian, R. De Leone, *Parallel gradient projection successive overrelaxation for symmetric linear complementarity problems and linear programs*, Annals of Operations Research 14 (1988), 41–59.

[25] P. Marcotte, J.H. Wu, *On the convergence of projection methods: application to the decomposition of affine variational inequalities*, Journal of Optimization Theory and Applications 85 (1995), 347–362.

[26] J.J. Morè, S.J. Wright, *Optimization Software Guide*, SIAM, Philadelphia, PA, 1993.

[27] W. Murray, *Sequential quadratic programming methods for large-scale problems*, Computational Optimization and Applications 7 (1997), 127–142.

[28] B.A. Murtagh, M.A. Saunders, *MINOS 5.4 User's Guide*, Report SOL 83-20R. Department of Operations Research, Stanford University, 1995.

[29] *NAG Fortran Library Manual*, Mark 18, 1998.

[30] Y. Nesterov, A. Nemirovskii, *Interior–Point Polynomial Algorithms in Convex Programming*, SIAM, Philadelphia, PA, 1994.

[31] S.N. Nielsen, S.A. Zenios, *Massively parallel algorithms for single constrained convex programs*, ORSA Journal on Computing 4 (1992), 166–181.

[32] J.M. Ortega, W.C. Rheinboldt, *Iterative Solution of Nonlinear Equations in Several Variables*, Academic Press, New York, 1970.

[33] J.S. Pang, D. Chan, *Iterative methods for variational and complementarity problems*, Mathematical Programming 24 (1982), 284–313.

[34] P.M. Pardalos, N. Kovoor, *An algorithm for a single constrained class of quadratic programs subject to upper and lower bounds*, Mathematical Programming 46 (1990), 321–328.

[35] P.M. Pardalos, Y. Ye, C.G. Han, *Computational aspects of an interior point algorithm for quadratic problems with box constraints*, In *Large–Scale Numerical Optimization*, T. Coleman and Y. Li eds., SIAM Philadelphia, 1990.

[36] M. Patriksson, *The Traffic Assignment Problem: Models and Methods*, Topics in Transportation, VSP, Utrecht, 1994.

[37] M. Patriksson, *Nonlinear Programming and Variational Inequality Problems: A Unified Approach*. Kluwer Academic Publisher, Dordrecht, 1999.

[38] V. Ruggiero, L. Zanni, *Modified projection algorithm for large strictly convex quadratic programs* Journal of Optimization Theory and Applications 104 (2000), 281–299.

[39] V. Ruggiero, L. Zanni, *On the efficiency of splitting and projection methods for large strictly convex quadratic programs*, In *Nonlinear Optimization and Related Topics*, G. Di Pillo ang F. Giannessi eds., Kluwer Academic Publishers, Dordrecht, 1999.

[40] M.V. Solodov, P. Tseng, *Modified projection-type methods for monotone variational inequalities* SIAM Journal on Control and Optimization 34 (1996), 1814–1830.

[41] G.W. Stewart, *The efficient generation of random orthogonal matrices with an application to condition estimators*, SIAM Journal on Numerical Analysis 17 (1980), 403–409.

[42] R.J. Vanderbei, T.J. Carpenter, *Symmetric indefinite systems for interior point methods*, Mathematical Programming 58 (1993), 1–32.

[43] D.L. Zhu, P. Marcotte, *An extended descent framework for variational inequalities*, Journal of Optimization Theory and Applications 80 (1994), 349–366.

Nonconvex Optimization and Its Applications

Nonconvex Optimization and Its Applications

Nonconvex Optimization and Its Applications

44. A. Rubinov: *Abstract Convexity and Global Optimization.* 2000
ISBN 0-7923-6323-X
45. R.G. Strongin and Y.D. Sergeyev: *Global Optimization with Non-Convex Constraints.* 2000 ISBN 0-7923-6490-2
46. X.-S. Zhang: *Neural Networks in Optimization.* 2000 ISBN 0-7923-6515-1
47. H. Jongen, P. Jonker and F. Twilt: *Nonlinear Optimization in Finite Dimensions.* Morse Theory, Chebyshev Approximation, Transversability, Flows, Parametric Aspects. 2000 ISBN 0-7923-6561-5
48. R. Horst, P.M. Pardalos and N.V. Thoai: *Introduction to Global Optimization.* 2nd Edition. 2000 ISBN 0-7923-6574-7
49. S.P. Uryasev (ed.): *Probabilistic Constrained Optimization.* Methodology and Applications. 2000 ISBN 0-7923-6644-1
50. D.Y. Gao, R.W. Ogden and G.E. Stavroulakis (eds.): *Nonsmooth/Nonconvex Mechanics.* Modeling, Analysis and Numerical Methods. 2001 ISBN 0-7923-6786-3
51. A. Atkinson, B. Bogacka and A. Zhigljavsky (eds.): *Optimum Design 2000.* 2001 ISBN 0-7923-6798-7
52. M. do Rosário Grossinho and S.A. Tersian: *An Introduction to Minimax Theorems and Their Applications to Differential Equations.* 2001 ISBN 0-7923-6832-0
53. A. Migdalas, P.M. Pardalos and P. Värbrand (eds.): *From Local to Global Optimization.* 2001 ISBN 0-7923-6883-5
54. N. Hadjisavvas and P.M. Pardalos (eds.): *Advances in Convex Analysis and Global Optimization.* Honoring the Memory of C. Caratheodory (1873-1950). 2001
ISBN 0-7923-6942-4
55. R.P. Gilbert, P.D. Panagiotopoulos[†] and P.M. Pardalos (eds.): *From Convexity to Nonconvexity.* 2001 ISBN 0-7923-7144-5
56. D.-Z. Du, P.M. Pardalos and W. Wu: *Mathematical Theory of Optimization.* 2001 ISBN 1-4020-0015-4
57. M.A. Goberna and M.A. López (eds.): *Semi-Infinite Programming. Recent Advances.* 2001 ISBN 1-4020-0032-4
58. F. Giannessi, A. Maugeri and P.M. Pardalos (eds.): *Equilibrium Problems: Nonsmooth Optimization and Variational Inequality Models.* 2001 ISBN 1-4020-0161-4

KLUWER ACADEMIC PUBLISHERS – DORDRECHT / BOSTON / LONDON